TEXTS AND READINGS IN MATHEMATICS **36**

Measure Theory

Texts and Readings in Mathematics

Measure Theory

Krishna B. Athreya
Soumendra N. Lahiri
Iowa State University
Ames, Iowa

HINDUSTAN
BOOK AGENCY

Published in India by

Hindustan Book Agency (India)
P 19 Green Park Extension
New Delhi 110 016
India

email: hba@vsnl.com
http://www.hindbook.com

Reprinted 2014

ISBN 978-81-85931-60-9

Digitally Printed at Replika Press Pvt. Ltd.

Preface

This book arose out of a course on measure and integration that one of us (KA) has taught over the last several years.

The traditional approach to a first course in measure theory, such as in Royden (1988), is to teach the Lebesgue measure on the real line, then the differentiation theorems of Lebesgue, L^p spaces on $\mathbb{R}$, and do general measure at the end of the course with one main application to the construction of product measures. This approach does have the pedagogic advantage of seeing one concrete case first before going to the general one. But this also has the disadvantage of making many students' perspective on measure theory somewhat narrow. It leads them to think mostly in terms of the Lebesgue measure and to believe that measure theory is intimately tied to the topology of the real line. As students of statistics, probability, physics, engineering, economics and biology know very well, there are mass distributions that are typically nonuniform and hence it is useful to gain a more general perspective.

This book attempts to provide that general perspective right from the beginning. After an opening chapter on an informal introduction to measure and integration theory which shows that the notions of σ-algebra of sets and countable additivity of a set function are dictated by certain very natural approximation procedures from practical applications and they are not just some abstract ideas, the general extension theorem of Caratheodory is presented in Chapter 1. As immediate examples, constructions of the large class of Lebesgue-Stieltjes measures on the real line and Euclidean spaces are discussed, as are measures on finite and countable spaces. Concrete examples such as the classical Lebesgue measure and various probability distributions on the real line are provided. This is further developed in Chapter 6, leading to the construction of measures on sequence spaces (i.e. sequences of random variables) via Kolmogorov's consistency theorem.

This book presents the basic material on measure and integration theory in a self contained and step by step manner. It is hoped that students will find it accessible, informative, and useful and also that they will be motivated to master the details by carefully working out the text material as well as some of the large number of exercises. The authors hope that the presentation here is found to be clear and comprehensive without being intimidating.

Here is a quick summary of the various chapters of the book. As mentioned earlier, Chapters 0 and 1 focus on the construction of measures starting with set functions on a small class of sets while integration theory is taken up in Chapter 2, where all basic convergence theorems including the monotone convergence theorem (MCT), Fatou lemma, dominated convergence theorem (DCT), bounded convergence theorem (BCT), Egorov's theorem and Scheffe's theorem are proved. Included here are also the notion of uniform integrability and the classical approximation theorem of Lusin and its use in L^p approximation by smooth functions. The third chapter presents basic inequalities for L^p spaces, the Riesz-Fischer theorem and elementary theory of Banach and Hilbert spaces. Chapter 4 deals with the Radon-Nikodym theorem via the Riesz representation on L^2 spaces and its application to differentiation theorems on the real line as well as to signed measures. Chapter 5 deals with product measures and the Fubini-Tonelli theorems. Two constructions of the product measure are presented – one using the extension theorem and another via iterated integrals. This is followed by a discussion on Convolutions, Laplace, and Fourier transforms. Kolmogorov's consistency theorem for the construction of stochastic processes is taken up in Chapter 6. There is an appendix that reviews elementary set theory and advanced calculus including metric spaces.

A word on the numbering system used in the book. Statements of results (i.e., Theorems, Corollaries, Lemmas, and Propositions) are numbered consecutively within each section, in the format $a.b.c$, where a is the chapter number, b is the section number, and c is the counter. Definitions, Examples, and Remarks have separate counters, also of the form $a.b.c$ as above. Equation numbers appear on the right, in the form $(b.c)$, where b is the section number and c is the equation number. Equations in a given chapter a are referred to as $(b.c)$ within the chapter, but as $(a.b.c)$ outside chapter a. Problems are listed at the end of each chapter in the form $a.c$, where a is the chapter number and c is the problem number.

In the writing of this book, material from existing books such as Royden (1988) and Rudin (1976, 1987) has been freely used. The authors owe a great debt to these books. The authors have used earlier drafts of the present book in courses taught over several years and have benefited greatly from suggestions for improvement from the students and colleagues at Iowa

State University, Cornell University, the Indian Institute of Science, and the Indian Statistical Institute. We are grateful to them.

Our special thanks go to Dean Isaacson, Kenneth Koehler, and Justin Peters at Iowa State University for their administrative support of this long project. Krishna Athreya would also like to thank Cornell University and the Indian Institute of Science for their support.

We are most indebted to Sharon Shepard who typed and retyped several times this book, patiently putting up with our never ending "final" versions. Without her patient and generous help, this book could not have been written.

John Kimmel of Springer-Verlag and Rajendra Bhatia of the Indian Statistical Institute have been very supportive of this endeavor. John Kimmel got the book reviewed at various stages and the referees' reports were very helpful and encouraging. We express our grateful thanks to these anonymous referees. Rajendra Bhatia was kind to accept to publish it in India to make it available for students in India.

We have tried hard to make this book free of mathematical and typographical errors and misleading or ambiguous statements. But we are aware that there will still be many such remaining that we have not caught. We will be most grateful to receive such corrections and suggestions for improvement. They can be emailed to us at kba@iastate.edu or snlahiri@iastate.edu.

On a personal note we would like to thank our families for their patience and support especially our wives Krishna Siddhanta and Pubali Banerjee. Krishna Athreya would like to record his profound gratitude to his maternal granduncle, the late Shri K. Venkatarama Iyer who opened the door to mathematical reasoning for him at a crucial stage in high school in the small village of Pattamadai in Tamil Nadu.

Contents

0 Measures and Integration: An Informal Introduction **1**
0.1 Introduction 1
0.2 What is measure theory? 1
0.3 What is integration theory? 5

1 Measures **9**
1.1 Classes of sets 9
1.2 Measures 14
1.3 The extension theorems and Lebesgue-Stieltjes measures . . 19
1.3.1 Caratheodory extension of measures 19
1.3.2 Lebesgue-Stieltjes measures on $\mathbb{R}$ 25
1.3.3 Lebesgue-Stieltjes measures on $\mathbb{R}^2$ 26
1.3.4 More on extension of measures 27
1.4 Completeness of measures 29
1.5 Problems 31

2 Integration **39**
2.1 Measurable transformations 39
2.2 Induced measures, distribution functions 44
2.2.1 Generalizations to higher dimensions 47
2.3 Integration 48
2.4 Riemann and Lebesgue integrals 59
2.5 More on convergence 61
2.6 Problems 72

3 L^p-Spaces **83**

3.1 Inequalities . . . 83

3.2 L^p Spaces . . . 89

3.2.1 Basic properties . . . 89

3.2.2 Dual Spaces . . . 93

3.3 Banach and Hilbert spaces . . . 94

3.3.1 Banach spaces . . . 94

3.3.2 Linear transformations . . . 96

3.3.3 Dual spaces . . . 97

3.3.4 Hilbert space . . . 98

3.4 Problems . . . 102

4 Differentiation **113**

4.1 The Lebesgue-Radon-Nikodym theorem . . . 113

4.2 Signed measures . . . 119

4.3 Functions of bounded variation . . . 125

4.4 Absolutely continuous functions on $\mathbb{R}$ . . . 128

4.5 Singular distributions . . . 133

4.5.1 Decomposition of a cdf . . . 133

4.5.2 Cantor ternary set . . . 134

4.5.3 Cantor ternary function . . . 136

4.6 Problems . . . 137

5 Product Measures, Convolutions, and Transforms **147**

5.1 Product spaces and product measures . . . 147

5.2 Fubini-Tonelli theorems . . . 152

5.3 Extensions to products of higher orders . . . 157

5.4 Convolutions . . . 160

5.4.1 Convolution of measures on $\big(\mathbb{R}, \mathcal{B}(\mathbb{R})\big)$ . . . 160

5.4.2 Convolution of sequences . . . 162

5.4.3 Convolution of functions in $L^1(\mathbb{R})$ . . . 162

5.4.4 Convolution of functions and measures . . . 164

5.5 Generating functions and Laplace transforms . . . 164

5.6 Fourier series . . . 166

5.7 Fourier transforms on $\mathbb{R}$ . . . 173

5.8 Plancherel transform . . . 178

5.9 Problems . . . 181

6 Probability Spaces **189**

6.1 Kolmogorov's probability model . . . 189

6.2 Random variables and random vectors . . . 191

6.3 Kolmogorov's consistency theorem . . . 199

6.4 Problems . . . 212

A Advanced Calculus: A Review **219**

A.1 Elementary set theory 219
A.1.1 Set operations 220
A.1.2 The principle of induction 223
A.1.3 Equivalence relations 223
A.2 Real numbers, continuity, differentiability and integration . 224
A.2.1 Real numbers 224
A.2.2 Sequences, series, limits, limsup, liminf 226
A.2.3 Continuity and differentiability 228
A.2.4 Riemann integration 230
A.3 Complex numbers, exponential and trigonometric functions 232
A.4 Metric spaces 236
A.4.1 Basic definitions 236
A.4.2 Continuous functions 238
A.4.3 Compactness 238
A.4.4 Sequences of functions and uniform convergence . . 239
A.5 Problems 240

B List of Symbols and Abbreviations **245**
B.1 Symbols 245
B.2 Abbreviations 246

References **249**

Subject Index **251**

Measures and Integration: An Informal Introduction

0.1 Introduction

For many students, who are learning measure and integration theory for the first time, the notions of a σ-algebra of subsets of a set Ω, countable additivity of a set function λ, measurability of a function, the definition of an integral, and the interchange of limits and integration are not easy to understand and often seem not so intuitive. The goals of this informal introduction to this subject are: 1) to show that the notions of σ-algebra and countable additivity are logical consequences of certain natural approximation procedures; 2) the dividends for the assumption of these two properties are great, and they lead to a nice and natural theory that is also very powerful for the handling of limits. Of course, as the saying goes, the devil is in the details. After this informal introduction the necessary details are given in the next few sections. It is hoped that after this heuristic explanation of the subject, the motivation for and the process of mastering the details on the part of the students will be forthcoming.

0.2 What is measure theory?

A simple answer is that it is a theory about the distribution of mass over a set $\mathbb{S}$. If the mass is uniformly distributed and $\mathbb{S}$ is an Euclidean space $\mathbb{R}^k$, it is the theory of Lebesgue measure on $\mathbb{R}^k$ (i.e., length in $\mathbb{R}$, area in $\mathbb{R}^2$, volume in $\mathbb{R}^3$, etc.). Probability theory is concerned with the case when

$\mathbb{S}$ is the sample space of a random experiment and the total mass is one. Consider the following example.

Imagine a cricket field $\mathbb{S}$ being prepared for a match. Suppose the field is sprayed with some chemical K to help the growth of grass. Suppose one wants to measure the amount of chemical K in as many of the subsets of $\mathbb{S}$ as possible. Suppose now that one has the tools to measure the amount of chemical K exactly on a class of subsets, such as – triangles, rectangles, circular shapes, elliptic shapes, etc., no matter how small. It is natural to try to approximate oddly-shaped regions by combinations of these "standard shapes," and then use a limiting process to obtain a measure for the oddly-shaped regions and reach some limit for such sets. Let $\mathcal{B}$ denote the class of subsets of $\mathbb{S}$ whose measure is obtained this way and let $\lambda(B)$ denote the amount of chemical K in each $B \in \mathcal{B}$. Call $\mathcal{B}$ the class of all (chemical K) measurable sets and $\lambda(B)$ the measure (of chemical K) on B for each $B \in \mathcal{B}$. It is reasonable to expect that the following properties of $\mathcal{B}$ and $\lambda(\cdot)$ hold:

Properties of $\mathcal{B}$

(i) $A \in \mathcal{B} \Rightarrow A^c \in \mathcal{B}$ (i.e., if one can measure the amount of chemical K on A and knows the total amount on $\mathbb{S}$, then one knows the amount on A^c).

(ii) $A_1, A_2 \in \mathcal{B} \Rightarrow A_1 \cup A_2 \in \mathcal{B}$ (i.e., if one can measure the amount of chemical K on A_1 and A_2, then one can do the same for $A_1 \cup A_2$).

(iii) If $\{A_n : n \geq 1\} \subset \mathcal{B}$, and $A_n \subset A_{n+1}$ for all $n \geq 1$, then $\lim_{n\to\infty} A_n \equiv \cup_{n=1}^{\infty} A_n \in \mathcal{B}$ (i.e., if one can measure the amount of chemical K on A_n for each $n \geq 1$ on an *increasing sequence* of sets then one can do so on the limit of A_n).

(iv) $\mathcal{C} \subset \mathcal{B}$ where $\mathcal{C}$ is the class of nice sets such as triangles, squares, etc., that one started with.

Properties of $\lambda(\cdot)$

(i) $\lambda(A) \geq 0$ for $A \in \mathcal{B}$ (i.e., the amount of chemical K on any set is nonnegative!)

(ii) If $A_1, A_2 \in B, A_1 \cap A_2 = \emptyset, \lambda(A_1 \cup A_2) = \lambda(A_1) + \lambda(A_2)$ (i.e., the amounts of chemical K on two disjoint sets simply add up! This property of λ is referred to as *finite additivity*).

(iii) If $\{A_n : n \geq 1\} \subset \mathcal{B}$, are such that $A_n \subset A_{n+1}$ for all n, then $\lambda(\lim_{n\to\infty} A_n) = \lambda(\cup_{n=1}^{\infty} A_n) = \lim_{n\to\infty} \lambda(A_n)$ (i.e., if we can approximate a set A by an increase sequence of sets $\{A_n\}_{n\geq 1}$ from $\mathcal{B}$, then $\lambda(A) = \lim_{n\to\infty} \lambda(A_n)$. This property of λ is referred to as *monotone continuity from below* or *m.c.f.b.* in short).

This last assumption (iii) is what guarantees that different approximations lead to consistent limits. Thus, if there are two increasing sequences $\{A'_n\}_{n\geq 1}$ and $\{A''_n\}_{n\geq 1}$ having the same limit A but $\{\lambda(A'_n)\}_{n\geq 1}$ and $\{\lambda(A''_n)\}_{n\geq 1}$ have different limits, then the approximating procedures are not consistent.

It turns out that the above set of reasonable and natural assumptions lead to a very rich and powerful theory that is widely applicable.

A triplet $(\mathbb{S}, \mathcal{B}, \lambda)$ that satisfies the above two sets of assumptions is called a *measure space*. The assumptions on $\mathcal{B}$ and λ are *equivalent* to the following:

On $\mathcal{B}$

$\mathcal{B}(\text{i})'$: $\emptyset$, the empty set, lies in $\mathcal{B}$

$\mathcal{B}(\text{ii})'$: $A \in \mathcal{B} \Rightarrow A^c \in \mathcal{B}$ (same as (i) before)

$\mathcal{B}(\text{iii})'$: $A_1, A_2, \ldots \in \mathcal{B} \Rightarrow \cup_i A_i \in \mathcal{B}$ (combines (ii) and (iii) above) (*Closure under countable unions*).

On λ

$\lambda(\text{i})'$: $\lambda(\cdot) \geq 0$ (same as (i) before) and $\lambda(\emptyset) = 0$.

$\lambda(\text{ii})'$: $\lambda(\cup_{n\geq 1} A_n) = \sum_{n=1}^{\infty} \lambda(A_n)$ if $\{A_n\}_{n\geq 1} \subset \mathcal{B}$ are *pairwise disjoint*, i.e. $A_i \cap A_j = \emptyset$ for $i \neq j$ (*Countable additivity*).

Any collection $\mathcal{B}$ of subsets of $\mathbb{S}$ that satisfies $\mathcal{B}(\text{i})'$, $\mathcal{B}(\text{ii})'$, $\mathcal{B}(\text{iii})'$ above is called a *σ-algebra*. Any set function λ on a σ-algebra $\mathcal{B}$ that satisfies $\lambda(\text{i})'$ and $\lambda(\text{ii})'$ above is called a *measure*. Thus, a *measure space* is a triplet $(\mathbb{S}, \mathcal{B}, \lambda)$ where $\mathbb{S}$ is a nonempty set, $\mathcal{B}$ is a *σ-algebra* of subsets of $\mathbb{S}$ and λ is a *measure* on $\mathcal{B}$. Notice that the σ-algebra structure on $\mathcal{B}$ and the countable additivity of λ are necessary consequences of the very natural assumptions (i), (ii), and (iii) on $\mathcal{B}$ and λ defined at the beginning.

It is not often the case that one is given $\mathcal{B}$ and λ explicitly. Typically, one starts with a small collection $\mathcal{C}$ of subsets of $\mathbb{S}$ that have properties resembling intervals or rectangles and a set function λ on $\mathcal{C}$. Then, $\mathcal{B}$ is the smallest σ-algebra containing $\mathcal{C}$ obtained from $\mathcal{C}$ by various operations such as countable unions, intersections, and their limits. The key properties on $\mathcal{C}$ that one needs are:

(i) $A, B \in \mathcal{C} \Rightarrow A \cap B \in \mathcal{C}$ (e.g., intersection of intervals is an interval).

(ii) $A \in \mathcal{C} \Rightarrow A^c$ is a finite union of sets from $\mathcal{C}$ (e.g., the complement of an interval is the union of two intervals or an interval itself).

A collection $\mathcal{C}$ satisfying (i) and (ii) is called a *semialgebra*. The function λ on $\mathcal{B}$ is an extension of λ on $\mathcal{C}$. For this extension to be a measure on $\mathcal{B}$, the conditions needed are:

(i) $\lambda(A) \geq 0$ for all $A \in \mathcal{C}$

(ii) If $A_1, A_2, \ldots \in \mathcal{C}$ are pairwise disjoint and $A = \cup_{n\geq 1} A_n \in \mathcal{C}$, then $\lambda(A) = \sum_{n=1}^{\infty} \lambda(A_n)$.

There is a result, known as the *extension theorem*, that says that given such a pair $(\mathcal{C}, \lambda)$, it is possible to extend λ to $\mathcal{B}$, the smallest σ-algebra containing $\mathcal{C}$, such that $(\mathbb{S}, \mathcal{B}, \lambda)$ is a measure space. Actually, it does more. It constructs a σ-algebra $\mathcal{B}^*$ larger than $\mathcal{B}$ and a measure λ^* on $\mathcal{B}^*$ such that $(\mathbb{S}, \mathcal{B}^*, \lambda^*)$ is a larger measure space, λ^* coincides with λ on $\mathcal{C}$ and it provides nice approximation theorems. For example, the following approximation result is available:

If $B \in \mathcal{B}^*$ with $\lambda^*(B) < \infty$, then for every $\epsilon > 0$, B can be approximated by a finite union of sets from $\mathcal{C}$, i.e. there exist sets $A_1, \ldots, A_k \in \mathcal{C}$ with $k < \infty$ such that $\lambda^*(A \triangle B) < \epsilon$ where $A \equiv \cup_{i=1}^{k} A_i$ and $A \triangle B = (A \cap B^c) \cup (A^c \cap B)$, the *symmetric difference* between A and B.

That is, in principle, every (measurable) set B of finite measure (i.e., B belonging to $\mathcal{B}^*$ with $\lambda^*(B) < \infty$) is nearly a finite union of (elementary) sets that belong to $\mathcal{C}$. For example if $\mathbb{S} = \mathbb{R}$ and $\mathcal{C}$ is the class of intervals, then every measurable set of finite measure is nearly a finite union of disjoint bounded open intervals.

The following are some concrete examples of the above extension procedure.

Theorem 0.2.1: *(Lebesgue-Stieltjes measures). Let $F : \mathbb{R} \to \mathbb{R}$ satisfy*

(i) $x_1 < x_2 \Rightarrow F(x_1) \leq F(x_2)$ (nondecreasing);

(ii) $F(x) = F(x+) \equiv \lim_{y \downarrow x} F(y)$ for all $x \in \mathbb{R}$ (i.e., $F(\cdot)$ is right continuous).

Let $\mathcal{C}$ be the class of sets of the form $(a, b]$, or (b, ∞), $-\infty \leq a < b < \infty$. Then, there exists a measure μ_F defined on $\mathcal{B} \equiv \mathcal{B}(\mathbb{R})$, the smallest σ-algebra generated by $\mathcal{C}$ such that

$$\mu_F((a, b]) = F(b) - F(a) \quad \textit{for all} \quad -\infty < a < b < \infty.$$

The σ-algebra $\mathcal{B} \equiv \mathcal{B}(\mathbb{R})$ is called the *Borel σ-algebra of* $\mathbb{R}$.

Corollary 0.2.2: *There exists a measure m on $\mathcal{B}(\mathbb{R})$ such that $m(I) =$ the length of I, for any interval I.*

Proof: Take $F(x) \equiv x$ in Theorem 0.2.1.
This measure is called the *Lebesgue measure on* $\mathbb{R}$. □

Corollary 0.2.3: *There exists a measure λ on $\mathcal{B}(\mathbb{R})$ such that*

$$\lambda((a, b]) = \frac{1}{\sqrt{2\pi}} \int_a^b e^{-x^2/2} dx.$$

Proof: Take $F(x) = \int_{-\infty}^{x} \frac{1}{\sqrt{2\pi}} e^{-u^2/2} du$, $x \in \mathbb{R}$.

This measure is called the *standard normal probability measure* on $\mathbb{R}$. □

Theorem 0.2.4: *(Lebesgue-Stieltjes measures in $\mathbb{R}^2$). Let $F : \mathbb{R}^2 \to \mathbb{R}$ be a function satisfying the following:*

(i) (Monotonicity) For $x = (x_1, x_2)', y = (y_1, y_2)'$ with $x_i \leq y_i$ for $i = 1, 2$, $(\Delta F)(x, y) \equiv F(y_1, y_2) - F(x_1, y_2) - F(y_1, x_2) + F(x_1, x_2) \geq 0$.

(ii) (Continuity from above) $F(x) = \lim_{y_i \downarrow x_i, i=1,2} F(y)$ for all $x \in \mathbb{R}^2$.

Let $\mathcal{C}$ be the class of all rectangles of the form $(a, b] \equiv (a_1, b_1] \times (a_2, b_2]$ with $a = (a_1, a_2)', b = (b_1, b_2)' \in \mathbb{R}^2$. Then there exists a measure μ_F, defined on the σ-algebra $\mathcal{B} \equiv \mathcal{B}(\mathbb{R}^2)$, generated by $\mathcal{C}$, such that

$$\mu_F((a, b]) = \Big(\Delta F\Big)(a, b).$$

Theorems 0.2.1 and 0.2.4 have a converse that says that every measure on $(\mathbb{R}^k, \mathcal{B}(\mathbb{R}^k))$ that is finite on bounded sets, arises from some function F (called a *distribution function*) and is, therefore, a *Lebesgue-Stieltjes measure*.

Here is another simple example of a measure space (with discrete $\mathbb{S}$).

Example 0.2.1:: Let $\mathbb{S} = \{s_1, s_2, \ldots, s_k\}, k \leq \infty$, and let $\mathcal{B} = \mathcal{P}(\mathbb{S})$, the power set of $\mathbb{S}$, i.e. the collection of all possible subsets of $\mathbb{S}$. Let $p_1, p_2, \ldots$ be nonnegative numbers. Let

$$\lambda(A) \equiv \sum_{1 \leq i \leq k} p_i I_A(s_i),$$

where I_A is the indicator function of the set A, defined by $I_A(s) = 1$ if $s \in A$ and 0 otherwise. It is easy to verify that $(\mathbb{S}, \mathcal{B}, \lambda)$ is a measure space and also that every measure λ on $\mathcal{B}$ arises this way.

0.3 What is integration theory?

In short, it is a theory about weighted sums of functions on a set $\mathbb{S}$ when the weights are specified by a mass distribution λ. Here is a more detailed answer.

Let $(\mathbb{S}, \mathcal{B}, \lambda)$ be a measure space. Suppose $f : \mathbb{S} \to \mathbb{R}$ is a *simple function*, i.e., f is such that $f(\mathbb{S})$ is a finite set $\{a_1, a_2, \ldots, a_k\}$. It is reasonable to define the weighted sum of f with respect to λ as $\sum_{i=1}^{k} a_i \lambda(A_i)$ where $A_i = f^{-1}\{a_i\}$. Of course, for this to be well defined, one needs A_i to be in $\mathcal{B}$ and $\lambda(A_i) < \infty$ for all i such that $a_i \neq 0$.

Notice that the quantity $\sum_{i=1}^{k} a_i\lambda(A_i)$ remains the same whether the a_i's are distinct or not. Call this the integral of f with respect to λ and denote this by $\int f d\lambda$. If f and g are simple, then for $\alpha, \beta \in \mathbb{R}$, $\int(\alpha f + \beta g)d\lambda = \alpha \int f dx + \beta \int g d\lambda$. Now how should one define $\int f d\lambda$ (integral of f with respect to λ) for a nonsimple f? The answer, of course, is to "approximate" by simple functions. Let f be a nonnegative function. To define the integral of f, one would like to approximate f by *simple* functions. It turns out that a necessary and sufficient condition for this is that for any $a \in \mathbb{R}$, the set $\{s : f(s) \leq a\}$ is in $\mathcal{B}$. Such a function f is called *measurable* with respect to $\mathcal{B}$ or $\mathcal{B}$-*measurable* or simply, *measurable* (if $\mathcal{B}$ is kept fixed throughout). Let f be a nonnegative $\mathcal{B}$ measurable function. Then there exists a sequence $\{f_n\}_{n\geq 1}$ of *simple nonnegative* functions such that for each $s \in \mathbb{S}$, $\{f_n(s)\}_{n\geq 1}$ is a nondecreasing sequence converging to $f(s)$. It is now natural to define the weighted sum of f with respect to λ, i.e. the *integral of f with respect to* λ, denoted by $\int f d\lambda$, as

$$\int f d\lambda = \lim_{n\to\infty} \int f_n d\lambda.$$

An immediate question is: *Is the right side the same for all such approximating sequences $\{f_n\}_{n\geq 1}$?* The answer is a *yes*; it is guaranteed by the very natural assumption imposed on λ that it is finitely additive and monotone continuous from below, i.e. λ(ii) and λ(iii) (or equivalently, that λ is countably additive, i.e. $\lambda(\text{ii})'$).

One can strengthen this to a stronger result known as the *monotone convergence theorem*, a key result which in turn leads to two other major convergence results.

The monotone convergence theorem (MCT): *Let $(\mathbb{S}, \mathcal{B}, \lambda)$ be a measure space and let $f_n : \mathbb{S} \to \mathbb{R}_+$, $n \geq 1$ be a sequence of nonnegative $\mathcal{B}$-measurable functions (not necessarily simple) such that for all $s \in \mathbb{S}$,*

(i) $f_n(s) \leq f_{n+1}(s)$, *for all* $n \geq 1$, *and*

(ii) $\lim_{n\to\infty} f_n(s) = f(s)$.

Then f is $\mathcal{B}$-measurable and $\int f d\lambda = \lim_{n\to\infty} \int f_n d\lambda$.

This says that the integral and the limit can be interchanged for monotone nondecreasing nonnegative $\mathcal{B}$-measurable functions. Note that if $f_n = I_{A_n}$, the indicator function of a set A_n and if $A_n \subset A_{n+1}$ for each n, then the MCT is the same as m.c.f.b. (cf. property λ(ii)). *Thus, the very natural assumption of m.c.f.b. yields a basic convergence result that makes the integration theory so elegant and powerful.*

To extend the definition of $\int f d\lambda$ to a real valued, $\mathcal{B}$-measurable function $f : \mathbb{S} \to \mathbb{R}$, one uses the simple idea that f can be decomposed as $f = f^+ - f^-$ where $f^+(s) = \max\{f(s), 0\}$ and $f^-(s) = \max\{-f(s), 0\}$, $s \in \mathbb{S}$.

Since both f^+ and f^- are nonnegative and $\mathcal{B}$-measurable, $\int f^+ d\lambda$ and $\int f^- d\lambda$ are both well defined. Now set

$$\int f d\lambda = \int f^+ d\lambda - \int f^- d\lambda,$$

provided at least one of the two terms on the right is finite. The function f is said to be *integrable* with respect to (w.r.t.) λ if both $\int f^+ d\lambda$ and $f^- d\lambda$ are finite or, equivalently, if $\int |f| d\lambda < \infty$. The following is a consequence of the MCT.

Fatou's lemma: *Let $\{f_n\}_{n\geq 1}$ be a sequence of nonnegative $\mathcal{B}$-measurable functions on a measure space $(\mathbb{S}, \mathcal{B}, \lambda)$. Then*

$$\int \liminf_{n\to\infty} f_n d\lambda \leq \liminf_{n\to\infty} \int f_n d\lambda.$$

This in turn leads to

(Lebesgue's) dominated convergence theorem (DCT): *Let $\{f_n\}_{n\geq 1}$ be a sequence of $\mathcal{B}$-measurable functions from a measure space $(\mathbb{S}, \mathcal{B}, \lambda)$ to $\mathbb{R}$ and let g be a $\mathcal{B}$-measurable nonnegative integrable function on $(\mathbb{S}, \mathcal{B}, \lambda)$. Suppose that for each s in $\mathbb{S}$,*

(i) $|f_n(s)| \leq g(s)$ *for all* $n \geq 1$ *and*

(ii) $\lim_{n\to\infty} f_n(s) = f(s)$.

Then, f is integrable and

$$\lim_{n\to\infty} \int f_n d\lambda = \int f d\lambda = \int \lim_{n\to\infty} f_n d\lambda.$$

Thus some very natural assumptions on $\mathcal{B}$ and λ lead to an interesting measure and integration theory that is quite general and that allows the interchange of limits and integrals under fairly general conditions. A systematic treatment of the measure and integration theory is given in the next five chapters.

1 Measures

Section 1.1 deals with algebraic operations on subsets of a given nonempty set Ω. Section 1.2 treats nonnegative set functions on classes of sets and defines the notion of a measure on an algebra. Section 1.3 treats the extension theorem while Section 1.4 deals with completeness of measures.

1.1 Classes of sets

Let Ω be a nonempty set and $\mathcal{P}(\Omega) \equiv \{A : A \subset \Omega\}$ be the *power set of* Ω, i.e., the class of all subsets of Ω.

Definition 1.1.1: A collection of sets $\mathcal{F} \subset \mathcal{P}(\Omega)$ is called an *algebra* if (a) $\Omega \in \mathcal{F}$, (b) $A \in \mathcal{F}$ implies $A^c \in \mathcal{F}$, and (c) $A, B \in \mathcal{F}$ implies $A \cup B \in \mathcal{F}$ (i.e., closure under pairwise unions).

Thus, an algebra is a class of sets containing Ω that is closed under complementation and pairwise (and hence finite) unions. It is easy to see that one can equivalently define an algebra by requiring that properties (a), (b) hold and that the property

$$(c)' \quad A, B \in \mathcal{F} \Rightarrow A \cap B \in \mathcal{F}$$

holds (i.e. closure under finite intersections).

Definition 1.1.2: A class $\mathcal{F} \subset \mathcal{P}(\Omega)$ is called a σ-*algebra* if it is an algebra and if it satisfies:
(d) $A_n \in \mathcal{F}$ for $n \geq 1 \Rightarrow \cup_{n\geq 1} A_n \in \mathcal{F}$.

Thus, a σ-algebra is a class of subsets of Ω that contains Ω and is closed under complementation and countable unions. As pointed out in Chapter 0, a σ-algebra can be alternatively defined as an algebra that is closed under monotone unions as the following shows.

Proposition 1.1.1: *Let $\mathcal{F} \subset \mathcal{P}(\Omega)$. Then $\mathcal{F}$ is a σ-algebra if and only if $\mathcal{F}$ is an algebra and satisfies*

$$A_n \in \mathcal{F}, A_n \subset A_{n+1} \quad \textit{for all } n \Rightarrow \bigcup_{n\geq 1} A_n \in \mathcal{F}.$$

Proof: The 'only if' part is obvious. For the 'if' part, let $\{B_n\}_{n=1}^{\infty} \subset \mathcal{F}$. Then, since $\mathcal{F}$ is an algebra, $A_n \equiv \cup_{j=1}^{n} B_j \in \mathcal{F}$ for all n. Further, $A_n \subset A_{n+1}$ for all n and $\cup_{n\geq 1} B_n = \cup_{n\geq 1} A_n$. Since by hypothesis $\cup_n A_n \in \mathcal{F}$, $\cup_n B_n \in \mathcal{F}$. □

Here are some examples of algebras and σ-algebras.

Example 1.1.1: Let $\Omega = \{a, b, c, d\}$. Consider the classes

$$\mathcal{F}_1 = \{\Omega, \emptyset, \{a\}\}$$

and

$$\mathcal{F}_2 = \{\Omega, \emptyset, \{a\}, \{b, c, d\}\}.$$

Then, $\mathcal{F}_2$ is an algebra (and also a σ-algebra), but $\mathcal{F}_1$ is not an algebra, since $\{a\}^c \notin \mathcal{F}_1$.

Example 1.1.2: Let Ω be any nonempty set and let

$$\mathcal{F}_3 = \mathcal{P}(\Omega) \equiv \{A : A \subset \Omega\}, \quad \text{the power set of} \quad \Omega$$

and

$$\mathcal{F}_4 = \{\Omega, \emptyset\}.$$

Then, it is easy to check that both $\mathcal{F}_3$ and $\mathcal{F}_4$ are σ-algebras. The latter σ-algebra is often called the *trivial σ-algebra* on Ω (Problem 1.1).

From the definition it is clear that any σ-algebra is also an algebra and thus $\mathcal{F}_2, \mathcal{F}_3, \mathcal{F}_4$ are examples of algebras, too. The following is an example of an algebra that is not a σ-algebra.

Example 1.1.3: Let Ω be a nonempty set, and let $|A|$ denote the number of elements of a set $A \subset \Omega$. Define.

$$\mathcal{F}_5 = \{A \subset \Omega : \text{ either } |A| \text{ is finite or } |A^c| \text{ is finite}\}.$$

Then, note that (i) $\Omega \in \mathcal{F}_5$ (since $|\Omega^c| = |\emptyset| = 0$)), (ii) $A \in \mathcal{F}_5$ implies $A^c \in \mathcal{F}_5$ (if $|A| < \infty$, then $|(A^c)^c| = |A| < \infty$ and if $|A^c| < \infty$, then

$A^c \in \mathcal{F}_5$ trivially). Next, suppose that $A, B \in \mathcal{F}_5$. If either $|A| < \infty$ or $|B| < \infty$, then

$$|A \cap B| \leq \min\{|A|, |B|\} < \infty,$$

so that $A \cap B \in \mathcal{F}_5$. On the other hand, if both $|A^c| < \infty$ and $|B^c| < \infty$, then

$$|(A \cap B)^c| = |A^c \cup B^c| \leq |A^c| + |B^c| < \infty,$$

implying that $A \cap B \in \mathcal{F}_5$. Thus, property (iii)$'$ holds, and $\mathcal{F}_5$ is an algebra. However, if $|\Omega| = \infty$, then $\mathcal{F}_5$ is not a σ-algebra. To see this, suppose that $|\Omega| = \infty$ and $\{\omega_1, \omega_2, \ldots\} \subset \Omega$. Then, by definition, $A_i = \{\omega_i\} \in \mathcal{F}_5$ for all $i \geq 1$, but $A \equiv \cup_{i=1}^{\infty} A_{2i-1} = \{\omega_1, \omega_3, \ldots\} \notin \mathcal{F}_5$, since $|A| = |A^c| = \infty$.

Example 1.1.4: Let Ω be a nonempty set and let

$$\mathcal{F}_6 = \{A \subset \Omega : A \text{ is countable or } A^c \text{ is countable}\}.$$

Then, it is easy to show that $\mathcal{F}_6$ is a σ-algebra (Problem 1.3).

Suppose $\{\mathcal{F}_\theta : \theta \in \Theta\}$ is a family of σ-algebras on Ω. From the definition, it follows that the intersection $\cap_{\theta \in \Theta} \mathcal{F}_\theta$ is a σ-algebra, no matter how large the index set Θ is (Problem 1.4). However, the union of two σ-algebras may not even be an algebra (Problem 1.5). For the development of measure theory and probability theory, the concept of a σ-algebra plays a crucial role. In many instances, given an arbitrary collection of subsets of Ω, one would like to extend it to a possibly larger class that is a σ-algebra. This leads to the following definition.

Definition 1.1.3: If $\mathcal{A}$ is a class of subsets of Ω, then the σ-algebra generated by $\mathcal{A}$, denoted by $\sigma\langle\mathcal{A}\rangle$, is defined as

$$\sigma\langle\mathcal{A}\rangle = \bigcap_{\mathcal{F} \in \mathcal{I}(\mathcal{A})} \mathcal{F},$$

where $\mathcal{I}(\mathcal{A}) \equiv \{\mathcal{F} : \mathcal{A} \subset \mathcal{F} \text{ and } \mathcal{F} \text{ is a } \sigma\text{-algebra on } \Omega\}$ is the collection of all σ-algebras containing the class $\mathcal{A}$.

Note that since the power set $\mathcal{P}(\Omega)$ contains $\mathcal{A}$ and is itself a σ-algebra, the collection $\mathcal{I}(\mathcal{A})$ is not empty and hence, the intersection in the above definition is well defined.

Example 1.1.5: In the set up of Example 1.1.1, $\sigma\langle\mathcal{F}_1\rangle = \mathcal{F}_2$ (why?).

A particularly useful class of σ-algebras are those generated by open sets of a topological space. These are called Borel σ-algebras. A *topological space* is a pair $(\mathbb{S}, \mathcal{T})$ where $\mathbb{S}$ is a nonempty set and $\mathcal{T}$ is a collection of subsets of $\mathbb{S}$ such that (i) $\mathbb{S} \in \mathcal{T}$, (ii) $\mathcal{O}_1, \mathcal{O}_2 \in \mathcal{T} \Rightarrow \mathcal{O}_1 \cap \mathcal{O}_2 \in \mathcal{T}$, and (iii) $\{\mathcal{O}_\alpha : \alpha \in I\} \subset \mathcal{T} \Rightarrow \cup_{\alpha \in I} \mathcal{O}_\alpha \in \mathcal{T}$. Elements of $\mathcal{T}$ are called *open sets*.

A *metric space* is a pair $(\mathbb{S}, d)$ where $\mathbb{S}$ is a nonempty set and d is a function from $\mathbb{S} \times \mathbb{S}$ to $\mathbb{R}^+$ satisfying (i) $d(x, y) = d(y, x)$ for all x, y in $\mathbb{S}$,

(ii) $d(x, y) = 0$ iff $x = y$, and (iii) $d(x, z) \leq d(x, y) + d(y, z)$ for all x, y, z in $\mathbb{S}$. Property (iii) is called the triangle inequality. The function d is called a metric on $\mathbb{S}$ (cf. see A.4).

Any Euclidean space $\mathbb{R}^n$ $(1 \leq n < \infty)$ is a metric space under any one of the following metrics:

(a) For $1 \leq p < \infty$, $d_p(x, y) = \Big(\sum_{i=1}^{n} |x_i - y_i|^p \Big)^{1/p}$.

(b) $d_\infty(x, y) = \max_{1 \leq i \leq n} |x_i - y_i|$.

(c) For $0 < p < 1$, $d_p(x, y) = \Big(\sum_{i=1}^{n} |x_i - y_i|^p \Big)$.

A metric space $(\mathbb{S}, d)$ is a topological space where a set $\mathcal{O}$ is open if for all $x \in \mathcal{O}$, there is an $\epsilon > 0$ such that $B(x, \epsilon) \equiv \{y : d(y, x) < \epsilon\} \subset \mathcal{O}$.

Definition 1.1.4: The *Borel σ-algebra* on a topological space $\mathbb{S}$ (in particular, on a metric space or an Euclidean space) is defined as the σ-algebra generated by the collection of open sets in $\mathbb{S}$.

Example 1.1.6: Let $\mathcal{B}(\mathbb{R}^k)$ denote the Borel σ-algebra on $\mathbb{R}^k$, $1 \leq k < \infty$. Then,

$$\mathcal{B}(\mathbb{R}^k) \equiv \sigma\langle\{A : A \text{ is an open subset of } \mathbb{R}^k\}\rangle$$

is also generated by each of the following classes of sets

$$\begin{aligned} \mathcal{O}_1 &= \{(a_1, b_1) \times \ldots \times (a_k, b_k) : -\infty \leq a_i < b_i \leq \infty, 1 \leq i \leq k\}; \\ \mathcal{O}_2 &= \{(-\infty, x_i) \times \cdots \times (-\infty, x_k) : x_1, \ldots, x_k \in \mathbb{R}\}; \\ \mathcal{O}_3 &= \{(a_1, b_1) \times \ldots \times (a_k, b_k) : a_i, b_i \in \mathcal{Q}, a_i < b_i, 1 \leq i \leq k\}; \\ \mathcal{O}_4 &= \{(-\infty, x_1) \times \ldots \times (-\infty, x_k) : x_1, \ldots, x_k \in \mathcal{Q}\}, \end{aligned}$$

where $\mathcal{Q}$ denotes the set of all rational numbers.

To show this, note that $\sigma\langle\mathcal{O}_i\rangle \subset \mathcal{B}(\mathbb{R}^k)$ for $i = 1, 2, 3, 4$, and hence, it is enough to show that $\sigma\langle\mathcal{O}_i\rangle \supset \mathcal{B}(\mathbb{R}^k)$. Let $\mathcal{G}$ be a σ-algebra containing $\mathcal{O}_3$. Observe that given any open set $A \subset \mathbb{R}^k$, there exist a sequence of sets $\{B_n\}_{n \geq 1}$ in $\mathcal{O}_3$ such that $A = \cup_{n \geq 1} B_n$ (Problem 1.9). Since $\mathcal{G}$ is a σ-algebra and $B_n \in \mathcal{G}$ for all $n \geq 1$, $A \in \mathcal{G}$. Thus, $\mathcal{G}$ is a σ-algebra containing all open subsets of $\mathbb{R}^k$, and hence $\mathcal{G} \supset \mathcal{B}(\mathbb{R}^k)$. Hence, it follows that

$$\mathcal{B}(\mathbb{R}^k) \supset \sigma\langle\mathcal{O}_1\rangle \supset \sigma\langle\mathcal{O}_3\rangle = \bigcap_{\mathcal{G}:\mathcal{G} \supset \mathcal{O}_3} \mathcal{G} \supset \mathcal{B}(\mathbb{R}^k).$$

Next note that any interval $(a, b) \subset \mathbb{R}$ can be expressed in terms of half spaces of the form $(-\infty, x)$, $x \in \mathbb{R}$ as

$$(a, b) = \bigcup_{n=1}^{\infty} [(-\infty, b) \setminus (-\infty, a + n^{-1})],$$

where for any two sets A and B, $A \backslash B = \{x : x \in A, x \notin B\}$. It is not difficult to show that this implies that $\sigma\langle \mathcal{O}_i \rangle = \mathcal{B}(\mathbb{R}^k)$ for $i = 2, 4$ (Problem 1.10).

Example 1.1.7: Let Ω be a nonempty set with $|\Omega| = \infty$ and $\mathcal{F}_5$ and $\mathcal{F}_6$ be as in Examples 1.1.3 and 1.1.4. Then $\mathcal{F}_6 = \sigma\langle \mathcal{F}_5 \rangle$. To see this, note that $\mathcal{F}_6$ is a σ-algebra containing $\mathcal{F}_5$, so that $\sigma\langle \mathcal{F}_5 \rangle \subset \mathcal{F}_6$. To prove the reverse inclusion, let $\mathcal{G}$ be a σ-algebra containing $\mathcal{F}_5$. It is enough to show that $\mathcal{F}_6 \subset \mathcal{G}$. Let $A \in \mathcal{F}_6$. If A is countable, say $A = \{\omega_1, \omega_2, \ldots\}$, then $A_i \equiv \{\omega_i\} \in \mathcal{F}_5 \subset \mathcal{G}$ for all $i \geq 1$ and hence, $A = \cup_{i=1}^{\infty} A_i \in \mathcal{G}$. On the other hand, if A^c is countable, then by the above argument, $A^c \in \mathcal{G} \Rightarrow A \in \mathcal{G}$.

Definition 1.1.5: A class $\mathcal{C}$ of subsets of Ω is a *π-system* or a *π-class* if $A, B \in \mathcal{C} \Rightarrow A \cap B \in \mathcal{C}$.

Example 1.1.8: The class $\mathcal{C}$ of intervals in $\mathbb{R}$ is a π-system whereas the class of all open discs in $\mathbb{R}^2$ is not.

Definition 1.1.6: A class $\mathcal{L}$ of subsets of Ω is a *λ-system* or a *λ-class* if (i) $\Omega \in \mathcal{L}$, (ii) $A, B \in \mathcal{L}, A \subset B \Rightarrow B \setminus A \in \mathcal{L}$, and (iii) $A_n \in \mathcal{L}, A_n \subset A_{n+1}$ for all $n \geq 1 \Rightarrow \cup_{n\geq 1} A_n \in \mathcal{L}$.

Example 1.1.9: Every σ-algebra $\mathcal{F}$ is a λ-system. But an algebra need not be a λ-system.

It is easily checked that if $\mathcal{L}_1$ and $\mathcal{L}_2$ are λ-systems, then $\mathcal{L}_1 \cap \mathcal{L}_2$ is also a λ-system. Recall that $\sigma\langle \mathcal{B} \rangle$, the σ-algebra generated by $\mathcal{B}$, is the intersection of all σ-algebras containing $\mathcal{B}$ and is also the smallest σ-algebra containing $\mathcal{B}$. Similarly, for any $\mathcal{B} \subset \mathcal{P}(\Omega)$, the λ-system generated by $\mathcal{B}$, denoted by $\lambda\langle \mathcal{B} \rangle$, is defined as the intersection of all λ-systems containing $\mathcal{B}$. It is the smallest λ-system containing $\mathcal{B}$.

Theorem 1.1.2: (*The π-λ theorem*). *If $\mathcal{C}$ is a π-system, then $\lambda\langle \mathcal{C} \rangle = \sigma\langle \mathcal{C} \rangle$.*

Proof: For any $\mathcal{C}$, $\sigma\langle \mathcal{C} \rangle$ is a λ-system and $\sigma\langle \mathcal{C} \rangle$ contains $\mathcal{C}$. Thus, $\lambda\langle \mathcal{C} \rangle) \subset \sigma\langle \mathcal{C} \rangle$ for any $\mathcal{C}$. Hence, it suffices to show that if $\mathcal{C}$ is a π-system, then $\lambda\langle \mathcal{C} \rangle$ is a σ-algebra . Since $\lambda\langle \mathcal{C} \rangle$ is a λ-system, it is closed under complementation and monotone increasing unions. By Proposition 1.1.1, it is enough to show that it is closed under intersection. Let $\lambda_1(\mathcal{C}) \equiv \{A : A \in \lambda\langle \mathcal{C} \rangle, A \cap B \in \lambda\langle \mathcal{C} \rangle$ for all $B \in \mathcal{C}\}$. Then, $\lambda_1(\mathcal{C})$ is a λ-system and $\mathcal{C}$ being a π-system, $\lambda_1(\mathcal{C}) \supset \mathcal{C}$. Therefore, $\lambda_1(\mathcal{C}) \supset \lambda\langle \mathcal{C} \rangle$. But $\lambda_1(\mathcal{C}) \subset \lambda\langle \mathcal{C} \rangle$. So $\lambda_1(\mathcal{C}) = \lambda\langle \mathcal{C} \rangle$.

Next, let $\lambda_2(\mathcal{C}) \equiv \{A : A \in \lambda\langle \mathcal{C} \rangle, A \cap B \in \lambda\langle \mathcal{C} \rangle$ for all $B \in \lambda\langle \mathcal{C} \rangle\}$. Then $\lambda_2(\mathcal{C})$ is a λ-system and by the previous step $\mathcal{C} \subset \lambda_2(\mathcal{C}) \subset \lambda\langle \mathcal{C} \rangle$. Hence, it follows that $\lambda_2(\mathcal{C}) = \lambda\langle \mathcal{C} \rangle$, i.e., $\lambda\langle \mathcal{C} \rangle$ is closed under intersection. This completes the proof of the theorem. □

Corollary 1.1.3: *If $\mathcal{C}$ is a π-system and $\mathcal{L}$ is a λ-system containing $\mathcal{C}$, then $\mathcal{L} \supset \sigma\langle \mathcal{C} \rangle$.*

Remark 1.1.1: There are several equivalent definitions of λ-systems; see, for example, Billingsley (1995). A closely related concept is that of a monotone class; see, for example, Chung (1974).

1.2 Measures

A *set function* is an extended real valued function defined on a class of subsets of a set Ω. Measures are nonnegative set functions that, intuitively speaking, measure the content of a subset of Ω. As explained in Section 0.2, a measure has to satisfy certain natural requirements, such as the measure of the union of a countable collection of *disjoint* sets is the *sum* of the measures of the individual sets. Formally, one has the following definition.

Definition 1.2.1: Let Ω be a nonempty set and $\mathcal{F}$ be an algebra on Ω. Then, a set function μ on $\mathcal{F}$ is called a *measure* if

(a) $\mu(A) \in [0, \infty]$ for all $A \in \mathcal{F}$;

(b) $\mu(\emptyset) = 0$;

(c) for any disjoint collection of sets $A_1, A_2, \ldots, \in \mathcal{F}$ with $\cup_{n\geq 1} A_n \in \mathcal{F}$,

$$\mu\Big(\bigcup_{n\geq 1} A_n\Big) = \sum_{n=1}^{\infty} \mu(A_n).$$

As discussed in Section 0.2, these conditions on μ are equivalent to finite additivity and monotone continuity from below.

Proposition 1.2.1: *Let Ω be a nonempty set and $\mathcal{F}$ be an algebra of subsets of Ω and μ be a set function on $\mathcal{F}$ with values in $[0, \infty]$ and with $\mu(\emptyset) = 0$. Then, μ is a measure iff μ satisfies*

$(iii)'_a$: (finite additivity) for all $A_1, A_2 \in \mathcal{F}$ with $A_1 \cap A_2 = \emptyset$, $\mu(A_1 \cup A_2) = \mu(A_1) + \mu(A_2)$, and

$(iii)'_b$: (monotone continuity from below or, m.c.f.b., in short) for any collection $\{A_n\}_{n\geq 1}$ of sets in $\mathcal{F}$ such that $A_n \subset A_{n+1}$ for all $n \geq 1$ and $\cup_{n\geq 1} A_n \in \mathcal{F}$,

$$\mu\Big(\bigcup_{n\geq 1} A_n\Big) = \lim_{n\to\infty} \mu(A_n).$$

Proof: Let μ be a measure on $\mathcal{F}$. Since μ satisfies (iii), taking $A_3, A_4, \ldots$ to be $\emptyset$ yields (iii)$'_a$. This implies that for A and B in $\mathcal{F}$, $A \subset B \Rightarrow \mu(B) = \mu(A) + \mu(B \setminus A) \geq \mu(A)$, i.e., μ is *monotone*. To establish (iii)$'_b$, note that if $\mu(A_n) = \infty$ for some $n = n_0$, then $\mu(A_n) = \infty$ for all $n \geq n_0$ and $\mu(\cup_{n\geq 1} A_n) = \infty$ and (iii)$'_b$ holds in this case. Hence, suppose that $\mu(A_n) < \infty$ for all $n \geq 1$. Setting $B_n = A_n \setminus A_{n-1}$ for $n \geq 1$ (with $A_0 = \emptyset$), by (iii)$'_a$, $\mu(B_n) = \mu(A_n) - \mu(A_{n-1})$. Since $\{B_n\}_{n\geq 1}$ is a *disjoint* collection of sets in $\mathcal{F}$ with $\cup_{n\geq 1} B_n = \cup_{n\geq 1} A_n$, by (iii)

$$\begin{aligned}\mu\Big(\bigcup_{n\geq 1} A_n\Big) &= \mu\Big(\bigcup_{n\geq 1} B_n\Big) = \sum_{n=1}^{\infty} \mu(B_n) = \lim_{N\to\infty} \sum_{n=1}^{N} [\mu(A_n) - \mu(A_{n-1})] \\ &= \lim_{N\to\infty} \mu(A_N),\end{aligned}$$

and so (iii)$'_b$ holds also in this case.

Conversely, let μ satisfy $\mu(\emptyset) = 0$ and (iii)$'_a$ and (iii)$'_b$. Let $\{A_n\}_{n\geq 1}$ be a disjoint collection of sets in $\mathcal{F}$ with $\cup_{i\geq 1} A_i \in \mathcal{F}$. Let $C_n = \cup_{j=1}^{n} A_j$ for $n \geq 1$. Since $\mathcal{F}$ is an algebra, $C_n \in \mathcal{F}$ for all $n \geq 1$. Also, $C_n \subset C_{n+1}$ for all $n \geq 1$. Hence, $\cup_{n\geq 1} C_n = \cup_{j\geq 1} A_j$. By (iii)$'_b$,

$$\begin{aligned}\mu\Big(\bigcup_{j\geq 1} A_j\Big) &= \mu\Big(\bigcup_{n\geq 1} C_n\Big) = \lim_{n\to\infty} \mu(C_n) \\ &= \lim_{n\to\infty} \sum_{j=1}^{n} \mu(A_j) \quad (\text{by (iii)}'_a) \\ &= \sum_{j=1}^{\infty} \mu(A_j).\end{aligned}$$

Thus, (iii) holds. □

Remark 1.2.1: The definition of a measure given in Definition 1.2.1 is valid when $\mathcal{F}$ is a σ-algebra. However, very often one may start with a measure on an algebra $\mathcal{A}$ and then extend it to a measure on the σ-algebra $\sigma\langle\mathcal{A}\rangle$. This is why the definition of a measure on an algebra is given here. In the same vein, one may begin with a definition of a measure on a class of subsets of Ω that form only a semialgebra (cf. Definition 1.3.1). As described in Chapter 0, such preliminary collections of sets are "nice" sets for which the measure may be defined easily, and the extension to a σ-algebra containing these sets may be necessary if one is interested in more general sets. This topic is discussed in greater detail in the next section.

Definition 1.2.2: A measure μ is called *finite* or *infinite* according as $\mu(\Omega) < \infty$ or $\mu(\Omega) = \infty$. A finite measure with $\mu(\Omega) = 1$ is called a *probability* measure. A measure μ on a σ-algebra $\mathcal{F}$ is called *σ-finite* if there exist a countable collection of sets $A_1, A_2, \ldots, \in \mathcal{F}$, not necessarily disjoint, such that

$$\text{(a)} \bigcup_{n\geq 1} A_n = \Omega \quad \text{and} \quad \text{(b)}\ \mu(A_n) < \infty \quad \text{for all} \quad n \geq 1.$$

Here are some examples of measures.

Example 1.2.1: (*The counting measure*). Let Ω be a nonempty set and $\mathcal{F}_3 = \mathcal{P}(\Omega)$ be the set of all subsets of Ω (cf. Example 1.1.2). Define

$$\mu(A) = |A|, \quad A \in \mathcal{F}_3,$$

where $|A|$ denotes the number of elements in A. It is easy to check that μ satisfies the requirements (a)–(c) of a measure. This measure μ is called the *counting measure* on Ω. Note that μ is finite iff Ω is finite and it is σ-finite if Ω is countably infinite.

Example 1.2.2: (*Discrete probability measures*). Let $\omega_1, \omega_2, \ldots, \in \Omega$ and $p_1, p_2, \ldots \in [0,1]$ be such that $\sum_{i=1}^{\infty} p_i = 1$. Define for any $A \subset \Omega$

$$P(A) = \sum_{i=1}^{\infty} p_i I_A(\omega_i),$$

where $I_A(\cdot)$ denotes the indicator function of a set A, defined by $I_A(\omega) = 0$ or 1 according as $\omega \notin A$ or $\omega \in A$. For any disjoint collection of sets $A_1, A_2, \ldots \in \mathcal{P}(\Omega)$,

$$\begin{aligned} P\Big(\bigcup_{i=1}^{\infty} A_i\Big) &= \sum_{j=1}^{\infty} p_j I_{\cup_{i=1}^{\infty} A_i}(\omega_j) \\ &= \sum_{j=1}^{\infty} p_j \Big(\sum_{i=1}^{\infty} I_{A_i}(\omega_j)\Big) \\ &= \sum_{i=1}^{\infty} \Big(\sum_{j=1}^{\infty} p_j I_{A_i}(\omega_j)\Big) \\ &= \sum_{i=1}^{\infty} P(A_i), \end{aligned}$$

where interchanging the order of summation is permissible since the summands are nonnegative. This shows that P is a probability measure on $\mathcal{P}(\Omega)$.

Example 1.2.3: (*Lebesgue-Stieltjes measures on* $\mathbb{R}$). As mentioned earlier (cf. Section 0.2), a large class of measures on the Borel σ-algebra $\mathcal{B}(\mathbb{R})$ of subsets of $\mathbb{R}$, known as the *Lebesgue-Stieltjes measures*, arise from nondecreasing right continuous functions $F : \mathbb{R} \to \mathbb{R}$. For each such F, the corresponding measure μ_F satisfies $\mu_F((a,b]) = F(b) - F(a)$ for all $-\infty < a < b < \infty$. The construction of these μ_F's via the extension theorem will be discussed in the next section. Also, note that if $A_n = (-n, n)$, $n = 1, 2, \ldots$, then $\mathbb{R} = \cup_{n \geq 1} A_n$ and $\mu_F(A_n) < \infty$ for each $n \geq 1$ (such measures are called Radon measures) and thus, μ_F is necessarily σ-finite.

Proposition 1.2.2: *Let μ be a measure on an algebra $\mathcal{F}$, and let $A, B, A_1, \ldots, A_k \in \mathcal{F}, 1 \leq k < \infty$. Then,*

(i) (Monotonicity) $\mu(A) \leq \mu(B)$ if $A \subset B$;

(ii) (Finite subadditivity) $\mu(A_1 \cup \ldots \cup A_k) \leq \mu(A_1) + \ldots + \mu(A_k)$;

(iii) (Inclusion-exclusion formula) If $\mu(A_i) < \infty$ for all $i = 1, \ldots, k$, then

$$\begin{aligned}\mu(A_1 \cup \ldots \cup A_k) &= \sum_{i=1}^{k} \mu(A_i) - \sum_{1 \leq i < j < k} \mu(A_i \cap A_j) \\ &\quad + \ldots + (-1)^{k-1} \mu(A_1 \cap \ldots \cap A_k).\end{aligned}$$

Proof: $\mu(B) = \mu(A \cup (A^c \cap B)) = \mu(A) + \mu(B \setminus A) \geq \mu(A)$, by (a) and (c) of Definition 1.2.1. This proves (i).

To prove (ii), note that if either $\mu(A)$ or $\mu(B)$ is finite, then $\mu(A \cap B) < \infty$, by (i). Hence, using the countable additivity property (c), we have

$$\begin{aligned}\mu(A \cup B) &= \mu(A) + \mu(B \setminus A) \\ &= \mu(A) + [\mu(B \setminus A) + \mu(A \cap B)] - \mu(A \cap B) \\ &= \mu(A) + \mu(B) - \mu(A \cap B). \end{aligned} \tag{2.1}$$

Hence, (ii) follows from (2.1) and by induction.

To prove (iii), note that the case $k = 2$ follows from (2.1). Next, suppose that (iii) holds for all sets $A_1, \ldots, A_k \in \mathcal{F}$ with $\mu(A_i) < \infty$ for all $i = 1, \ldots, k$ for some $k = n$, $n \in \mathbb{N}$. To show that it holds for $k = n + 1$, note that by (2.1),

$$\begin{aligned}&\mu\Big(\bigcup_{i=1}^{n+1} A_i\Big) \\ &= \mu\Big(\bigcup_{i=1}^{n} A_i\Big) + \mu(A_{n+1}) - \mu\Big(\bigcup_{i=1}^{n} (A_i \cap A_{n+1})\Big) \\ &= \Big\{\sum_{i=1}^{n} \mu(A_i) - \sum_{1<i<j\leq n} \mu(A_i \cap A_j) + \cdots + (-1)^{n-1} \mu(A_1 \cap \ldots \cap A_n)\Big\} \\ &\quad + \mu(A_{n+1}) - \Big[\sum_{i=1}^{n} \mu(A_i \cap A_{n+1}) - \sum_{1 \leq i < j \leq n} \mu(A_i \cap A_j \cap A_{n+1}) \\ &\qquad\qquad + \cdots + (-1)^{n-1} \mu(A_1 \cap \ldots \cap A_{n+1})\Big] \\ &= \sum_{i=1}^{n+1} \mu(A_i) - \sum_{1 \leq i < j \leq n+1} \mu(A_i \cap A_j) + \cdots + (-1)^{n} \mu\Big(\bigcap_{j=1}^{n+1} A_j\Big).\end{aligned}$$

By induction, this completes the proof of Proposition 1.2.2. □

In Proposition 1.2.1, it was shown that a set function μ on an algebra $\mathcal{F}$ is a measure iff it is finitely additive and monotone continuous from below. A natural question is: if μ is a measure on $\mathcal{F}$ and $\{A_n\}_{n\geq 1}$ is a collection of decreasing sets in $\mathcal{F}$ with $A \equiv \cap_{n\geq 1} A_n$ also in $\mathcal{F}$, does the relation $\mu(A) = \lim_{n\to\infty} \mu(A_n)$ hold, i.e., does *monotone continuity from above* hold? The answer is positive under the assumption $\mu(A_{n_0}) < \infty$ for some $n_0 \in \mathbb{N}$. It turns out that, in general, this assumption cannot be dropped (Problem 1.18).

Proposition 1.2.3: *Let μ be a measure on an algebra $\mathcal{F}$.*

(i) (Monotone continuity from above) Let $\{A_n\}_{n\geq 1}$ be a sequence of sets in $\mathcal{F}$ such that $A_{n+1} \subset A_n$ for all $n \geq 1$ and $A \equiv \cap_{n\geq 1} A_n \in \mathcal{F}$. Also, let $\mu(A_{n_0}) < \infty$ for some $n_0 \in \mathbb{N}$. Then,

$$\lim_{n\to\infty} \mu(A_n) = \mu(A).$$

(ii) (Countable subadditivity) If $\{A_n\}_{n\geq 1}$ is a sequence of sets in $\mathcal{F}$ such that $\cup_{n\geq 1} A_n \in \mathcal{F}$, then

$$\mu\Big(\bigcup_{n=1}^{\infty} A_n\Big) \leq \sum_{n=1}^{\infty} \mu(A_n).$$

Proof: To prove part (i), without loss of generality (w.l.o.g.), assume that $n_0 = 1$, i.e., $\mu(A_1) < \infty$. Let $C_n = A_1 \setminus A_n$ for $n \geq 1$, and $C_\infty = A_1 \setminus A$. Then C_n and C_∞ belong to $\mathcal{F}$ and $C_n \uparrow C_\infty$. By Proposition 1.2.1 $(\text{iii})'_b$, (i.e., by the m.c.f.b. property), $\mu(C_n) \uparrow \mu(C_\infty)$ and by $(\text{iii})'_a$, (i.e., finite additivity), $\mu(C_n) = \mu(A_1) - \mu(A_n)$ for all $1 \leq n < \infty$, due to the fact $\mu(A_1) < \infty$. This proves (i).

To prove part (ii), let $D_n = \cup_{i=1}^n A_i, n \geq 1$. Then, $D_n \uparrow D \equiv \cup_{i\geq 1} A_i$. Hence, by m.c.f.b. and finite subadditivity,

$$\mu(D) = \lim_{n\to\infty} \mu(D_n) \leq \lim_{n\to\infty} \sum_{i=1}^{n} \mu(A_i) = \sum_{n=1}^{\infty} \mu(A_n).$$ □

Theorem 1.2.4: (*Uniqueness of measures*). *Let μ_1 and μ_2 be two finite measures on a measurable space $(\Omega, \mathcal{F})$. Let $\mathcal{C} \subset \mathcal{F}$ be a π-system such that $\mathcal{F} = \sigma\langle\mathcal{C}\rangle$. If $\mu_1(C) = \mu_2(C)$ for all $C \in \mathcal{C}$ and $\mu_1(\Omega) = \mu_2(\Omega)$, then $\mu_1(A) = \mu_2(A)$ for all $A \in \mathcal{F}$.*

Proof: Let $\mathcal{L} \equiv \{A : A \in \mathcal{F},\ \mu_1(A) = \mu_2(A)\}$. It is easy to verify that $\mathcal{L}$ is a λ-system. Since $\mathcal{C} \subset \mathcal{L}$, by Theorem 1.1.2, $\mathcal{L} = \sigma\langle\mathcal{C}\rangle = \mathcal{F}$. □

1.3 The extension theorems and Lebesgue-Stieltjes measures

As discussed in Section 1.0, in many situations, one starts with a given set function μ defined on a small class $\mathcal{C}$ of subsets of a set Ω, and then, wants to extend μ to a larger class $\mathcal{M}$ by some approximation procedure. In this section, a general result in this direction, known as the *extension theorem*, is established. This is then applied to the construction of Lebesgue-Stieltjes measures on Euclidean spaces. For another application, see Chapter 6.

1.3.1 Caratheodory extension of measures

Definition 1.3.1: Let Ω be a nonempty set and let $\mathcal{P}(\Omega)$ be the power set of Ω. A class $\mathcal{C} \subset \mathcal{P}(\Omega)$ is called a *semialgebra* if (i) $A, B \in \mathcal{C} \Rightarrow A \cap B \in \mathcal{C}$ and (ii) for any $A \in \mathcal{C}$, there exist sets $B_1, B_2, \ldots, B_k \in \mathcal{C}$, for some $1 \leq k < \infty$, such that $B_i \cap B_j = \emptyset$ for $i \neq j$, and $A^c = \cup_{i=1}^k B_i$.

Example 1.3.1: $\Omega = \mathbb{R}$, $\mathcal{C} \equiv \{(a, b], (b, \infty) : -\infty \leq a, b < \infty\}$.

Example 1.3.2: $\Omega = \mathbb{R}$, $\mathcal{C} \equiv \{I : I \text{ is an interval}\}$. An interval I in $\mathbb{R}$ is a set in $\mathbb{R}$ such that $a, b \in I$, $a < b \Rightarrow (a, b) \subset I$.

Example 1.3.3: $\Omega = \mathbb{R}^k$, $\mathcal{C} \equiv \{I_1 \times I_2 \times \ \times I_k : I_j$ is an interval in $\mathbb{R}$ for $1 \leq j \leq k\}$.

Recall that a collection $\mathcal{A} \subset \mathcal{P}(\Omega)$ is an algebra if it is closed under finite union and complementation. It is easily verified (Problem 1.19) that the smallest algebra containing a semialgebra $\mathcal{C}$ is $\mathcal{A}(\mathcal{C}) \equiv \{A : A = \cup_{i=1}^k B_i, B_i \in \mathcal{C}$ for $i = 1, \ldots, k, k < \infty,\}$, i.e., the class of finite unions of sets from $\mathcal{C}$.

Definition 1.3.2: A set function μ on a semialgebra $\mathcal{C}$, taking values in $\bar{\mathbb{R}}_+ \equiv [0, \infty]$, is called a *measure* if (i) $\mu(\emptyset) = 0$ and (ii) for any sequence of sets $\{A_n\}_{n\geq 1} \subset \mathcal{C}$ with $\cup_{n\geq 1} A_n \in \mathcal{C}$, and $A_i \cap A_j = \emptyset$ for $i \neq j$, $\mu(\cup_{n\geq 1} A_n) = \sum_{n=1}^\infty \mu(A_n)$.

Proposition 1.3.1: *Let μ be a measure on a semialgebra $\mathcal{C}$. Let $\mathcal{A} \equiv \mathcal{A}(\mathcal{C})$ be the smallest algebra generated by $\mathcal{C}$. For each $A \in \mathcal{A}$, set*

$$\bar{\mu}(A) = \sum_{i=1}^k \mu(B_i),$$

if the set A has the representation $A = \cup_{i=1}^k B_i$ for some $B_1, \ldots, B_k \in \mathcal{C}, k < \infty$ with $B_i \cap B_j = \emptyset$ for $i \neq j$. Then,

(i) $\bar{\mu}$ is independent of the representation of A as $A = \cup_{i=1}^k B_i$;

(ii) $\bar{\mu}$ *is finitely additive on* $\mathcal{A}$*, i.e.,* $A, B \in \mathcal{A}, A \cap B = \emptyset \Rightarrow \bar{\mu}(A \cup B) = \bar{\mu}(A) + \bar{\mu}(B)$*; and*

(iii) $\bar{\mu}$ *is countably additive on* $\mathcal{A}$*, i.e., if* $A_n \in \mathcal{A}$ *for all* $n \geq 1$*,* $A_i \cap A_j = \emptyset$ *for all* $i = j$*, and* $\cup_{n\geq 1} A_n \in \mathcal{A}$*, then*

$$\bar{\mu}\Big(\bigcup_{n\geq 1} A_n\Big) = \sum_{n=1}^{\infty} \bar{\mu}(A_n).$$

Proof: Parts (i) and (ii) are easy to verify. Turning to part (iii), let each $n \geq 1$, $A_n = \cup_{j=1}^{k_n} B_{nj}$, $B_{nj} \in \mathcal{C}$, $\{B_{nj}\}_{j=1}^{k_n}$ disjoint. Since $\cup_{n\geq 1} A_n \in \mathcal{A}$ then exist disjoint sets $\{B_i\}_{i=1}^{k} \subset \mathcal{C}$ such that $\cup_{n\geq 1} A_n = \cup_{i=1}^{k} B_i$. Now

$$\begin{aligned} B_i &= B_i \cap \Big(\bigcup_{n\geq 1} A_n\Big) = \bigcup_{n\geq 1} (B_i \cap A_n) \\ &= \bigcup_{n\geq 1} \bigcup_{j=1}^{k_n} (B_i \cap B_{nj}). \end{aligned}$$

Since for all i, $B_i \in \mathcal{C}$, $B_i \cap B_{nj} \in \mathcal{C}$ for all j, n and μ is a measure on $\mathcal{C}$

$$\mu(B_i) = \sum_{n\geq 1} \sum_{j=1}^{k_n} \mu(B_i \cap B_{nj}).$$

Thus,

$$\begin{aligned} \bar{\mu}\Big(\bigcup_{n\geq 1} A_n\Big) &= \sum_{i=1}^{k} \mu(B_i) = \sum_{i=1}^{k} \sum_{n\geq 1} \Big(\sum_{j=1}^{k_n} \mu(B_i \cap B_{nj})\Big) \\ &= \sum_{n\geq 1} \Big(\sum_{i=1}^{k} \sum_{j=1}^{k_n} \mu(B_i \cap B_{nj})\Big) \\ &= \sum_{n\geq 1} \bar{\mu}(A_n), \end{aligned}$$

since

$$\begin{aligned} A_n &= A_n \cap \bigcup_{i=1}^{k} B_i \\ &= \bigcup_{i=1}^{k} \bigcup_{j=1}^{k_n} (B_i \cap B_{nj}). \end{aligned}$$

□

Thus, the set function $\bar{\mu}$ defined above is a measure on $\mathcal{A}$. To go beyond $\mathcal{A} = \mathcal{A}(\mathcal{C})$ to $\sigma\langle\mathcal{A}\rangle$, which is also the same as $\sigma\langle\mathcal{C}\rangle$ (see Problem 1.19 (ii)), the first step involves using the given set function μ on $\mathcal{C}$ to define a set function μ^* on the class of all subsets of Ω, i.e., on $\mathcal{P}(\Omega)$, where for any $A \in \mathcal{P}(\Omega)$, $\mu^*(A)$ is an estimate 'from above' of the μ-content of A.

Definition 1.3.3: Given a measure μ on a semialgebra $\mathcal{C}$, the *outer measure induced by* μ is the set function μ^*, defined on $\mathcal{P}(\Omega)$, as

$$\mu^*(A) \equiv \inf\left\{ \sum_{n=1}^{\infty} \mu(A_n) : \{A_n\}_{n\geq 1} \subset \mathcal{C},\ A \subset \cup_{n\geq 1} A_n \right\}. \tag{3.1}$$

Thus, a given set A is covered by countable unions of sets from $\mathcal{C}$ and the sums of the measures on such covers are computed and $\mu^*(A)$ is the greatest lower bound one can get in this way. It is not difficult to show that on $\mathcal{C}$ and $\mathcal{A}$, this is not an overestimate. That is, $\mu^* = \mu$ on $\mathcal{C}$ and on $\mathcal{A}$, $\mu^* = \bar{\mu}$. Now suppose $\mu(\Omega) < \infty$. Let $A \subset \Omega$ be a set such that the overestimates of the μ-contents of A and A^c, namely, $\mu^*(A)$ and $\mu^*(A^c)$, add up to $\mu(\Omega)$, i.e.,

$$\mu(\Omega) = \mu^*(\Omega) = \mu^*(A) + \mu^*(A^c). \tag{3.2}$$

Then, there is no room for error and the estimates $\mu^*(A)$ and $\mu^*(A^c)$ are in fact not overestimates at all. In this case, the set A may be considered 'exactly measurable' by this approximation procedure. This observation motivates the following definition.

Definition 1.3.4: A set A is said to be μ^*-*measurable* if

$$\mu^*(E) = \mu^*(E \cap A) + \mu^*(E \cap A^c) \text{ for all } E \subset \Omega. \tag{3.3}$$

In other words, an analog of (3.2) should hold in every portion E of Ω for a set A to be μ^*-measurable.

It can be shown (Problem 1.20) that μ^* defined in (3.1) satisfies:

$$\mu^*(\emptyset) = 0, \tag{3.4}$$

$$A \subset B \Rightarrow \mu^*(A) \leq \mu^*(B), \tag{3.5}$$

and for any $\{A_n\}_{n\geq 1} \subset \mathcal{P}(\Omega)$,

$$\mu^*\Big(\bigcup_{n\geq 1} A_n\Big) \leq \sum_{n=1}^{\infty} \mu^*(A_n). \tag{3.6}$$

Definition 1.3.5: Any set function $\mu^* : \mathcal{P}(\Omega) \to \bar{\mathbb{R}}_+ \equiv [0, \infty]$ satisfying (3.4)–(3.6) is called an *outer measure on* Ω.

The following result (due to C. Caratheodory) yields a measure space on Ω starting from a general outer measure μ^* that need not arise from a measure μ as in (3.1).

Theorem 1.3.2: *Let μ^* be an outer measure on Ω, i.e., it satisfies (3.4)–(3.6). Let $\mathcal{M} \equiv \mathcal{M}_{\mu^*} \equiv \{A : A$ is μ^*-measurable, i.e., A satisfies (3.3)$\}$. Then*

(i) *$\mathcal{M}$ is a σ-algebra,*

(ii) *μ^* restricted to $\mathcal{M}$ is a measure, and*

(iii) *$\mu^*(A) = 0 \Rightarrow \mathcal{P}(A) \subset \mathcal{M}$.*

Proof: From (3.3), it follows that $\emptyset \in \mathcal{M}$ and that $A \in \mathcal{M} \Rightarrow A^c \in \mathcal{M}$. Next, it will be shown that $\mathcal{M}$ is closed under finite unions. Let $A_1, A_2 \in \mathcal{M}$. Then, for any $E \subset \Omega$,

$$\begin{aligned} \mu^*(E) &= \mu^*(E \cap A_1) + \mu^*(E \cap A_1^c) \quad \text{(since } A_1 \in \mathcal{M}) \\ &= \mu^*(E \cap A_1 \cap A_2) + \mu^*(E \cap A_1 \cap A_2^c) \\ &\quad + \mu^*(E \cap A_1^c \cap A_2) + \mu^*(E \cap A_1^c \cap A_2^c) \quad \text{(since } A_2 \in \mathcal{M}). \end{aligned}$$

But $(A_1 \cap A_2) \cup (A_1 \cap A_2^c) \cup (A_1^c \cap A_2) = A_1 \cup A_2$. Since μ^* is subadditive, it follows that

$$\mu^*(E \cap (A_1 \cup A_2)) \leq \mu^*(E \cap A_1 \cap A_2) + \mu^*(E \cap A_1 \cap A_2^c) + \mu^*(E \cap A_1^c \cap A_2).$$

Thus

$$\mu^*(E) \geq \mu^*(E \cap (A_1 \cup A_2)) + \mu^*(E \cap (A_1 \cup A_2)^c).$$

The subadditivity of μ^* yields the opposite inequality and so, $A_1 \cup A_2 \in \mathcal{M}$ and hence, $\mathcal{M}$ is an algebra. To show that $\mathcal{M}$ is a σ-algebra, it suffices to show that $\mathcal{M}$ is closed under monotone unions, i.e., $A_n \in \mathcal{M}, A_n \subset A_{n+1}$ for all $n \geq 1 \Rightarrow A \equiv \cup_{n \geq 1} A_n \in \mathcal{M}$. Let $B_1 = A_1$ and $B_n = A_n \cap A_{n-1}^c$ for all $n \geq 2$. Then, for all $n \geq 1$, $B_n \in \mathcal{M}$ (since $\mathcal{M}$ is an algebra), $\cup_{j=1}^n B_j = A_n$, and $\cup_{j=1}^\infty B_j = A$. Hence, for any $E \subset \Omega$,

$$\begin{aligned} \mu^*(E) &= \mu^*(E \cap A_n) + \mu^*(E \cap A_n^c) \\ &= \mu^*(E \cap A_n \cap B_n) + \mu^*(E \cap A_n \cap B_n^c) + \mu^*(E \cap A_n^c) \\ &\qquad\qquad \text{(since } B_n \in \mathcal{M}) \\ &= \mu^*(E \cap B_n) + \mu^*(E \cap A_{n-1}) + \mu^*(E \cap A_n^c) \\ &= \sum_{j=1}^n \mu^*(E \cap B_j) + \mu^*(E \cap A_n^c) \quad \text{(by iteration)} \\ &\geq \sum_{j=1}^n \mu^*(E \cap B_j) + \mu^*(E \cap A^c) \quad \text{(by monotonicity).} \end{aligned}$$

Now letting $n \to \infty$, and using the subadditivity of μ^* and the fact that $\cup_{j=1}^\infty B_j = A$, one gets

$$\mu^*(E) \geq \sum_{j=1}^\infty \mu^*(E \cap B_j) + \mu^*(E \cap A^c)$$

$$\geq \quad \mu^*(E \cap A) + \mu^*(E \cap A^c).$$

This completes the proof of part (i).

To prove part (ii), let $\{B_n\}_{n\geq 1} \subset \mathcal{M}$ and $B_i \cap B_j = \emptyset$ for $i \neq j$. Let $A_j = \cup_{i=j}^{\infty} B_i$, $j \in \mathbb{N}$. Then, by (i), $A_j \in \mathcal{M}$ for all $j \in \mathbb{N}$ and so

$$\begin{aligned}
\mu^*(A_1) &= \mu^*(A_1 \cap B_1) + \mu^*(A_1 \cap B_1^c) \quad (\text{since } B_1 \in \mathcal{M}) \\
&= \mu^*(B_1) + \mu^*(A_2) \\
&= \mu^*(B_1) + \mu^*(B_2) + \mu^*(A_3) \quad (\text{by iteration}) \\
&= \sum_{i=1}^{n} \mu^*(B_i) + \mu^*(A_{n+1}) \quad (\text{by iteration}) \\
&\geq \sum_{i=1}^{n} \mu^*(B_i) \quad \text{for all} \quad n \geq 1.
\end{aligned}$$

Now letting $n \to \infty$, one has $\mu^*(A_1) \geq \sum_{i=1}^{\infty} \mu^*(B_i)$. By subadditivity of μ^*, the opposite inequality holds and so

$$\mu^*(A_1) = \mu^*\left(\bigcup_{i=1}^{\infty} B_i\right) = \sum_{i=1}^{\infty} \mu^*(B_i),$$

proving (ii).

As for (iii), note that by monotonicity, $\mu^*(A) = 0$, $B \subset A \Rightarrow \mu^*(B) = 0$ and hence, for any E, $\mu^*(E \cap B) = 0$. Since $\mu^*(E) \geq \mu^*(E \cap B^c)$, this implies $\mu^*(E) \geq \mu^*(E \cap B^c) + \mu^*(E \cap B)$. The opposite inequality holds by the subadditivity of μ^*. So $B \in \mathcal{M}$, and (iii) is proved. □

Definition 1.3.6: A measure space $(\Omega, \mathcal{F}, \nu)$ is called *complete* if for any $A \in \mathcal{F}$ with $\nu(A) = 0 \Rightarrow \mathcal{P}(A) \subset \mathcal{F}$.

Thus, by part (iii) of the above theorem, $(\Omega, \mathcal{M}_{\mu^*}, \mu^*)$ is a *complete measure space*. Now the above theorem is applied to a μ^* that is generated by a given measure μ on a semialgebra $\mathcal{C}$ via (3.1).

Theorem 1.3.3: (*Caratheodory's extension theorem*). *Let μ be a measure on a semialgebra $\mathcal{C}$ and let μ^* be the set function induced by μ as defined by (3.1). Then,*

(i) μ^* is an outer measure,

(ii) $\mathcal{C} \subset \mathcal{M}_{\mu^*}$, and

(iii) $\mu^* = \mu$ on $\mathcal{C}$, where $\mathcal{M}_{\mu^*}$ is as in Theorem 1.3.2.

Proof: The proof of (i) involves verifying (3.4)–(3.6) which is left as an exercise (Problem 1.20). To prove (ii), let $A \in \mathcal{C}$. Let $E \subset \Omega$ and $\{A_n\}_{n\geq 1} \subset$

$\mathcal{C}$ be such that $E \subset \cup_{n\geq 1} A_n$. Then, for all $i \in \mathbb{N}$, $A_i = (A_i \cap A) \cup (A_i \cap B_1) \cup \ldots \cup (A_i \cap B_k)$ where $B_1, \ldots, B_k$ are disjoint sets in $\mathcal{C}$ such that $\cup_{j=1}^k B_j = A^c$. Since μ is finitely additive on $\mathcal{C}$,

$$\begin{aligned} \mu(A_i) &= \mu(A_i \cap A) + \sum_{j=1}^{k} \mu(A_i \cap B_j) \\ \Rightarrow \quad \sum_{n=1}^{\infty} \mu(A_n) &= \sum_{n=1}^{\infty} \mu(A_n \cap A) + \sum_{n=1}^{\infty} \sum_{j=1}^{k} \mu(A_n \cap B_j) \\ &\geq \mu^*(E \cap A) + \mu^*(E \cap A^c), \end{aligned}$$

since $\{A_n \cap A\}_{n\geq 1}$ and $\{A_n \cap B_j : 1 \leq j \leq k, n \geq 1\}$ are both countable subcollections of $\mathcal{C}$ whose unions cover $E \cap A$ and $E \cap A^c$, respectively. From the definition of $\mu^*(E)$, it now follows that

$$\mu^*(E) \geq \mu^*(E \cap A) + \mu^*(E \cap A^c).$$

Now the subadditivity of μ^* completes the proof of part (ii).

To prove (iii), let $A \in \mathcal{C}$. Then, by definition, $\mu^*(A) \leq \mu(A)$. If $\mu^*(A) = \infty$, then $\mu(A) = \infty = \mu^*(A)$. If $\mu^*(A) < \infty$, then by the definition of 'infimum', for any $\epsilon > 0$, there exists $\{A_n\}_{n\geq 1} \subset \mathcal{C}$ such that $A \subset \cup_{n\geq 1} A_n$ and

$$\mu^*(A) \leq \sum_{n=1}^{\infty} \mu(A_n) \leq \mu^*(A) + \epsilon.$$

But $A = A \cap (\cup_{n\geq 1} A_n) = \cup_{n\geq 1}(A \cap A_n)$. Note that the set function $\bar{\mu}$ defined in Proposition 1.3.1 is a measure on $\mathcal{A}(\mathcal{C})$ and it coincides with μ on $\mathcal{C}$. Since $A, A \cap A_n \in \mathcal{C}$ for all $n \geq 1$, by Proposition 1.2.3 (b) applied to $\bar{\mu}$,

$$\mu(A) = \bar{\mu}(A) \leq \sum_{n=1}^{\infty} \bar{\mu}(A \cap A_n) \leq \sum_{n=1}^{\infty} \bar{\mu}(A_n) = \sum_{n=1}^{\infty} \mu(A_n) \leq \mu^*(A) + \epsilon.$$

(Alternately, w.l.o.g., assume that $\{A_n\}_{n\geq 1}$ are disjoint. Then $A = A \cap \cup_{n\geq 1} A_n = \cup_{n\geq 1}(A \cap A_n)$. Since $A \cap A_n \in \mathcal{C}$ for all n, by the countable additivity of μ on $\mathcal{C}$, $\mu(A) = \sum_{n=1}^{\infty} \mu(A \cap A_n) \leq \sum_{n=1}^{\infty} \mu(A_n) = \mu^*(A) + \epsilon$.) Since $\epsilon > 0$ is arbitrary, this yields $\mu(A) \leq \mu^*(A)$. $\square$

Thus, given a measure μ on a semialgebra $\mathcal{C} \subset \mathcal{P}(\Omega)$, there is a complete measure space $(\Omega, \mathcal{M}_{\mu^*}, \mu^*)$ such that $\mathcal{M}_{\mu^*} \supset \mathcal{C}$ and μ^* restricted to $\mathcal{C}$ equals μ. For this reason, μ^* is called an *extension* of μ. The measure space $(\Omega, \mathcal{M}_{\mu^*}, \mu^*)$ is called the *Caratheodory extension* of μ. Since $\mathcal{M}_{\mu^*}$ is a σ-algebra and contains $\mathcal{C}$, $\mathcal{M}_{\mu^*}$ must contain $\sigma\langle\mathcal{C}\rangle$, the σ-algebra generated by $\mathcal{C}$, and thus, $(\Omega, \sigma\langle\mathcal{C}\rangle, \mu^*)$ is also a measure space. However, the latter may *not* be *complete* (see Section 1.4).

Now the above method is applied to the construction of Lebesgue-Stieltjes measures on $\mathbb{R}$ and $\mathbb{R}^2$.

1.3.2 Lebesgue-Stieltjes measures on $\mathbb{R}$

Let $F : \mathbb{R} \to \mathbb{R}$ be nondecreasing. For $x \in \mathbb{R}$, let $F(x+) \equiv \lim_{y \downarrow x} F(y)$, and $F(x-) \equiv \lim_{y \uparrow x} F(y)$. Set $F(\infty) = \lim_{x \uparrow \infty} F(x)$ and $F(-\infty) = \lim_{x \downarrow -\infty} F(y)$. Let

$$\mathcal{C} \equiv \Big\{(a, b] : -\infty \leq a \leq b < \infty\Big\} \cup \Big\{(a, \infty) : -\infty \leq a < \infty\Big\}. \tag{3.7}$$

Define

$$\begin{aligned} \mu_F((a, b]) &= F(b+) - F(a+), \\ \mu_F((a, \infty)) &= F(\infty) - F(a+). \end{aligned} \tag{3.8}$$

Then, it may be verified that

(i) $\mathcal{C}$ is a semialgebra;

(ii) μ_F is a measure on $\mathcal{C}$. (For (ii), one needs to use the Heine-Borel theorem. See Problems 1.22 and 1.23.)

Let $(\mathbb{R}, \mathcal{M}_{\mu_F^*}, \mu_F^*)$ be the Caratheodory extension of μ_F, i.e., the measure space constructed as in the above two theorems.

Definition 1.3.7: Let $F : \mathbb{R} \to \mathbb{R}$ be nondecreasing. The (measure) space $(\mathbb{R}, \mathcal{M}_{\mu_F^*}, \mu_F^*)$ is called a *Lebesgue-Stieltjes measure space* and μ_F^* is the *Lebesgue-Stieltjes measure* generated by F.

Since $\sigma\langle\mathcal{C}\rangle = \mathcal{B}(\mathbb{R})$, the class of all Borel sets of $\mathbb{R}$, every Lebesgue-Stieltjes measure μ_F^* is also a measure on $(\mathbb{R}, \mathcal{B}(\mathbb{R}))$. Note also that μ_F^* is finite on bounded intervals.

Conversely, given any *Radon measure* μ on $(\mathbb{R}, \mathcal{B}(\mathbb{R}))$, i.e., a measure on $(\mathbb{R}, \mathcal{B}(\mathbb{R}))$ that is finite on bounded intervals, set

$$F(x) = \begin{cases} \mu((0, x]) & \text{if } x > 0 \\ 0 & \text{if } x = 0 \\ -\mu((x, 0]) & \text{if } x \leq 0. \end{cases}$$

Then $\mu_F = \mu$ on $\mathcal{C}$. By the uniqueness of the extension (discussed later in this section, see also Theorem 1.2.4), it follows that μ_F^* coincides with μ on $\mathcal{B}(\mathbb{R})$. Thus, every Radon measure on $(\mathbb{R}, \mathcal{B}(\mathbb{R}))$ is necessarily a Lebesgue-Stieltjes measure.

Lebesgue Measure on $\mathbb{R}$

Definition 1.3.8: When $F(x) \equiv x, x \in \mathbb{R}$, the measure μ_F^* is called the *Lebesgue measure* and the σ-algebra $\mathcal{M}_{\mu_F^*}$ is called the class of *Lebesgue measurable sets.*

The Lebesgue measure will be denoted by $m(\cdot)$ or $\mu_L(\cdot)$. Given below are some important results on $m(\cdot)$.

(i) It follows from equation (3.1), that $\mu_F^*(\{x\}) = F(x+) - F(x-)$ and hence $= 0$ if F is continuous at x. Thus $m(\{x\}) \equiv 0$ on $\mathbb{R}$.

(ii) By countable additivity of $m(\cdot)$, $m(A) = 0$ for any countable set A.

(iii) (*Cantor set*). There exists an uncountable set C such that $m(C) = 0$. An example is the *Cantor set* constructed as follows: Start with $I_0 = [0, 1]$. Delete the open middle third, i.e., $\left(\frac{1}{3}, \frac{2}{3}\right)$. Next from the closed intervals $I_{11} = [0, \frac{1}{3}]$ and $I_{12} = [\frac{2}{3}, 1]$ delete the open middle thirds, i.e., $\left(\frac{1}{9}, \frac{2}{9}\right)$ and $\left(\frac{7}{9}, \frac{2}{9}\right)$ respectively. Repeat this process of deleting the middle third from each of the remaining closed intervals. Thus at stage n there will be 2^{n-1} new closed intervals and 2^{n-1} deleted open intervals, each of length $\frac{1}{3^n}$. Let U_n denote the union of all the deleted open intervals at the nth stage. Then $\{U_n\}_{n \geq 1}$ are disjoint open sets. Let $U \equiv \cup_{n \geq 1} U_n$. By countable additivity

$$m(U) = \sum_{n=1}^{\infty} m(U_n) = \sum_{n=1}^{\infty} \frac{2^{n-1}}{3^n} = 1.$$

Let $C \equiv [0, 1] - U$. Since U is open and [0,1] is closed, C is nonempty. It can be shown that $C \equiv \{x : x = \sum_1^{\infty} \frac{a_i}{3^i}, a_i = 0 \text{ or } 2\}$ (Problem 1.33). Thus C can be mapped in (1–1) manner on to the set of all sequences $\{\delta_i\}_{i \geq 1}$ such that $\delta_i = 0$ or 2. But this set is uncountable. Since $m([0, 1]) = 1$, it follows that $m(C) = 0$. For more properties of the Cantor set, see Rudin (1976) and Chapter 4.

(iv) $m(\cdot)$ is invariant under reflection and translation. That is, for any E in $\mathcal{B}(\mathbb{R})$,

$$m(-E) = m(E) \quad \text{and} \quad m(E + c) = m(E)$$

for all c in $\mathbb{R}$ where $-E = \{-x : x \in E\}$ and $E + c = \{y : y = x + c, x \in E\}$. This follows from Theorem 1.2.4 and the fact that the claim holds for intervals (cf. Problem 2.48).

(v) There exists a subset $A \subset \mathbb{R}$ such that $A \notin \mathcal{M}_m$. That is, there exists a *non-Lebesgue measurable set*. The proof of this requires the use of the axiom of choice (cf. A.1). For a proof see Royden (1988).

1.3.3 Lebesgue-Stieltjes measures on $\mathbb{R}^2$

Let $F : \mathbb{R}^2 \to \mathbb{R}$ satisfy

$$F(a_2, b_2) - F(a_2, b_1) - F(a_1, b_2) + F(a_1, b_1) \geq 0, \tag{3.9}$$

and

$$F(a_2, b_2) - F(a_1, b_1) \geq 0, \tag{3.10}$$

for all $a_1 \le a_2$, $b_1 \le b_2$. Extend F to $\bar{\mathbb{R}}^2$ by appropriate limiting procedure. Let

$$\mathcal{C}_2 \equiv \{I_1 \times I_2 : I_1, I_2 \in \mathcal{C}\}, \tag{3.11}$$

where $\mathcal{C}$ is as in (3.7). Next, for $I_1 = (a_1, a_2]$, $I_2 = (b_1, b_2]$, $-\infty < a_1, a_2, b_1, b_2 < \infty$, set

$$\mu_F(I_1 \times I_2) \equiv F(a_2+, b_2+) - F(a_2+, b_1+) - (F(a_1+, b_2+) + F(a_1+, b_1+)), \tag{3.12}$$

where for any $a, b \in \mathbb{R}$, $F(a+, b+)$ is defined as

$$F(a+, b+) \equiv \lim_{a' \downarrow a, b' \downarrow b} F(a', b').$$

Note that by (3.10), the limit exists and hence, $F(a+, b+)$ is well defined. Further, by (3.9), the right side of (3.12) is nonnegative. Next extend the definition of μ_F to unbounded sets in $\mathcal{C}_2$ by the limiting procedure:

$$\mu_F(I_1 \times I_2) = \lim_{n \to \infty} \mu_F\big((I_1 \times I_2) \cap J_n\big), \tag{3.13}$$

where $J_n = (-n, n] \times (-n, n]$. Then it may be verified (Problems 1.24, 1.25) that

(i) $\mathcal{C}_2$ is a semialgebra

(ii) μ_F is a measure on $\mathcal{C}_2$.

Let $(\mathbb{R}^2, \mathcal{M}_{\mu_F^*}, \mu_F^*)$ be the measure space constructed as in the above two theorems. The measure μ_F^* is called the *Lebesgue-Stieltjes measure* generated by F and $\mathcal{M}_{\mu_F^*}$ a *Lebesgue-Stieltjes σ-algebra.* Again, in this case, $\mathcal{M}_{\mu^*}$ includes the σ-algebra $\sigma\langle\mathcal{C}\rangle \equiv \mathcal{B}(\mathbb{R}^2)$ and so $(\mathbb{R}^2, \mathcal{B}(\mathbb{R}^2), \mu_F^*)$ is also a measure space. If $F(a, b) = ab$, then μ_F is called the *Lebesgue measure on* $\mathbb{R}^2$.

A similar construction holds for any $\mathbb{R}^k$, $k < \infty$.

1.3.4 More on extension of measures

Next the uniqueness of the extension μ^* and approximation of the μ^*-measure of a set in $\mathcal{M}_{\mu^*}$ by that of a set from the algebra $\mathcal{A} \equiv \mathcal{A}(\mathcal{C})$ are considered. As in the case of measures defined on an algebra, a measure μ on a semialgebra $\mathcal{C} \subset \mathcal{P}(\Omega)$ is said to be *σ-finite* if there exists a *countable* collection $\{A_n\}_{n \ge 1} \subset \mathcal{C}$ such that (i) $\mu(A_n) < \infty$ for each $n \ge 1$ and (ii) $\cup_{n \ge 1} A_n = \Omega$. The following approximation result holds.

Theorem 1.3.4: *Let $A \in \mathcal{M}_{\mu^*}$ and $\mu^*(A) < \infty$. Then, for each $\epsilon > 0$, there exist $B_1, B_2, \ldots, B_k \in \mathcal{C}$, $k < \infty$ with $B_i \cap B_j = \emptyset$ for $1 \le i \ne j \le k$, such that*

$$\mu^*\left(A \,\triangle\, \bigcup_{j=1}^{k} B_j\right) < \epsilon,$$

where for any two sets E_1 and E_2, $E_1 \triangle E_2$ is the symmetric difference of E_1 and E_2, defined by $E_1 \triangle E_2 \equiv (E_1 \cap E_2^c) \cup (E_1^c \cap E_2)$.

Proof: By definition of μ^*, $\mu^*(A) < \infty$ implies that for every $\epsilon > 0$, there exist $\{B_n\}_{n\geq 1} \subset \mathcal{C}$ such that $A \subset \cup_{n\geq 1} B_n$ and

$$\mu^*(A) \leq \sum_{n=1}^{\infty} \mu(B_n) \leq \mu^*(A) + \epsilon/2 < \infty.$$

Since $B_n \in \mathcal{C}$, B_n^c is a finite union of disjoint sets from $\mathcal{C}$. W.l.o.g., it can be assumed that $\{B_n\}_{n\geq 1}$ are disjoint. (Otherwise, one can consider the sequence $B_1, B_2 \cap B_1^c, B_3 \cap B_2^c \cap B_1^c, \ldots$.) Next, $\sum_{n=1}^{\infty} \mu(B_n) < \infty$ implies that for every $\epsilon > 0$, there exists $k \in \mathbb{N}$ such that $\sum_{n=k+1}^{\infty} \mu(B_n) < \frac{\epsilon}{2}$. Since both A and $\cup_{j=1}^{k} B_j$ are subsets of $\cup_{n\geq 1} B_n$,

$$\mu^*\left(A \triangle \left[\bigcup_{j=1}^{k} B_j\right]\right) \leq \mu^*\left(A^c \cap \left[\bigcup_{j=1}^{\infty} B_j\right]\right) + \mu^*\left(\bigcup_{j=k+1}^{\infty} B_j\right).$$

But since μ^* is a measure on $(\Omega, \mathcal{M}_{\mu^*})$, $\mu^*(\cup_{n\geq 1} B_n) = \mu^*(A) + \mu^*((\cup_{n\geq 1} B_n) \cap A^c)$. Further, since $\mu^*(A) < \infty$,

$$\begin{aligned}
&\mu^*\left(\left[\bigcup_{n\geq 1} B_n\right] \cap A^c\right) \\
&= \mu^*\left(\bigcup_{n\geq 1} B_n\right) - \mu^*(A)) \\
&\leq \sum_{n=1}^{\infty} \mu^*(B_n) - \mu^*(A), \quad \text{(since } \mu^* \text{ is countably subadditive)} \\
&= \sum_{n=1}^{\infty} \mu(B_n) - \mu^*(A), \quad \text{(since } \mu^* = \mu \text{ on } \mathcal{C}) \\
&< \frac{\epsilon}{2} \quad \text{(by the choice of } \{B_n\}_{n\geq 1}).
\end{aligned}$$

Also, by the definition of k,

$$\mu^*\left(\bigcup_{j=k+1}^{\infty} B_j\right) \leq \sum_{j=k+1}^{\infty} \mu^*(B_j) = \sum_{j=k+1}^{\infty} \mu(B_j) < \frac{\epsilon}{2}.$$

Thus, $\mu^*(A \triangle [\cup_{j=1}^{k} B_j]) < \epsilon$. This completes the proof of the theorem. □

Thus, every μ^*-measurable set of finite measure is nearly a finite union of disjoint elements from the semialgebra $\mathcal{C}$. This was enunciated by J.E. Littlewood as the first of his three principles of approximation (the other two being: every Lebesgue measurable function is nearly continuous (cf. Theorem 2.5.12) and every almost everywhere convergent sequence of functions on a finite measure space is nearly uniformly convergent (Egorov's theorem) cf. Theorem 2.5.11).

One may strengthen Theorem 1.3.4 to prove the following result on regularity of *Radon* measures on $\big(\mathbb{R}^k, \mathcal{B}(\mathbb{R}^k)\big)$ (cf. Problem 1.32). See also Rudin (1987), Chapter 2.

Corollary 1.3.5: *(Regularity of measures). Let μ be a Radon measure on $(\mathbb{R}^k, \mathcal{B}(\mathbb{R}^k))$ for some $k \in \mathbb{N}$, i.e., $\mu(A) < \infty$ for all bounded sets $A \in \mathcal{B}(\mathbb{R}^k)$. Let $A \in \mathcal{B}(\mathbb{R}^k)$ be such that $\mu(A) < \infty$. Then, for each $\epsilon > 0$, there exist a compact set K and an open set G such that $K \subset A \subset G$ and $\mu(G \setminus K) < \epsilon$.*

The following uniqueness result can be established by an application of the above approximation theorem (Theorem 1.3.4) or by applying the π-λ theorem (Corollary 1.1.3), as in Theorem 1.2.4 (Problem 1.26).

Theorem 1.3.6: *(Uniqueness). Let μ be a σ-finite measure on a semi-algebra $\mathcal{C}$. Let ν be a measure on the measurable space $(\Omega, \sigma\langle\mathcal{C}\rangle)$ such that $\nu = \mu$ on $\mathcal{C}$. Then $\nu = \mu^*$ on $\sigma\langle\mathcal{C}\rangle$.*

An application of Theorem 1.3.4 yields an useful approximation result known as Lusin's theorem (see Theorem 2.1.3 in Section 2) for approximating Borel measurable functions by continuous functions.

1.4 Completeness of measures

Recall from Definition 1.3.6 that a measure space $(\Omega, \mathcal{F}, \mu)$ is called *complete* if for any $A \in \mathcal{F}$, $\mu(A) = 0$, $B \subset A \;\Rightarrow B \in \mathcal{F}$. That is, for any set A in $\mathcal{F}$ whose μ measure is zero, all subsets of a set A are also in $\mathcal{F}$. For example, the very construction of the Lebesgue-Stieltjes measure for a nondecreasing F on $\mathbb{R}$, discussed in Section 1.3, yields a complete measure space $(\mathbb{R}, \mathcal{M}_{\mu_F^*}, \mu_F^*)$. The Borel σ-algebra $\mathcal{B}(\mathbb{R})$ is a sub-σ-algebra of $\mathcal{M}_{\mu_F^*}$ and $(\mathbb{R}, \mathcal{B}(\mathbb{R}), \mu_F^*)$ need not be complete. For example, if μ_F is the Lebesgue measure, then the Cantor set C (Section 1.3.2) has measure 0 and hence $\mathcal{M} \equiv$ the Lebesgue σ-algebra, contains the power set of C and hence, has cardinality larger than that of $\mathbb{R}$. It can be shown that the cardinality of $\mathcal{B}(\mathbb{R})$ is the same as that of $\mathbb{R}$ (see Hewitt and Stromberg (1965)). For another example, if F is a degenerate distribution at 0, i.e., $F(x) = 0$ for $x < 0$ and $F(x) = 1$ for $x \geq 0$, then $\mathcal{M}_{\mu_F^*} = \mathcal{P}(\mathbb{R})$, the power set of $\mathbb{R}$ (Problem 1.28), and hence, $(\mathbb{R}, \mathcal{B}(\mathbb{R}), \mu_F)$ is not complete. The same is true for any discrete distribution function. However, it is always possible to complete an incomplete measure space $(\Omega, \mathcal{F}, \mu)$ by adding new sets to $\mathcal{F}$. This procedure is discussed below.

Theorem 1.4.1: *Let $(\Omega, \mathcal{F}, \mu)$ be a measure space. Let $\tilde{\mathcal{F}} \equiv \{A : B_1 \subset A \subset B_2$ for some $B_1, B_2 \in \mathcal{F}$ satisfying $\mu(B_2 \setminus B_1) = 0\}$. For any $A \in \tilde{\mathcal{F}}$, set*

$\tilde{\mu}(A) = \mu(B_1) = \mu(B_2)$ for any pair of sets $B_1, B_2 \in \mathcal{F}$ with $B_1 \subset A \subset B_2$ and $\mu(B_2 \setminus B_1) = 0$. Then

(i) $\tilde{\mathcal{F}}$ is a σ-algebra and $\mathcal{F} \subset \tilde{\mathcal{F}}$,

(ii) $\tilde{\mu}$ is well defined,

(iii) $(\Omega, \tilde{\mathcal{F}}, \tilde{\mu})$ is a complete measure space and $\tilde{\mu} = \mu$ on $\mathcal{F}$.

Proof:

(i) Since $A \in \tilde{\mathcal{F}}$, there exist $B_1, B_2 \in \mathcal{F}$, $B_1 \subset A \subset B_2$ and $\mu(B_2 \setminus B_1) = 0$. Clearly, $B_2^c \subset A^c \subset B_1^c$, and $B_1^c, B_2^c \in \mathcal{F}$ and $\mu(B_1^c \setminus B_2^c) = \mu(B_2 \setminus B_1) = 0$ and so $A^c \in \tilde{\mathcal{F}}$. Next, let $\{A_n\}_{n=1}^\infty \subset \tilde{\mathcal{F}}$ and $A = \cup_{n \geq 1} A_n$. Then, for each n there exist B_{1n} and B_{2n} in $\mathcal{F}$ such that $B_{1n} \subset A_n \subset B_{2n}$ and $\mu(B_{2n} \setminus B_{1n}) = 0$. Let $B_1 = \cup_{n \geq 1} B_{1n}$ and $B_2 = \cup_{n \geq 1} B_{2n}$. Then $B_1 \subset A \subset B_2$, $B_1, B_2 \in \mathcal{F}$ and $B_2 \setminus B_1 \subset \cup_{n \geq 1}(B_{2n} \setminus B_{1n})$ and hence, $\mu(B_2 \setminus B_1) \leq \sum_{n=1}^\infty \mu(B_{2n} \setminus B_{1n}) = 0$. Thus, $A \in \tilde{\mathcal{F}}$ and hence $\tilde{\mathcal{F}}$ is a σ-algebra. Clearly, $\mathcal{F} \subset \tilde{\mathcal{F}}$ since for every $A \in \mathcal{F}$, one may take $B_1 = B_2 = A$.

(ii) Let $B_1 \subset A \subset B_2$, $B_1' \subset A \subset B_2'$, $B_1, B_1', B_2, B_2' \in \mathcal{F}$ and $\mu(B_2 \setminus B_1) = 0 = \mu(B_2' \setminus B_1')$. Then $B_1 \cup B_1' \subset A \subset B_2 \cap B_2'$ and $(B_2 \cap B_2') \setminus (B_1 \cup B_1') = (B_2 \cap B_2') \cap (B_1^c \cap B_1'^c) \subset B_2 \cap B_1^c$. Thus

$$\mu([B_2 \cap B_2'] \setminus [B_1 \cup B_1']) = 0.$$

Hence, $\mu(B_2) = \mu(B_1) + \mu(B_2 \setminus B_1) = \mu(B_1) \leq \mu(B_1 \cup B_1') = \mu(B_2 \cap B_2') \leq \mu(B_2')$. By symmetry $\mu(B_2') \leq \mu(B_2)$ and so $\mu(B_2) = \mu(B_2')$. But $\mu(B_2) = \mu(B_1)$ and $\mu(B_2') = \mu(B_1')$ and also all four quantities agree.

(iii) It remains to show that $\tilde{\mu}$ is countably additive and complete on $\tilde{\mathcal{F}}$. Let $\{A_n\}_{n=1}^\infty$ be a disjoint sequence of sets from $\tilde{\mathcal{F}}$ and let $A = \cup_{n \geq 1} A_n$. Let $\{B_{1n}\}_{n \geq 1}, \{B_{2n}\}_{n \geq 1}, B_1, B_2$ be as in the proof of (i). Then, the fact that $\{A_n\}_{n=1}^\infty$ are disjoint implies $\{B_{1n}\}_{n=1}^\infty$ are also disjoint. And since $B_1 = \cup_{n \geq 1} B_{1n}$ and μ is a measure on $(\Omega, \mathcal{F})$,

$$\mu(B_1) \equiv \sum_{n=1}^\infty (B_{1n}).$$

Also, by definition of B_{1n}'s, $\mu(B_{1n}) = \tilde{\mu}(A_n)$ for all $n \geq 1$, and by (i), $\tilde{\mu}(A) = \mu(B_1)$. Thus,

$$\tilde{\mu}(A) = \mu(B_1) = \sum_{n=1}^\infty (B_{1n}) = \sum_{n=1}^\infty \tilde{\mu}(A_n),$$

establishing the countable additivity of $\tilde{\mu}$.

Next, suppose that $A \in \tilde{\mathcal{F}}$ and $\tilde{\mu}(A) = 0$. Then there exist $B_1, B_2 \in \mathcal{F}$ such that $B_1 \subset A \subset B_2$ and $\mu(B_2 \setminus B_1) = 0$. Further, by definition of $\tilde{\mu}$, $\mu(B_2) = \tilde{\mu}(A) = 0$. If $D \subset A$, then $\emptyset \subset D \subset B_2$ and $\mu(B_2 \setminus \emptyset) = 0$. Therefore, $D \in \tilde{\mathcal{F}}$ and hence, $(\Omega, \tilde{\mathcal{F}}, \tilde{\mu})$ is complete.

Finally, if $A \in \mathcal{F}$, then take $B_1 = B_2 = A$ and so, $\tilde{\mu}(A) = \mu(B_1) = \mu(A)$, and thus, $\tilde{\mu} = \mu$ on $\mathcal{F}$. Hence, the proof of the theorem is complete. □

1.5 Problems

1.1 Let Ω be a nonempty set. Show that $\mathcal{F} \equiv \{\Omega, \emptyset\}$ and $\mathcal{G} = \mathcal{P}(\Omega) \equiv \{A : A \subset \Omega\}$ are both σ-algebras.

1.2 Let Ω be a finite set, i.e., the number of elements in Ω is finite. Let $\mathcal{F} \subset \mathcal{P}(\Omega)$ be an algebra. Show that $\mathcal{F}$ is a σ-algebra.

1.3 Show that $\mathcal{F}_6$ in Example 1.1.4 is a σ-algebra.

1.4 Let $\{\mathcal{F}_\theta : \theta \in \Theta\}$ be a family of σ-algebras on Ω. Show that $\mathcal{G} \equiv \cap_{\theta \in \Theta} \mathcal{F}_\theta$ is also a σ-algebra.

1.5 Let $\Omega = \{1, 2, 3\}$, $\mathcal{F}_1 = \{\{1\}, \{2, 3\}, \Omega, \emptyset\}$, $\mathcal{F}_2 = \{\{1, 2\}, \{3\}, \Omega, \emptyset\}$. Verify that $\mathcal{F}_1$ and $\mathcal{F}_2$ are both algebras (in fact, σ-algebras) but $\mathcal{F}_1 \cup \mathcal{F}_2$ is not an algebra.

1.6 Let Ω be a nonempty set and let $\mathcal{A} \equiv \{A_i : i \in \mathbb{N}\}$ be a *partition* of Ω, i.e., $A_i \cap A_j = \emptyset$ for all $i \neq j$ and $\cup_{n \geq 1} A_n = \Omega$. Let $\mathcal{F} = \{\cup_{i \in J} A_i : J \subset \mathbb{N}\}$ where, for $J = \emptyset$, $\cup_{i \in J} A_i \equiv \emptyset$. Show that $\mathcal{F}$ is a σ-algebra .

1.7 Let Ω be a nonempty set and let $\mathcal{B} \equiv \{B_i : 1 \leq i \leq k < \infty\} \subset \mathcal{P}(\Omega)$, $\mathcal{B}$ not necessarily a partition. Find $\sigma\langle\mathcal{B}\rangle$.

(**Hint:** For each $\delta = (\delta_1, \delta_2, \ldots, \delta_k)$, $\delta_i \in \{0, 1\}$ let $B_\delta = \cap_{i=1}^k B_i(\delta_i)$, where $B_i(0) = B_i^c$ and $B_i(1) = B_i$, $i \geq 1$. Show that $\sigma(\mathcal{B}) = \{E : E = \cup_{\delta \in J} B_\delta,\ J \subset \{1, 2, \ldots, k\}\}$.)

1.8 Let $\Omega \equiv \{1, 2, \ldots\} = \mathbb{N}$ and $A_i \equiv \{j : j \in \mathbb{N}, j \geq i\}, i \in \mathbb{N}$. Show that $\sigma\langle\mathcal{A}\rangle = \mathcal{P}(\Omega)$ where $\mathcal{A} = \{A_i : i \in \mathbb{N}\}$.

1.9 (a) Show that every open set $A \subset \mathbb{R}$ is a countable union of open intervals.

(**Hint:** Use the fact that the set $\mathcal{Q}$ of all rational numbers is dense in $\mathbb{R}$.)

(b) Extend the above to $\mathbb{R}^k$ for any $k \in \mathbb{N}$.

(c) Strengthen (a) to assert that the open intervals in (a) can be chosen to be disjoint.

1.10 Show that in Example 1.1.6, $\mathcal{O}_j \subset \sigma\langle\mathcal{O}_i\rangle$ for all $1 \le i, j \le 4$.

1.11 For $k \in \mathbb{N}$, let $\mathcal{O}_5 \equiv \{(a_1, b_1] \times \ldots \times (a_k, b_k] : -\infty < a_i < b_i < \infty, 1 \le i \le k\}$ and $\mathcal{O}_6 \equiv \{[a_1, b_1] \times \ldots \times [a_k, b_k] : -\infty < a_i < b_i < \infty, 1 \le i \le k\}$. Show that $\sigma\langle\mathcal{O}_5\rangle = \sigma\langle\mathcal{O}_6\rangle = \mathcal{B}(\mathbb{R}^k)$.

1.12 Let $\mathcal{O}_S \equiv \{\{x\} : x \in \mathbb{R}^k\}$ be the class of all singletons in $\mathbb{R}^k$, $k \in \mathbb{N}$. Show that $\sigma\langle\mathcal{O}_S\rangle$ is *properly* contained in $\mathcal{B}(\mathbb{R}^k)$.

(**Hint:** Show that $\sigma\langle\mathcal{O}_S\rangle$ coincides with $\mathcal{F}_6$ of Example 1.1.4).

1.13 Let $\bar{\mathbb{R}} \equiv \mathbb{R} \cup \{+\infty\} \cup \{-\infty\}$ be the extended real line. The Borel σ-algebra on $\bar{\mathbb{R}}$, denoted by $\mathcal{B}(\bar{\mathbb{R}})$, is defined as the σ-algebra on $\bar{\mathbb{R}}$ generated by the collection $\mathcal{B}(\mathbb{R}) \cup \{\infty\} \cup \{-\infty\}$.

(a) Show that $\mathcal{B}(\bar{\mathbb{R}}) = \big\{A \cup B : A \in \mathcal{B}(\mathbb{R}), B \subset \{-\infty, \infty\}\big\}$.

(b) Show, however, that the σ-algebra on $\bar{\mathbb{R}}$ generated by $\mathcal{B}(\mathbb{R})$ is given by $\sigma\langle\mathcal{B}(\mathbb{R})\rangle = \big\{A \cup B : A \in \mathcal{B}(\mathbb{R}), B = \{-\infty, \infty\}$ or $B = \emptyset\big\}$.

1.14 Let $\mathcal{A}_1, \mathcal{A}_2$ and $\mathcal{A}_3$ denote, respectively, the class of triangles, discs, and pentagons in $\mathbb{R}^2$. Show that $\sigma\langle\mathcal{A}_i\rangle \equiv \mathcal{B}(\mathbb{R}^2)$. Thus, the σ-algebra $\mathcal{B}(\mathbb{R}^2)$ (and similarly $\mathcal{B}(\mathbb{R}^k)$) can be generated by starting with various classes of sets of different shapes and geometry.

1.15 Let Ω be a nonempty set and $\mathcal{B}$ be a σ-algebra on Ω. Let $A \subset \Omega$ and $\mathcal{B}_A \equiv \{B \cap A : B \in \mathcal{B}\}$. Show that $\mathcal{B}_A$ is a σ-algebra on A. The σ-algebra $\mathcal{B}_A$ is called the *trace σ-algebra of $\mathcal{B}$ on A*.

1.16 Let $\Omega = \mathbb{R}$ and $\mathcal{F}$ be the collection of all finite unions of disjoint intervals of the form $(a, b] \cap \mathbb{R}$, $-\infty \le a < b \le \infty$. Show that $\mathcal{F}$ is an algebra but not a σ-algebra.

1.17 Let Ω be a nonempty set and $\{A_i\}_{i \in \mathbb{N}}$ be a sequence of subsets of Ω such that $A_{i+1} \subset A_i$ for all $i \in \mathbb{N}$. Verify that $\mathcal{A} = \{A_i : i \in \mathbb{N}\}$ is a π-system and also determine $\lambda\langle\mathcal{A}\rangle$, the λ-system generated by $\mathcal{A}$.

1.18 Let $\Omega \equiv \mathbb{N}$, $\mathcal{F} = \mathcal{P}(\Omega)$, and $A_n = \{j : j \in \mathbb{N}, j \ge n\}, n \in \mathbb{N}$. Let μ be the counting measure on $(\Omega, \mathcal{F})$. Verify that $\lim_{n\to\infty} \mu(A_n) \ne \mu(\cap_{n \ge 1} A_n)$.

1.19 Let Ω be a nonempty set and let $\mathcal{C} \subset \mathcal{P}(\Omega)$ be a semialgebra. Let

$$\mathcal{A}(\mathcal{C}) \equiv \Big\{A : A = \bigcup_{i=1}^{k} B_i : B_i \in \mathcal{C}, i = 1, 2, \ldots, k, k \in \mathbb{N}\Big\}.$$

(a) Show that $\mathcal{A}(\mathcal{C})$ is the smallest algebra containing $\mathcal{C}$.

(b) Show also that $\sigma\langle\mathcal{C}\rangle = \sigma\langle\mathcal{A}(\mathcal{C})\rangle$.

1.20 Let μ^* be as in (3.1) of Section 1.3. Verify (3.4)–(3.6).

(**Hint:** Fix $0 < \epsilon < \infty$. If $\mu^*(A_n) < \infty$ for all $n \in \mathbb{N}$, then find, for each n, a cover $\{A_{nj}\}_{j\geq 1} \subset \mathcal{C}$ such that $\mu^*(A_n) \leq \sum_{j=1}^{\infty} \mu(A_{nj}) + \frac{\epsilon}{2^n}$.)

1.21 Prove Proposition 1.3.1.

1.22 Let $F : \mathbb{R} \to \mathbb{R}$ be nondecreasing. Let $(a, b], (a_n, b_n], n \subset \mathbb{N}$ be intervals in $\mathbb{R}$ such that $(a, b] = \cup_{n\geq 1}(a_n, b_n]$ and $\{(a_n, b_n] : n \geq 1\}$ are disjoint. Let $\mu_F(\cdot)$ be as in (3.8). Show that $\mu_F((a, b]) = \sum_{n=1}^{\infty} \mu_F((a_n, b_n])$ by completing the following steps:

(a) Let $G(x) \equiv F(x+)$ for all $x \in \mathbb{R}$ and let $G(\pm\infty) = F(\pm\infty)$. Verify that $G(\cdot)$ is nondecreasing and right continuous on $\mathbb{R}$ and that for any A in $\mathcal{C}$, $\mu_F(A) = \mu_G(A)$.

(b) In view of (a), assume w.l.o.g. that $F(\cdot)$ is right continuous. Show that for any $k \in \mathbb{N}$,

$$F(b) - F(a) \geq \sum_{i=1}^{k}(F(b_i) - F(a_i)),$$

and conclude that

$$F(b) - F(a) \geq \sum_{i=1}^{\infty}(F(b_i) - F(a_i)).$$

(c) To prove the reverse inequality, fix $\eta > 0$. Choose $c > a$ and $d_n > b_n$, $n \geq 1$ such that such that $F(c) - F(a) < \eta$, $[c, b] \subset \cup_{n\geq 1}(a_n, d_n)$ and $F(d_n) - F(b_n) < \eta/2^n$ for all $n \in \mathbb{N}$. Next, apply the Heine-Borel theorem to the interval $[c, b]$ and the open cover $\{(a_n, d_n)\}_{n\geq 1}$ and extract a finite cover $\{(a_i, d_i)\}_{i=1}^{k}$ for $[c, b]$. W.l.o.g., assume that $c \in (a_1, d_1)$ and $b \in (a_k, d_k)$. Now verify that

$$\begin{aligned} F(b) - F(a) &\leq \sum_{j=1}^{k}(F(b_j) - F(a_j)) + 2\eta \\ &\leq \sum_{j=1}^{\infty}(F(b_j) - F(a_j)) + 2\eta. \end{aligned}$$

1.23 Extend the above arguments to the case when $(a, b]$ and $(a_i, b_i]$, $i \geq 1$ are not necessarily bounded intervals.

1.24 Verify that $\mathcal{C}_2$, defined in (3.11), is a semialgebra.

1.25 (a) Verify that the limit in (3.13) exists.

(b) Extend the arguments in Problems 1.22 and 1.23 to verify that μ_F of (3.12) and (3.13) is a measure on $\mathcal{C}_2$.

1.26 Establish Theorem 1.3.6 by completing the following:

(a) Suppose first that $\nu(\Omega) < \infty$. Verify that $\mathcal{L} \equiv \{A : A \in \sigma\langle\mathcal{C}\rangle, \mu^*(A) = \nu(A)\}$ is a λ system and use the π-λ theorem.

(b) Extend the above to the σ-finite case.

1.27 Prove Corollary 1.3.5 for Lebesgue measure $m(\cdot)$.

1.28 Let F be a discrete distribution function, i.e., F is of the form

$$F(x) \equiv \sum_{j=1}^{\infty} a_j I(x_j \leq x), \quad x \in \mathbb{R},$$

where $0 < a_j < \infty$, $\sum_{j \geq 1} a_j = 1$, $x_j \in \mathbb{R}$, $j \geq 1$. Show that $\mathcal{M}_{\mu_F^*} = \mathcal{P}(\mathbb{R})$.

(**Hint:** Show that $\mu_F^*(A^c) = 0$, where $A \equiv \{x_j : j \geq 1\}$, and use the fact that for any $B \subset \mathbb{R}$, $B \cap A \in \mathcal{B}(\mathbb{R})$.)

1.29 Let

$$F(x) = \begin{cases} 0 & \text{for } x < 0 \\ x & \text{for } 0 \leq x \leq 1 \\ 1 & \text{for } x > 0. \end{cases}$$

Show that $\mathcal{M}_{\mu_F^*} \equiv \{A : A \in \mathcal{P}(\mathbb{R}), A \cap [0,1] \in \mathcal{M}\}$, where $\mathcal{M}$ is the σ-algebra of Lebesgue measurable sets as in Definition 1.3.7.

1.30 Let $F(\cdot) = \frac{1}{2}\Phi(\cdot) + \frac{1}{2}F_P(\cdot)$ where $\Phi(\cdot)$ is the standard normal cdf, i.e., $\Phi(x) \equiv \int_{-\infty}^{x} \frac{1}{\sqrt{2\pi}} e^{-u^2/2} Au$ and $F_P(x) \equiv \sum_{k=0}^{\infty} e^{-2} \frac{2^k}{k!} I_{(-\infty,x]}^{(k)}$, $x \in \mathbb{R}$. Let $F_1 = \Phi$, $F_2 = F_P$ and $F_3 = F$. Let $A_1 = (0,1)$, $A_2 = \{x : x \in \mathbb{R}, \sin x \in (0, \frac{1}{2})\}$, $A_3 = \{x :$ for some integers $a_0, a_1, \ldots, a_k$, $k < \infty$, $\sum_{i=0}^{k} a_i x^i = 0\}$, the set of all algebraic numbers. Compute $\mu_{F_i}(A_j)$, $1 \leq i, j \leq 3$.

1.31 Let μ be a measure on a semialgebra $\mathcal{C} \subset \mathcal{P}(\Omega)$ where Ω is a nonempty set. Let μ^* be the outer measure generated by μ and let $\mathcal{M}_{\mu^*}$ be the σ-algebra of μ^*-measurable sets as defined in Theorem 1.3.3.

(a) Show that for all $A \subset \Omega$, there exists a $B \in \sigma\langle\mathcal{C}\rangle$ such that $A \subset B$ and $\mu^*(A) = \mu^*(B)$.

(**Hint:** If $\mu^*(A) = \infty$, take B to be Ω. If $\mu^*(A) < \infty$, use the definition of μ^* to show that for each $n \geq 1$, there exists $\{B_{nj}\}_{j\geq 1} \subset \mathcal{C}$ such that $A \subset B_n \equiv \cup_{j\geq 1} B_{nj}$, $\mu^*(A) \leq \sum_{j=1}^{\infty} \mu(B_{nj}) < \mu^*(A) + \frac{1}{n}$. Take $B = \cap_{n\geq 1} B_n$.)

(b) Show that for all $A \in \mathcal{M}_{\mu^*}$ with $\mu^*(A) < \infty$, there exists $B \in \sigma\langle\mathcal{C}\rangle$ such that $A \subset B$ and $\mu^*(B \setminus A) = 0$.

(**Hint:** Use (a) and the relation $B = A \cup (B \setminus A)$ with A and $B \setminus A = B \cap A^c$ in $\mathcal{M}_{\mu^*}$.)

(c) Show that if μ is σ-finite (i.e., there exist sets $\Omega_n, n \geq 1$ in $\mathcal{C}$ with $\mu(\Omega_n) < \infty$ for all $n \geq 1$ and $\cup_{n\geq 1}\Omega_n = \Omega$), then in (b), the hypothesis that $\mu^*(A) < \infty$ can be dropped.

(**Hint:** Assume w.l.o.g. that $\{\Omega_n\}_{n\geq 1}$ are disjoint. Apply (b) to $\{A_n \equiv A \cap \Omega_n\}_{n\geq 1}$.)

(d) Show that if μ is σ-finite, then $A \in \mathcal{M}_{\mu^*}$ iff there exist sets $B_1, B_2 \in \sigma\langle\mathcal{C}\rangle$ such that $B_1 \subset A \subset B_2$ and $\mu^*(B_2 \setminus B_1) = 0$.

(**Hint:** Apply (c) to both A and A^c.)

This shows that $\mathcal{M}_{\mu^*}$ is the completion of $\sigma\langle\mathcal{C}\rangle$ w.r.t. μ^*.

1.32 (*An outline of a proof of Corollary 1.3.5*). Let $(\mathbb{R}, \mathcal{M}_{\mu_F^*}, \mu_F^*)$ be a Lebesgue Stieltjes measure space generated by a right continuous and nondecreasing function $F : \mathbb{R} \to \mathbb{R}$.

(a) Show that $A \in \mathcal{M}_{\mu_F^*}$ iff there exist Borel sets B_1 and $B_2 \in \mathcal{B}(\mathbb{R})$ such that $B_1 \subset A \subset B_2$ and $\mu_F^*(B_2 \setminus B_1) = 0$.

(**Hint:** Take $\mathcal{C}$ to be the semialgebra $\mathcal{C} = \{(a, b] : -\infty \leq a \leq b < \infty\} \cup \{(b, \infty) : -\infty \leq b < \infty\}$ and apply Problem 1.31 (d).)

(b) Let $A \in \mathcal{M}_{\mu_F^*}$ with $\mu_F^*(A) < \infty$. Show that for any $\epsilon > 0$, there exist a finite number of bounded open intervals $I_j, j = 1, 2, \ldots, k$ such that $\mu_F^*(A \triangle \cup_{j=1}^k I_j) < \epsilon$.

(**Outline:** *Claim: For any $B \in \mathcal{C}$ with $\mu_F(B) < \infty$, there exists an open interval I such that $\mu_F^*(I \triangle B) < \epsilon$.*

To see this, note that if $B = (a, b]$, $-\infty \leq a < b < \infty$, then one may choose $b' > b$ such that $F(b') - F(b) < \epsilon$. Now, with $I = (a, b')$, $\mu_F^*(I \triangle B) = \mu_F^*((b, b')) = \mu_F((b, b')) \leq F(b') - F(b) < \epsilon$. If $B = (b, \infty)$ and $\mu_F(B) < \infty$, then there exists $b' > b$ such that $F(\infty) - F(b'-) < \epsilon$. Hence, with $I = (b, b')$, $\mu_F^*(I \triangle B) = \mu_F^*([b', \infty)) = F(\infty) - F(b'-) < \epsilon$. This proves the claim.

Next, By Theorem 1.3.4, for all $\epsilon > 0$, there exist $B_1, B_2, \ldots, B_k \in \mathcal{C}$ such that $\mu_F^*(A \triangle \cup_{j=1}^k B_j) < \epsilon/2$. For each

B_j, find I_j, a bounded open interval such that $\mu_F^*(B_j \triangle I_j) < \frac{\epsilon}{2^j}$. Since $(A_1 \cup A_2) \triangle (C_1 \cup C_2) \subseteq (A_1 \triangle C_1) \cup (A_2 \triangle C_2)$ for any $A_1, A_2, C_1, C_2 \subset \Omega$, it follows that

$$\mu_F^*\Big(\Big[\bigcup_{j=1}^{k} B_j\Big] \triangle \Big[\bigcup_{j=1}^{k} I_j\Big]\Big) < \sum_{j=1}^{k} \mu_F^*(B_j \triangle I_j) < \frac{\epsilon}{2}.$$

Hence, $\mu_F^*(A \triangle [\cup_{j=1}^k I_j]) < \epsilon$.)

(c) Let $A \in \mathcal{M}_{\mu_F^*}$ with $\mu_F^*(A) < \infty$. Show that for every $\epsilon > 0$, there exists an open set O such that $A \subset O$ and $\mu_F^*(O \setminus A) < \epsilon$.

(**Hint:** By definition of μ_F^*, for every $\epsilon > 0$, there exist $\{B_j\}_{j\geq 1} \subset \mathcal{C}$ such that $A \subset \cup_{j\geq 1} B_j$ and $\mu_F^*(A) \leq \sum_{j=1}^{\infty} \mu_F(B_j) \leq \mu_F^*(A) + \epsilon$. Now as in (b), there exist open intervals I_j such that $B_j \subset I_j$ and $\mu_F^*(I_j \setminus B_j) < \epsilon/2^j$ for all $j \geq 1$. Then $A \subset \cup_{j=1}^{\infty} B_j \subset \cup_{j=1}^{\infty} I_j \equiv O$. Also, $\mu_F^*(O) = \mu_F^*(A) + \mu_F^*(O \setminus A) \Rightarrow \mu_F^*(O \setminus A) = \mu_F^*(O) - \mu^*(A) < 2\epsilon$, (since $\mu^*(O) \leq \sum_{j=1}^{\infty} \mu^*(I_j) = \sum_{j=1}^{\infty} \mu_F^*(B_j) + \epsilon \leq \mu_F^*(A) + 2\epsilon < \infty$).)

(d) Extend (c) to all $A \in \mathcal{M}_{\mu_F^*}$.

(**Hint:** Let $A_i = A \cap [i, i+1]$, $i \in \mathbb{Z}$. Apply (c) to A_i with $\epsilon_i = \frac{\epsilon}{2^{|i|+1}}$ and take unions.)

(e) Show that for all $A \in \mathcal{M}_{\mu_F^*}$ and for all $\epsilon > 0$, there exist a closed set C and an open set O such that $C \subset A \subset O$ and $\mu_F^*(O \setminus A) \leq \epsilon$, $\mu^*(A \setminus C) < \epsilon$.

(**Hint:** Apply (d) to A and A^c.)

(f) Show that for all $A \in \mathcal{M}_{\mu_F^*}$ with $\mu_F^*(A) < \infty$ and for all $\epsilon > 0$, there exist a closed and bounded set $F \subset A$ such that $\mu_F^*(A \setminus F) < \epsilon$ and an open set O with $A \subset O$ such that $\mu_F^*(O \setminus A) < \epsilon$.

(**Hint:** Apply (d) to $A \cap [-M, M]$ where M is chosen so that $\mu_F^*(A \cap [-M, M]^c) < \epsilon$. Why is this possible?)

Remark: Thus for all $A \in \mathcal{M}_{\mu_F^*}$ with $\mu_F^*(A) < \infty$ and for all $\epsilon > 0$, there exist a compact set $K \subset A$ and an open set $O \supset A$ such that $\mu_F^*(A \setminus K) < \epsilon$ and $\mu_F^*(O \setminus A) < \epsilon$. The first property is called *inner regularity of* μ_F^* and the second property is called *outer regularity* of μ_F^*.

(g) Show that for all $A \in \mathcal{M}_{\mu_F^*}$ with $\mu_F^*(A) < \infty$ and for all $\epsilon > 0$, there exists a continuous function g_ϵ with compact support (i.e., $g_\epsilon(x)$ is zero for $|x|$ large) such that

$$\mu_F^*(A \triangle\ g_\epsilon^{-1}\{1\}) < \epsilon.$$

(**Hint:** For any bounded open interval (a, b), let $\eta > 0$ be such that $\mu_F((a, a+\eta]) + \mu_F([b-\eta, b)) < \epsilon$. Next define

$$g_\epsilon(x) = \begin{cases} 1 & \text{if} \quad a+\eta \le x \le b-\eta \\ 0 & \text{if} \quad x \notin (a,b) \\ \text{linear} & \text{over} \quad [a, a+\eta] \cup [b-\eta, b]. \end{cases}$$

Then g_ϵ is continuous with compact support. Also, $g^{-1}\{1\} = [a+\eta, b-\eta]$ and $(a,b) \triangle g^{-1}\{1\} = (a, a+\eta) \cup (b-\eta, b)$. So $\mu_F\{(a,b) \triangle g_\epsilon^{-1}(1)\}| < \epsilon$, proving the claim for $A = (a, b)$. The general case follows from (b).)

(h) Show that for all $A \in \mathcal{M}_{\mu_F^*}$ and for all $\epsilon > 0$, there exists a continuous function g_ϵ (not necessarily with compact support) such that $\mu_F^*(A \triangle g_\epsilon^{-1}\{1\}) < \epsilon$ (i.e., drop the condition $\mu_F^*(A) < \infty$).

(**Hint:** Let $A_k = A \cap [k, k+1]$, $k \in \mathbb{Z}$. Find $g_k : \mathbb{R} \to \mathbb{R}$ continuous with support in $\left(k - \frac{1}{8}, k + \frac{9}{8}\right)$ such that $\mu_F^*(I_{A_k} \neq g_k) < \frac{\epsilon}{2^{|k|+1}}$. Let $g = \sum_{k \in \mathbb{Z}} g_k$. Note that for any $x \in \mathbb{R}$, atmost two $g_k(x) \neq 0$ and so g is continuous. Also, $\mu_F^*(I_A \neq g) \le \sum_{k \in \mathbb{Z}} \mu_F^*(I_{A_k} \neq g_k) < \epsilon$.)

1.33 Let C be the Cantor set in [0,1] as defined in Section 1.3.2.

(a) Show that

$$C = \left\{ x : x = \sum_{i=1}^{\infty} \frac{a_i}{3^i},\ a_i \in \{0, 2\} \right\}.$$

and hence that C is uncountable.

(b) Show that

$$C + C \equiv \{x + y : x, y \in C\} = [0, 2].$$

2

Integration

2.1 Measurable transformations

Oftentimes, one is not interested in the full details of a measure space $(\Omega, \mathcal{F}, \mu)$, but only in certain functions defined on Ω. For example, if Ω represents the outcomes of 10 tosses of a fair coin, one may only be interested in knowing the number of heads in the 10 tosses. It turns out that to assign measures (probabilities) to sets (events) involving such functions, one can allow only certain functions (called measurable functions) that satisfy some 'natural' restrictions, specified in the following definitions.

Definition 2.1.1: Let Ω be a nonempty set and let $\mathcal{F}$ be a σ-algebra on Ω. Then the pair $(\Omega, \mathcal{F})$ is called a *measurable space.* If μ is a measure on $(\Omega, \mathcal{F})$, then the triplet $(\Omega, \mathcal{F}, \mu)$ is called a *measure space.* If in addition, μ is a probability measure, then $(\Omega, \mathcal{F}, \mu)$ is called a *probability space.*

Definition 2.1.2:

(a) Let $(\Omega, \mathcal{F})$ be a measurable space. Then a function $f : \Omega$ to $\mathbb{R}$ is called $\langle \mathcal{F}, \mathcal{B}(\mathbb{R})\rangle$-*measurable* (or $\mathcal{F}$-*measurable*) if for each a in $\mathbb{R}$

$$f^{-1}\big((-\infty, a]\big) \equiv \{\omega : f(\omega) \leq a\} \in \mathcal{F}. \tag{1.1}$$

(b) Let $(\Omega, \mathcal{F}, P)$ be a probability space. Then a function $X : \Omega \to \mathbb{R}$ is called a *random variable*, if the event

$$X^{-1}((-\infty, a]) \equiv \{\omega : X(\omega) \leq a\} \in \mathcal{F}$$

for each a in $\mathbb{R}$, i.e., a random variable is a real valued $\mathcal{F}$-measurable function on a probability space $(\Omega, \mathcal{F}, P)$.

It will be shown later that condition (1.1) on f is equivalent to the stronger condition that $f^{-1}(A) \in \mathcal{F}$ for all Borel sets $A \in \mathcal{B}(\mathbb{R})$. Since for any Borel set $A \in \mathcal{B}(\mathbb{R})$, $f^{-1}(A)$ is a member of the underlying σ-algebra $\mathcal{F}$, one can assign a measure to the set $f^{-1}(A)$ using a measure μ on $(\Omega, \mathcal{F})$. Note that for an *arbitrary* function T from $\Omega \to \mathbb{R}$, $T^{-1}(A)$ need not be a member of $\mathcal{F}$ and hence, such an assignment may *not* be possible. Thus, condition (1.1) on real valued mappings is a 'natural' requirement while dealing with measure spaces.

The following definition generalizes (1.1) to maps between two measurable spaces.

Definition 2.1.3: Let $(\Omega_i, \mathcal{F}_i), i = 1, 2$ be measurable spaces. Then, a mapping $T : \Omega_1 \to \Omega_2$ is called *measurable with respect to the σ-algebras* $\langle \mathcal{F}_1, \mathcal{F}_2 \rangle$ (or $\langle \mathcal{F}_1, \mathcal{F}_2 \rangle$*-measurable*) if

$$T^{-1}(A) \in \mathcal{F}_1 \quad \text{for all} \quad A \in \mathcal{F}_2.$$

Thus, X is a random variable on a probability space $(\Omega, \mathcal{F}, P)$ iff X is $\langle \mathcal{F}, \mathcal{B}(\mathbb{R}) \rangle$-measurable. Some examples of measurable transformations are given below.

Example 2.1.1: Let $\Omega = \{a, b, c, d\}, \mathcal{F}_2 = \{\Omega, \emptyset, \{a\}, \{b, c, d\}\}$ and let $\mathcal{F}_3 =$ the set of all subsets of Ω. Define the mappings $T_i : \Omega \to \Omega, i = 1, 2$, by

$$T_1(\omega) \equiv a \text{ for } \omega \in \Omega$$

and

$$T_2(\omega) = \begin{cases} a & \text{if } \omega = a, b \\ c & \text{if } \omega = c, d. \end{cases}$$

Then, T_1 is $\langle \mathcal{F}_2, \mathcal{F}_3 \rangle$-measurable since for any $A \in \mathcal{F}_3$, $T_1^{-1}(A) = \Omega$ or $\emptyset$ according as $a \in A$ or $a \notin A$. By similar arguments, it follows that T_2 is $\langle \mathcal{F}_3, \mathcal{F}_2 \rangle$-measurable. However, T_2 is not $\langle \mathcal{F}_2, \mathcal{F}_3 \rangle$-measurable since $T_2^{-1}(\{a\}) = \{a, b\} \notin \mathcal{F}_2$.

As this simple example shows, measurability of a given mapping critically depends on the σ-algebras on its domain and range spaces. In general, if T is $\langle \mathcal{F}_1, \mathcal{F}_2 \rangle$-measurable, then T is $\langle \tilde{\mathcal{F}}_1, \mathcal{F}_2 \rangle$-measurable for any σ-algebra $\tilde{\mathcal{F}}_1 \supset \mathcal{F}_1$ and T is $\langle \mathcal{F}_1, \tilde{\mathcal{F}}_2 \rangle$-measurable for any $\tilde{\mathcal{F}}_2 \subset \mathcal{F}_2$.

Example 2.1.2: Let $T : \mathbb{R} \to \mathbb{R}$ be defined as

$$T(x) = \begin{cases} \sin 2x & \text{if } x > 0 \\ 1 + \cos x & \text{if } x \leq 0. \end{cases}$$

Is T measurable w.r.t. the Borel σ-algebras $\langle \mathcal{B}(\mathbb{R}), \mathcal{B}(\mathbb{R})\rangle$? If one is to apply the definition directly, one must check that $T^{-1}(A) \in \mathcal{B}(\mathbb{R})$ for all $A \in \mathcal{B}(\mathbb{R})$. However, finding $T^{-1}(A)$ for all Borel sets A is not an easy task. In many instances like this one, verification of the measurability property of a given mapping by directly using the definition can be difficult. In such situations, one may use some easy-to-verify sufficient conditions. Some results of this type are given below.

Proposition 2.1.1: *Let $(\Omega_i, \mathcal{F}_i)$, $i = 1, 2, 3$ be measurable spaces.*

(i) *Suppose that $\mathcal{F}_2 = \sigma\langle \mathcal{A}\rangle$ for some class of subsets $\mathcal{A}$ of Ω_2. If $T : \Omega_1 \to \Omega_2$ is such that $T^{-1}(A) \in \mathcal{F}_1$ for all $A \in \mathcal{A}$, then T is $\langle \mathcal{F}_1, \mathcal{F}_2\rangle$-measurable.*

(ii) *Suppose that $T_1 : \Omega_1 \to \Omega_2$ is $\langle \mathcal{F}_1, \mathcal{F}_2\rangle$-measurable and $T_2 : \Omega_2 \to \Omega_3$ is $\langle \mathcal{F}_2, \mathcal{F}_3\rangle$-measurable. Let $T = T_2 \circ T_1 : \Omega_1 \to \Omega_3$ denote the composition of T_1 and T_2, defined by $T(\omega_1) = T_2(T_1(\omega_1))$, $\omega_1 \in \Omega_1$. Then, T is $\langle \mathcal{F}_1, \mathcal{F}_3\rangle$-measurable.*

Proof:

(i) Define the collection of sets

$$\mathcal{F} = \{A \in \mathcal{F}_2 : T^{-1}(A) \in \mathcal{F}_1\}.$$

Then,

(a) $T^{-1}(\Omega_2) = \Omega_1 \in \mathcal{F}_1 \Rightarrow \Omega_2 \in \mathcal{F}$.

(b) If $A \in \mathcal{F}$, then $T^{-1}(A) \in \mathcal{F}_1 \Rightarrow (T^{-1}(A))^c \in \mathcal{F}_1 \Rightarrow T^{-1}(A^c) = (T^{-1}(A))^c \in \mathcal{F}_1$, implying $A^c \in \mathcal{F}$.

(c) If $A_1, A_2, \ldots, \in \mathcal{F}$, then, $T^{-1}(A_i) \in \mathcal{F}_1$ for all $i \geq 1$. Since $\mathcal{F}_1$ is a σ-algebra , $T^{-1}(\cup_{n\geq 1} A_n) = \cup_{n\geq 1} T^{-1}(A_n) \in \mathcal{F}_1$. Thus, $\cup_{n\geq 1} A_n \in \mathcal{F}$. (See also Problem 2.1 on de Morgan's laws.)

Hence, by (a), (b), (c), $\mathcal{F}$ is a σ-algebra and by hypothesis $\mathcal{A} \subset \mathcal{F}$. Hence, $\mathcal{F}_2 = \sigma\langle \mathcal{A}\rangle \subset \mathcal{F} \subset \mathcal{F}_2$. Thus, $\mathcal{F} = \mathcal{F}_2$ and T is $\langle \mathcal{F}_1, \mathcal{F}_2\rangle$-measurable.

(ii) Let $A \in \mathcal{F}_3$. Then, $T_2^{-1}(A) \in \mathcal{F}_2$, since T_2 is $\langle \mathcal{F}_2, \mathcal{F}_3\rangle$-measurable. Also, by the $\langle \mathcal{F}_1, \mathcal{F}_2\rangle$-measurability of T_1, $T^{-1}(A) = T_1^{-1}(T_2^{-1}(A)) \in \mathcal{F}_1$, showing that T is $\langle \mathcal{F}_1, \mathcal{F}_3\rangle$-measurable. □

Proposition 2.1.2: *For any $k, p \in \mathbb{N}$, if $f : \mathbb{R}^p \to \mathbb{R}^k$ is continuous, then f is $\langle \mathcal{B}(\mathbb{R}^p), \mathcal{B}(\mathbb{R}^k)\rangle$-measurable.*

Proof: Let $\mathcal{A} = \{A : A$ is an open set in $\mathbb{R}^k\}$. Then, by the continuity of f, $f^{-1}(A)$ is open and hence, is in $\mathcal{B}(\mathbb{R}^p)$ (cf. Section A.4). Thus, $f^{-1}(A) \in$

$\mathbb{R}^p$ for all $A \in \mathcal{A}$. Since $\mathcal{B}(\mathbb{R}^k) = \sigma\langle\mathcal{A}\rangle$, by Proposition 2.1.1 (a), f is $\langle\mathcal{B}(\mathbb{R}^p), \mathcal{B}(\mathbb{R}^k)\rangle$-measurable. □

While the converse to the above proposition is not true a result due to Lusin says that except on a set of small measure f coincides with a continuous function. This is stronger than the statement that except on set of small measure f is close to a continuous function. For the statement and proof of Lusin's theorem, see Theorem 2.5.12.

Proposition 2.1.3: *Let $f_1, \ldots, f_k$ $(k \in \mathbb{N})$ be $\langle\mathcal{F}, \mathcal{B}(\mathbb{R})\rangle$-measurable transformations from Ω to $\mathbb{R}$. Then,*

(i) $f = (f_1, \ldots, f_k)$ is $\langle\mathcal{F}, \mathcal{B}(\mathbb{R}^k)\rangle$-measurable.

(ii) $g = f_1 + \ldots + f_k$ is $\langle\mathcal{F}, \mathcal{B}(\mathbb{R})\rangle$-measurable.

(iii) $h \equiv \prod_{i=1}^k f_i$ is $\langle\mathcal{F}, \mathcal{B}(\mathbb{R})\rangle$-measurable.

(iv) Let $p \in \mathbb{N}$ and let $\psi : \mathbb{R}^k \to \mathbb{R}^p$ be continuous. Then, $\xi \equiv \psi \circ f$ is $\langle\mathcal{F}, \mathcal{B}(\mathbb{R}^p)\rangle$-measurable, where $f = (f_1, \ldots, f_k)$.

Proof: To prove (i), note that for any rectangle $R = (a_1, b_1) \times \ldots \times (a_k, b_k)$,

$$\begin{aligned} f^{-1}(R) &= \{\omega \in \Omega : a_1 < f_1(\omega) < b_1, \ldots, a_k < f_k(\omega) < b_k\} \\ &= \bigcap_{i=1}^k \{\omega \in \Omega : a_i < f_1(\omega) < b_i\} \\ &= \bigcap_{i=1}^k f_i^{-1}(a_i, b_i) \in \mathcal{F}, \end{aligned}$$

since each f_i is $\langle\mathcal{F}, \mathcal{B}(\mathbb{R})\rangle$-measurable. Hence, by Proposition 2.1.1 (i), f is $\langle\mathcal{F}, \mathcal{B}(\mathbb{R}^k)\rangle$-measurable. To prove (ii), note that the function $g_1(x) \equiv x_1 + \ldots + x_k$, $x = (x_1, \ldots, x_k) \in \mathbb{R}^k$ is continuous on $\mathbb{R}^k$, and hence, by Proposition 2.1.2, is $\langle\mathcal{B}(\mathbb{R}^k), \mathcal{B}(\mathbb{R})\rangle$-measurable. Since $g = g_1 \circ f$, g is $\langle\mathcal{F}, \mathcal{B}(\mathbb{R})\rangle$-measurable, by Proposition 2.1.1 (ii). The proofs of (iii) and (iv) are similar to that of (ii) and hence, are omitted. □

Corollary 2.1.4: *The collection of $\langle\mathcal{F}, \mathcal{B}(\mathbb{R})\rangle$-measurable functions from Ω to $\mathbb{R}$ is closed under pointwise addition and multiplication as well as under scalar multiplication.*

The proof of Corollary 2.1.4 is omitted.

In view of the above, writing the function T of Example 2.1.2 as

$$T(x) = (\sin 2x) I_{(0,\infty)}(x) + (1 + \cos x) I_{(-\infty,0]}(x),$$

$x \in \mathbb{R}$, the $\langle\mathcal{B}(\mathbb{R}), \mathcal{B}(\mathbb{R})\rangle$-measurability of T follows. Note that here T is not continuous over $\mathbb{R}$ but only piecewise continuous (see also Problem 2.2).

Next, measurability of the limit of a sequence of measurable functions is considered. Let $\bar{\mathbb{R}} = \mathbb{R} \cup \{+\infty, -\infty\}$ denote the extended real line and let $\mathcal{B}(\bar{\mathbb{R}}) \equiv \sigma\langle \mathcal{B}(\mathbb{R}) \cup \{+\infty\} \cup \{-\infty\}\rangle$ denote the extended Borel σ-algebra on $\bar{\mathbb{R}}$.

Proposition 2.1.5: *For each $n \in \mathbb{N}$, let $f_n : \Omega \to \bar{\mathbb{R}}$ be a $\langle \mathcal{F}, \mathcal{B}(\bar{\mathbb{R}})\rangle$-measurable function.*

(i) Then, each of the functions $\sup_{n\in\mathbb{N}} f_n$, $\inf_{n\in\mathbb{N}} f_n$, $\limsup_{n\to\infty} f_n$, and $\liminf_{n\to\infty} f_n$ is $\langle \mathcal{F}, \mathcal{B}(\bar{\mathbb{R}})\rangle$-measurable.

(ii) The set $A \equiv \{\omega : \lim_{n\to\infty} f_n(\omega)$ exists and is finite$\}$ lies in $\mathcal{F}$ and the function $h \equiv (\lim_{n\to\infty} f_n) \cdot I_A$ is $\langle \mathcal{F}, \mathcal{B}(\mathbb{R})\rangle$-measurable.

Proof:

(i) Let $g = \sup_{n\geq 1} f_n$. To show that g is $\langle \mathcal{F}, \mathcal{B}(\bar{\mathbb{R}})\rangle$-measurable, it is enough to show that $\{\omega : g(\omega) \leq r\} \in \mathcal{F}$ for all $r \in \mathbb{R}$ (cf. Problem 2.4). Now, for any $r \in \mathbb{R}$,

$$\begin{aligned}\{\omega : g(\omega) \leq r\} &= \bigcap_{n=1}^{\infty} \{\omega : f_n(\omega) \leq r\} \\ &= \bigcap_{n=1}^{\infty} f_n^{-1}((-\infty, r]) \in \mathcal{F},\end{aligned}$$

since $f_n^{-1}((-\infty, r]) \in \mathcal{F}$ for all $n \geq 1$, by the measurability of f_n.

Next note that $\inf_{n\geq 1} f_n = -\sup_{n\geq 1}(-f_n)$ and hence, $\inf_{n\geq 1} f_n$ is $\langle \mathcal{F}, \mathcal{B}(\bar{\mathbb{R}})\rangle$-measurable. To prove the measurability of $\limsup_{n\to\infty} f_n$, define the functions $g_m = \sup_{n\geq m} f_n$, $m \geq 1$. Then, g_m is $\langle \mathcal{F}, \mathcal{B}(\bar{\mathbb{R}})\rangle$-measurable for each $m \geq 1$ and since g_m is nonincreasing in m, $\inf_{m\geq 1} g_m \equiv \limsup_{n\to\infty} f_n$ is also $\langle \mathcal{F}, \mathcal{B}(\bar{\mathbb{R}})\rangle$-measurable. A similar argument works for $\liminf_{n\to\infty} f_n$.

(ii) Let $h_1 = \limsup_{n\to\infty} f_n$ and $h_2 = \liminf_{n\to\infty} f_n$, and define $\tilde{h}_i = h_i I_{\mathbb{R}}(h_i)$, $i = 1, 2$. Note that $\tilde{h}_1 - \tilde{h}_2$ is $\langle \mathcal{F}, \mathcal{B}(\mathbb{R})\rangle$-measurable. Hence,

$$\begin{aligned}&\{\omega : \lim_{n\to\infty} f_n(\omega) \quad \text{exists and is finite}\} \\ = \;& \{\omega : -\infty < \limsup_{n\to\infty} f_n(\omega) = \liminf_{n\to\infty} f_n(\omega) < \infty\} \\ = \;& \{\omega : -\infty < h_2(\omega) = h_1(\omega) < \infty\} \\ = \;& \{\omega : \tilde{h}_1(\omega) = \tilde{h}_2(\omega)\} \cap \{\omega : h_1(\omega) < \infty, h_2(\omega) > -\infty\} \\ = \;& (\tilde{h}_1 - \tilde{h}_2)^{-1}(\{0\}) \cap \{\omega : h_1(\omega) < \infty, h_2(\omega) > -\infty\} \in \mathcal{F}.\end{aligned}$$

Finally, note that $h = h_1 I_A$. □

Definition 2.1.4: Let $\{f_\lambda : \lambda \in \Lambda\}$ be a family of mappings from Ω_1 into Ω_2 and let $\mathcal{F}_2$ be a σ-algebra on Ω_2. Then,

$$\sigma\langle\{f_\lambda^{-1}(A) : A \in \mathcal{F}_2, \lambda \in \Lambda\}\rangle$$

is called the *σ-algebra generated by* $\{f_\lambda : \lambda \in \Lambda\}$ (w.r.t. $\mathcal{F}_2$) and is denoted by $\sigma\langle\{f_\lambda : \lambda \in \Lambda\}\rangle$.

Note that $\sigma\langle\{f_\lambda : \lambda \in \Lambda\}\rangle$ is the smallest σ-algebra on Ω_1 that makes all f_λ's measurable w.r.t. $\mathcal{F}_2$ on Ω_2.

Example 2.1.3: Let $f = I_A$ for some set $A \subset \Omega_1$ and $\Omega_2 = \mathbb{R}$ and $\mathcal{F}_2 = \mathcal{B}(\mathbb{R})$. Then,

$$\sigma\langle\{f\}\rangle = \sigma\langle\{A\}\rangle = \{\Omega_1, \emptyset, A, A^c\}.$$

Example 2.1.4: Let $\Omega_1 = \mathbb{R}^k$, $\Omega_2 = \mathbb{R}$, $\mathcal{F}_2 = \mathcal{B}(\mathbb{R})$, and for $1 \leq i \leq k$, let $f_i : \Omega_1 \to \Omega_2$ be defined as

$$f_i(x_1, \ldots, x_k) = x_i, \qquad (x_1, \ldots, x_k) \in \Omega_1 = \mathbb{R}^k.$$

Then, $\sigma\langle\{f_i : 1 \leq i \leq k\}\rangle = \mathcal{B}(\mathbb{R}^k)$.

To show this, note that any measurable rectangle $A_1 \times \ldots \times A_k$ can be written as $A_1 \times \ldots \times A_k = \cap_{i=1}^k f_i^{-1}(A_i)$ and hence $A_1 \times \ldots \times A_k \in \sigma\langle\{f_i : 1 \leq i \leq k\}\rangle$ for all $A_1, \ldots, A_k \in \mathbb{R}$. Since $\mathbb{R}^k$ is generated by the collection of all measurable rectangles, $\mathcal{B}(\mathbb{R}^k) \subset \sigma\langle\{f_i : 1 \leq i \leq k\}\rangle$. Conversely, for any $A \in \mathcal{B}(\mathbb{R})$ and for any $1 \leq i \leq k$, $f_i^{-1}(A) = \mathbb{R} \times \ldots \times A \times \ldots \times \mathbb{R}$ (with A in the ith position) is in $\mathcal{B}(\mathbb{R}^k)$. Therefore, $\sigma\langle\{f_i : 1 \leq i \leq k\}\rangle = \sigma\langle\{f_i^{-1}(A) : A \in \mathbb{R}, 1 \leq i \leq k\}\rangle \subset \mathcal{B}(\mathbb{R}^k)$. Hence, $\sigma\langle\{f_i : 1 \leq i \leq k\}\rangle = \mathcal{B}(\mathbb{R}^k)$.

Proposition 2.1.6: *Let $\{f_\lambda : \lambda \in \Lambda\}$ be an uncountable collection of maps from Ω_1 to Ω_2. Then for any $B \in \sigma\langle\{f_\lambda : \lambda \in \Lambda\}\rangle$, there exists a countable set $\Lambda_B \subset \Lambda$ such that $B \in \sigma\langle\{f_\lambda : \lambda \in \Lambda_B\}\rangle$.*

Proof: The proof of this result is left as an exercise (Problem 2.5). □

2.2 Induced measures, distribution functions

Suppose X is a random variable defined on a probability space $(\Omega, \mathcal{F}, P)$. Then P governs the probabilities assigned to events like $X^{-1}([a, b])$, $-\infty < a < b < \infty$. Since X takes values in the real line, one should be able to express such probabilities only as a function of the set $[a, b]$. Clearly, since X is $\langle\mathcal{F}, \mathbb{R}\rangle$-measurable, $X^{-1}(A) \in \mathcal{F}$ for all $A \in \mathcal{B}(\mathbb{R})$ and the function

$$P_X(A) \equiv P(X^{-1}(A)) \tag{2.1}$$

is a set function defined on $\mathcal{B}(\mathbb{R})$. Is this a (probability) measure on $\mathcal{B}(\mathbb{R})$? The following proposition answers the question more generally.

Proposition 2.2.1: *Let $(\Omega_i, \mathcal{F}_i)$, $i = 1, 2$ be measurable spaces and let $T : \Omega_1 \to \Omega_2$ be a $\langle \mathcal{F}_1, \mathcal{F}_2 \rangle$-measurable mapping from Ω_1 to Ω_2. Then, for any measure μ on $(\Omega_1, \mathcal{F}_1)$, the set function μT^{-1}, defined by*

$$\mu T^{-1}(A) \equiv \mu(T^{-1}(A)), \quad A \in \mathcal{F}_2 \tag{2.2}$$

is a measure on $\mathcal{F}_2$.

Proof: It is easy to check that μT^{-1} satisfies the three conditions for being a measure. The details are left as an exercise (cf. Problem 2.9).

Definition 2.2.1: The measure μT^{-1} is called the *measure induced by T* (or the *induced measure of T*) on $\mathcal{F}_2$.

In particular, if $\mu(\Omega_1) = 1$, then $\mu T^{-1}(\Omega_2) = 1$. Hence, the set function P defined in (2.1) is indeed a probability measure on $(\mathbb{R}, \mathcal{B}(\mathbb{R}))$.

Definition 2.2.2: For a random variable X defined on a probability space $(\Omega, \mathcal{F}, P)$, the *probability distribution* of X (or the *law* of X), denoted by P_X (say), is the induced measure of X under P on $\mathbb{R}$, as defined in (2.2).

In introductory courses on probability and statistics, one defines probabilities of events like '$X \in [a, b]$' by using the probability mass function for discrete random variables and the probability density function for 'continuous' random variables. The measure-theoretic definition above allows one to treat both these cases as well as the case of 'mixed' distributions under a unified framework.

Definition 2.2.3: The *cumulative distribution function* (or *cdf* in short) *of a random variable* X is defined as

$$F_X(x) \equiv P_X((-\infty, x]), \quad x \in \mathbb{R}. \tag{2.3}$$

Proposition 2.2.2: *Let F be the cdf of a random variable X.*

(i) For $x_1 < x_2$, $F(x_1) \leq F(x_2)$ (i.e., F is nondecreasing on $\mathbb{R}$).

(ii) For x in $\mathbb{R}$, $F(x) = \lim_{y \downarrow x} F(y)$ (i.e., F is right continuous on $\mathbb{R}$).

(iii) $\lim_{x \to -\infty} F(x) = 0$ and $\lim_{x \to +\infty} F(x) = 1$.

Proof: For $x_1 < x_2, (-\infty, x_1] \subset (-\infty, x_2]$. Since P_X is a measure on $\mathcal{B}(\mathbb{R})$,

$$F(x_1) = P_X((-\infty, x_1]) \leq P_X((-\infty, x_2]) = F(x_2),$$

proving (i).

To prove (ii), let $x_n \downarrow x$. Then, the sets $(-\infty, x_n] \downarrow (-\infty, x]$, and $P_X((-\infty, x_1]) = P(X \leq x_1) \leq 1$. Hence, using the monotone continuity from above of the measure P_X (m.c.f.a.) (cf. Proposition 1.2.3), one gets

$$\lim_{n\to\infty} F(x_n) = \lim_{n\to\infty} P_X((-\infty, x_n]) = P_X((-\infty, x]) = F(x).$$

Next consider part (iii). Note that if $x_n \downarrow -\infty$ and $y_n \uparrow \infty$, then $(-\infty, x_n] \downarrow \emptyset$ and $(-\infty, y_n] \uparrow (-\infty, \infty)$. Hence, part (iii) follows the m.c.f.a. and the m.c f.b. properties of P_X (cf. Propositions 1.2.1 and 1.2.3). □

Definition 2.2.4: A function $F : \mathbb{R} \to \mathbb{R}$ satisfying (i), (ii), and (iii) of Proposition 2.2.2 is called a *cumulative distribution function* (or *cdf* for short).

Thus, given a random variable X, its cdf F_X satisfies properties (i), (ii), (iii) of the Proposition 2.2.2. Conversely, given a cdf F, one can construct a probability space $(\Omega, \mathcal{F}, P)$ and a random variable X on it such that the cdf of X is F. Indeed, given a cdf F, note that by Theorem 1.3.3 and Definition 1.3.7, there exists a (Lebesgue-Stieltjes) probability measure μ_F on $(\mathbb{R}, \mathcal{B}(\mathbb{R}))$ such that $\mu_F((-\infty, x]) = F(x)$. Now define X to be the identity map on $\mathbb{R}$, i.e., let $X(x) \equiv x$ for all $x \in \mathbb{R}$. Then, X is a random variable on the probability space $(\mathbb{R}, \mathcal{B}(\mathbb{R}), \mu_F)$ with probability distribution $P_X = \mu_F$ and cdf $F_X = F$.

In addition to (i), (ii) and (iii) of Proposition 2.2.2, it is easy to verify that for any x in $\mathbb{R}$,

$$P(X < x) = F_X(x-) \equiv \lim_{y \uparrow x} F_X(y)$$

and hence

$$P(X = x) = F_X(x) - F_X(x-). \tag{2.4}$$

Thus, the function $F_X(\cdot)$ has a jump at x iff $P(X = x) > 0$. Since a monotone function from $\mathbb{R}$ to $\mathbb{R}$ can have only jump discontinuities and only a countable number of them (cf. Problem 2.11), for any random variable X, the set $\{a \in \mathbb{R} : P(X = a) > 0\}$ is countable. This leads to the following definitions.

Definition 2.2.5:

(a) A random variable X is called *discrete* if there exists a countable set $A \subset \mathbb{R}$ such that $P(X \in A) = 1$.

(b) A random variable X is called *continuous* if $P(X = x) = 0$ for all $x \in \mathbb{R}$.

Note that X is continuous iff F_X is continuous on all of $\mathbb{R}$ and X is discrete iff the sum of all the jumps of its cdf F_X is one. It may also be noted that if F_X is a step function, then X is discrete but not conversely. For example, consider the case where the set A in the above definition is the set of all rational numbers.

It turns out that a given cdf may be written as a weighted sum of a discrete and a continuous cdfs. Let F be a cdf. Let $A \equiv \{x : p(x) \equiv F(x) - F(x-) > 0\}$. As remarked earlier, A is at most countable. Write $\alpha = \sum_{y \in A} p(y)$ and let $\tilde{F}_d(x) = \sum_{y \in A} p(y) I_{(-\infty, x]}(y)$, and $\tilde{F}_c(x) = F(x) - \tilde{F}_d(x)$. It is easy to verify that $\tilde{F}_c(\cdot)$ is continuous on $\mathbb{R}$. If $\alpha = 0$, then $F(x) = \tilde{F}_c(x)$ and F is continuous. If $\alpha = 1$, then $F = \tilde{F}_d(x)$ and F is discrete. If $0 < \alpha < 1$, $F(\cdot)$ can be written as

$$F(x) = \alpha F_d(x) + (1 - \alpha) F_c(x), \tag{2.5}$$

where $F_d(x) \equiv \alpha^{-1} \tilde{F}_d(x)$ and $F_c(x) \equiv (1 - \alpha)^{-1} \tilde{F}_c(x)$ are both cdfs, with F_d being discrete and F_c being continuous. For a further decomposition of $F_c(\cdot)$ into absolutely continuous and singular continuous components, see Chapter 4.

2.2.1 *Generalizations to higher dimensions*

Induced distributions of random vectors and the associated cdfs are briefly considered in this section. Let $X = (X_1, X_2, \ldots, X_k)$ be a k-dimensional random vector defined on a probability space $(\Omega, \mathcal{F}, P)$. The probability distribution P_X of X is the induced probability measure on $(\mathbb{R}^k, \mathcal{B}(\mathbb{R}^k))$, defined by (cf. (2.1))

$$P_X(A) \equiv P(X^{-1}(A)) \quad A \in \mathcal{B}(\mathbb{R}^k). \tag{2.6}$$

The cdf F_X of X is now defined by

$$F_X(x) = P(X \le x), \quad x \in \mathbb{R}^k, \tag{2.7}$$

where for any $x = (x_1, x_2, \ldots, x_k)$ and $y = (y_1, y_2, \ldots, y_k)$ in $\mathbb{R}^k$, $x \le y$ means that $x_i \le y_i$ for all $i = 1, \ldots, k$.

The extension of Proposition 2.2.2 to the k-dimensional case is notationally involved. Here, an analog of Proposition 2.2.2 for the bivariate case, i.e., for $k = 2$ is stated.

Proposition 2.2.3: *Let F be the cdf of a bivariate random vector $X = (X_1, X_2)$.*

(i) Then, for any $x = (x_1, x_2) \le y = (y_1, y_2)$,

$$F(y_1, y_2) - F(x_1, y_2) - F(y_1, x_2) + F(x_1, x_2) \ge 0. \tag{2.8}$$

(ii) For any $x = (x_1, x_2) \in \mathbb{R}^2$,

$$\lim_{y_1 \downarrow x_1, y_2 \downarrow x_2} F(y_1, y_2) = F(x_1, x_2),$$

i.e., F is right continuous on $\mathbb{R}^2$.

(iii) $\lim_{x_1 \to -\infty} F(x_1, a) = \lim_{x_2 \to -\infty} F(a, x_2) = 0$ *for all* $a \in \mathbb{R}$; $\lim_{x_1 \to \infty, x_2 \to \infty} F(x_1, x_2) = 1$.

(iv) For any $a \in \mathbb{R}$, $\lim_{y \uparrow \infty} F(a, y) = P(X_1 \le a)$ *and* $\lim_{y \uparrow \infty} F(y, a) = P(X_2 \le a)$.

Proof: Clearly,

$$\begin{aligned} 0 &\le P(X \in (x_1, y_1] \times (x_2, y_2]) \\ &= P(x_1 < X_1 \le y_1, x_2 < X_2 \le y_2) \\ &= P(X_1 \le y_1, x_2 < X_2 \le y_2) - P(X_1 \le x_1, x_2 < X_2 \le y_2) \\ &= P(X_1 \le y_1, X_2 \le y_2) - P(X_1 \le y_1, X_2 \le x_2) \\ &\qquad -[P(X_1 \le x_1, X_2 \le y_2) - P(X_1 \le x_1, X_2 \le x_2)] \\ &= F(y_1, y_2) - F(y_1, x_2) - F(x_1, y_2) + F(x_1, x_2). \end{aligned}$$

This proves (i).

To prove (ii), note that for any sequence $y_{in} \downarrow x_i, i = 1, 2$, the sets $A_n = (-\infty, y_{1n}] \times (-\infty, y_{2n}] \downarrow A \equiv (-\infty, x_1] \times (-\infty, x_2]$. Hence, by m.c.f.a property of a probability measure,

$$F(y_{1n}, y_{2n}) = P(X \in A_n) \downarrow P(A) = F(x_1, x_2).$$

For (iii), note that $(-\infty, x_{1n}] \times (-\infty, a] \downarrow \emptyset$ for any sequence $x_{1n} \downarrow -\infty$ and for any $a \in \mathbb{R}$. Hence, again by the m.c.f.a. property,

$$F(x_{1n}, a) \downarrow 0 \quad \text{as} \quad n \to \infty\,.$$

By similar arguments, $F(a, x_{2n}) \downarrow 0$ whenever $x_{2n} \downarrow -\infty$. To prove the last relation in (iii), apply the m.c.f.b. property to the sets $(-\infty, x_{1n}] \times (-\infty, x_{2n}] \uparrow \mathbb{R}^2$ for $x_{1n} \uparrow \infty, x_{2n} \uparrow \infty$.

The proof of part (iv) is similar. □

Note that any function satisfying properties (i), (ii), (iii) of Proposition 2.2.3 determines a probability measure uniquely. This follows from the discussions in Section 1.3, as (1.3.9) and (1.3.10) follow from (i) and (iii) (Problem 2.12). For a general $k \ge 1$, an analog of property (i) above is cumbersome to write down explicitly. Indeed, for any $x \le y$, now a sum involving 2^k-terms must be nonnegative. However, properties (ii), (iii), and (iv) can be extended in an obvious way to the k-dimensional case. See Problem 2.13 for a precise statement. Also, for a general $k \ge 1$ functions satisfying the properties listed in Problem 2.13 uniquely determine a probability measure on $(\mathbb{R}^k, \mathcal{B}(\mathbb{R}^k))$.

2.3 Integration

Let $(\Omega, \mathcal{F}, \mu)$ be a measure space and $f : \Omega \to \mathbb{R}$ be a measurable function. The goal of this section is to define the integral of f with respect to the measure μ and establish some basic convergence results. The integral of a nonnegative function taking only *finitely* many values is defined first, which is then extended to all nonnegative measurable functions by approximation from below. Finally, the integral of an arbitrary measurable function is defined using the decomposition of the function into its positive and negative parts.

Definition 2.3.1: A function $f : \Omega \to \bar{\mathbb{R}} \equiv [-\infty, \infty]$ is called *simple* if there exist a finite set (of distinct elements) $\{c_1, \ldots, c_k\} \in \bar{\mathbb{R}}$ and sets $A_1, \ldots, A_k \in \mathcal{F}$, $k \in \mathbb{N}$ such that f can be written as

$$f = \sum_{i=1}^{k} c_i I_{A_i} . \tag{3.1}$$

Definition 2.3.2: (*The integral of a simple nonnegative function*). Let $f : \Omega \to \bar{\mathbb{R}}_+ \equiv [0, \infty]$ be a simple nonnegative function on $(\Omega, \mathcal{F}, \mu)$ with the representation (3.1). The *integral* of f w.r.t. μ, denoted by $\int f d\mu$, is defined as

$$\int f d\mu \equiv \sum_{i=1}^{k} c_i \mu(A_i) . \tag{3.2}$$

Here and in the following, the relation

$$0 \cdot \infty = 0$$

is adopted as a convention.

It may be verified that the value of the integral in (3.2) does not depend on the representation of f. That is, if f can be expressed as $f = \sum_{j=1}^{l} d_j I_{B_j}$ for some $d_1, \ldots, d_l \in \bar{\mathbb{R}}_+$ (not necessarily distinct) and for some sets $B_1, \ldots, B_l \in \mathcal{F}$, then $\sum_{i=1}^{k} c_i \mu(A_i) = \sum_{j=1}^{l} d_j \mu(B_j)$, so that the value of the integral remains unchanged (Problem 2.17). Also note that for the f in Definition 2.3.2,

$$0 \le \int f d\mu \le \infty.$$

The following proposition is an easy consequence of the definition and the above remark.

Proposition 2.3.1: *Let f and g be two simple nonnegative functions on $(\Omega, \mathcal{F}, \mu)$. Then*

(i) *(Linearity) For $\alpha \ge 0, \beta \ge 0, \int (\alpha f + \beta g) d\mu = \alpha \int f d\mu + \beta \int g d\mu$.*

(ii) *(Monotonicity) If $f \ge g$ a.e. (μ), i.e., $\mu(\{\omega : \omega \in \Omega, f(\omega) < g(\omega)\}) = 0$, then $\int f d\mu \ge \int g d\mu$.*

(iii) If $f = g$ *a.e.* (μ)*, i.e., if* $\mu(\{\omega : \omega \in \Omega, f(\omega) \neq g(\omega)\}) = 0$*, then* $\int f d\mu = \int g d\mu$.

Definition 2.3.3: (*The integral of a nonnegative measurable function*). Let $f : \Omega \to \bar{\mathbb{R}}_+$ be a nonnegative measurable function on $(\Omega, \mathcal{F}, \mu)$. The integral of f with respect to μ, also denoted by $\int f d\mu$, is defined as

$$\int f d\mu = \lim_{n\to\infty} \int f_n d\mu, \tag{3.3}$$

where $\{f_n\}_{n\geq 1}$ is *any* sequence of nonnegative simple functions such that $f_n(\omega) \uparrow f(\omega)$ for all ω.

Note that by Proposition 2.3.1 (ii), the sequence $\{\int f_n d\mu\}_{n\geq 1}$ is nondecreasing and hence the right side of (3.3) is well defined. That the right side of (3.3) is the *same* for all such approximating sequences of functions needs to be established and is the content of the following proposition. The proof of this proposition exploits in a crucial way the m.c.f.b. and the finite additivity of the set function μ (or, equivalently, the countable additivity of μ).

Proposition 2.3.2: *Let* $\{f_n\}_{n\geq 1}$ *and* $\{g_n\}_{n\geq 1}$ *be two sequences of simple nonnegative measurable functions on* $(\Omega, \mathcal{F}, \mu)$ *to* $\bar{\mathbb{R}}_+$ *such that as* $n \to \infty$*,* $f_n(\omega) \uparrow f(\omega)$ *and* $g_n(\omega) \uparrow f(\omega)$ *for all* $\omega \in \Omega$*. Then*

$$\lim_{n\to\infty} \int f_n d\mu = \lim_{n\to\infty} \int g_n d\mu. \tag{3.4}$$

Proof: Fix $N \in \mathbb{N}$ and $0 < \rho < 1$. It will now be shown that

$$\lim_{n\to\infty} \int f_n d\mu \geq \rho \int g_N d\mu. \tag{3.5}$$

Suppose that g_N has the representation $g_N \equiv \sum_{i=1}^{k} d_i I_{B_i}$. Let $D_n = \{\omega \in \Omega : f_n(\omega) \geq \rho g_N(\omega)\}$, $n \geq 1$. Since $f_n(\omega) \uparrow f(\omega)$ for all ω, $D_n \uparrow D \equiv \{\omega : f(\omega) \geq \rho g_N(\omega)\}$ (Problem 2.18 (b)). Also since $g_N(\omega) \leq f(\omega)$ and $0 < \rho < 1, D = \Omega$. Now writing $f_n = f_n I_{D_n} + f_n I_{D_n^c}$, it follows from Proposition 2.3.1 that

$$\begin{aligned} \int f_n d\mu &\geq \int f_n I_{D_n} d\mu \geq \rho \int g_N I_{D_n} d\mu \\ &= \rho \sum_{i=1}^{k} d_i \mu(B_i \cap D_n). \end{aligned} \tag{3.6}$$

By the m.c.f.b. property, for each $i \in \mathbb{N}$, $\mu(B_i \cap D_n) \uparrow \mu(B_i \cap \Omega) = \mu(B_i)$ as $n \to \infty$. Since the sequence $\{\int f_n d\mu\}_{n\geq 1}$ is nondecreasing, taking limits in (3.6), yields (3.5). Next, letting $\rho \uparrow 1$ yields $\lim_{n\to\infty} \int f_n d\mu \geq \int g_N d\mu$ for each $N \in \mathbb{N}$ and hence,

$$\lim_{n\to\infty} \int f_n d\mu \geq \lim_{n\to\infty} \int g_n d\mu .$$

By symmetry, (3.4) follows and hence, the proof is complete. □

Remark 2.3.1: It is easy to verify that Proposition 2.3.2 remains valid if $\{f_n\}_{n\geq 1}$ and $\{g_n\}_{n\geq 1}$ increase to f a.e. (μ).

Given a nonnegative measurable function f, one can always construct a nondecreasing sequence $\{f_n\}_{n\geq 1}$ of nonnegative simple functions such that $f_n(\omega) \uparrow f(\omega)$ for all $\omega \in \Omega$ in the following manner. Let $\{\delta_n\}_{n\geq 1}$ be a sequence of positive real numbers and let $\{N_n\}_{n\geq 1}$ be a sequence of positive integers such that as $n \to \infty$, $\delta_n \downarrow 0, N_n \uparrow \infty$ and $N_n\delta_n \uparrow \infty$. Further, suppose that the sequence $P_n \equiv \{j\delta_n : j = 0, 1, 2, \ldots, N_n\}$ is nested, i.e., $P_n \subset P_{n+1}$ for each $n \geq 1$. Now set

$$f_n(\omega) = \begin{cases} j\delta_n & \text{if} \quad j\delta_n \leq f(\omega) < (j+1)\delta_n, \ j = 0, 1, 2, \ldots, (N_n - 1) \\ N_n\delta_n & \text{if} \quad f(\omega) \geq N_n\delta_n . \end{cases} \tag{3.7}$$

A specific choice of δ_n and N_n is given by $\delta_n = 2^{-n}, N_n = n2^n$.

Thus, with the above choice of $\{f_n\}_{n\geq 1}$ in the definition of the *Lebesgue integral* $\int f d\mu$ in (3.3), the *range* of f is subdivided into intervals of decreasing lengths. This is in contrast to the definition of the *Riemann integral* of f over a bounded interval, which is defined via subdividing the *domain* of f into finer subintervals.

Remark 2.3.2: In some cases it may be more appropriate to choose the approximating sequence $\{f_n\}_{n\geq 1}$ in a manner different from (3.7). For example, let $\Omega = \{\omega_i : i \geq 1\}$ be a countable set, $\mathcal{F} = \mathcal{P}(\Omega)$, the power set of Ω, and let μ be a measure on $(\Omega, \mathcal{F})$. Then any function $f : \Omega \to \mathbb{R}_+ \equiv [0, \infty)$ is measurable and the integral $\int f d\mu$ coincides with the sum $\sum_{i=1}^{\infty} f(\omega_i)\mu(\{\omega_i\})$ as can be seen by choosing the approximating sequence $\{f_n\}_{n\geq 1}$ as

$$f_n(\omega_i) = \begin{cases} f(\omega_i) & \text{for} \quad i = 1, 2, \ldots, n \\ 0 & \text{for} \quad i > n. \end{cases}$$

Remark 2.3.3: The integral of a nonnegative measurable function can be alternatively defined as

$$\int f d\mu = \sup \Big\{ \int g d\mu : g \quad \text{nonnegative and simple,} \ g \leq f \Big\}.$$

The equivalence of this to (3.3) is seen as follows. Clearly the right side above, say, M is greater than or equal to $\int f d\mu$ as in (3.3). Conversely, there exist a sequence $\{g_n\}_{n\geq 1}$ of simple nonnegative functions with $g_n \leq f$

for all $n \geq 1$ such that $\lim_n \int g_n d\mu$ equals the supremum M defined above. Now set $h_n = \max\{g_j : 1 \leq j \leq n\}$, $n \geq 1$. Now it can be verified that for each $n \geq 1$, h_n is nonnegative, simple and satisfies $h_n \uparrow f$ and also that $\int h_n d\mu$ converges to M (Problem 2.19 (b)).

Corollary 2.3.3: *Let f and g be two nonnegative measurable functions on $(\Omega, \mathcal{F}, \mu)$. Then, the conclusions of Proposition 2.3.1 remain valid for such f and g.*

Proof: This follows from Proposition 2.3.1 for nonnegative simple functions and Definition 2.3.3. □

The definition of the integral $\int f d\mu$ of a nonnegative measurable function f in (3.3) makes it possible to interchange limits and integration in a fairly routine manner. In particular, the following key result is a direct consequence of the definition.

Theorem 2.3.4: (*The monotone convergence theorem or MCT*). *Let $\{f_n\}_{n\geq 1}$ and f be nonnegative measurable functions on $(\Omega, \mathcal{F}, \mu)$ such that $f_n \uparrow f$ a.e. (μ). Then*

$$\int f d\mu = \lim_{n\to\infty} \int f_n d\mu. \tag{3.8}$$

Remark 2.3.4: The important difference between (3.4) and (3.8) is that in (3.8), the f_n's need not be simple.

Proof: It is similar to the proof of Proposition 2.3.2. Let $\{g_n\}_{n\geq 1}$ be a sequence of nonnegative simple functions on $(\Omega, \mathcal{F}, \mu)$ such that $g_n(\omega) \uparrow f(\omega)$ for all ω. By hypothesis, there exists a set $A \in \mathcal{F}$ such that $\mu(A^c) = 0$ and for ω in A, $f_n(\omega) \uparrow f(\omega)$. Fix $k \in \mathbb{N}$ and $0 < \rho < 1$. Let $D_n = \{\omega : \omega \in A,\ f_n(\omega) \geq \rho g_k(\omega)\}$, $n \geq 1$. Then, $D_n \uparrow D \equiv \{\omega : \omega \in A, f(\omega) \geq \rho g_k(\omega)\}$. Since $g_k(\omega) \leq f(\omega)$ for all ω, it follows that $D = A$. Now, by Corollary 2.3.3,

$$\int f_n d\mu \geq \int f_n I_{D_n} d\mu \geq \rho \int g_k I_{D_n} d\mu \quad \text{for all} \quad n \geq 1.$$

By m.c.f.b, $\int g_k I_{D_n} d\mu \uparrow \int g_k I_A d\mu = \int g_k d\mu$ as $n \to \infty$, yielding

$$\liminf_{n\to\infty} \int f_n d\mu \geq \rho \int g_k d\mu$$

for all $0 < \rho < 1$ and all $k \in \mathbb{N}$. Letting $\rho \uparrow 1$ first and then $k \uparrow \infty$, from (3.3) one gets

$$\liminf_{n\to\infty} \int f_n d\mu \geq \int f d\mu.$$

By monotonicity (Corollary 2.3.3),

$$\int f_n d\mu \leq \int f d\mu \quad \text{for all} \quad n \geq 1$$

and so the proof is complete. □

Corollary 2.3.5: *Let $\{h_n\}_{n\geq 1}$ be a sequence of nonnegative measurable functions on a measure space $(\Omega, \mathcal{F}, \mu)$. Then*

$$\int \left(\sum_{n=1}^{\infty} h_n \right) d\mu = \sum_{n=1}^{\infty} \int h_n d\mu .$$

Proof: Let $f_n = \sum_{i=1}^{n} h_i, n \geq 1$, and let $f = \sum_{i=1}^{\infty} h_i$. Then, $0 \leq f_n \uparrow f$. By the MCT,

$$\int f_n d\mu \uparrow \int f d\mu.$$

But by Corollary 2.3.3,

$$\int f_n d\mu = \sum_{i=1}^{n} \int h_i d\mu.$$

Hence, the result follows. □

Corollary 2.3.6: *Let f be a nonnegative measurable function on a measurable space $(\Omega, \mathcal{F}, \mu)$. For $A \in \mathcal{F}$, let*

$$\nu(A) \equiv \int f I_A d\mu.$$

Then, ν is a measure on $(\Omega, \mathcal{F})$.

Proof: Let $\{A_n\}_{n\geq 1}$ be a sequence of disjoint sets in $\mathcal{F}$. Let $h_n = f I_{A_n}$ for $n \geq 1$. Then by Corollary 2.3.5,

$$\begin{aligned} \nu(\cup_{n\geq 1} A_n) &= \int f I_{[\cup_{n\geq 1} A_n]} d\mu = \int f \cdot \left[\sum_{n=1}^{\infty} I_{A_n} \right] d\mu \\ &= \int \left[\sum_{n=1}^{\infty} h_n \right] d\mu = \sum_{n=1}^{\infty} \int h_n d\mu = \sum_{n=1}^{\infty} \nu(A_n). \end{aligned}$$

□

Remark 2.3.5: Notice that $\mu(A) = 0 \Rightarrow \nu(A) = 0$. In this case ν is said to be *dominated by* or *absolutely continuous with respect to* μ. The Radon-Nikodym theorem (see Chapter 4) provides a converse to this. That is, if ν and μ are two measures on a measurable space $(\Omega, \mathcal{F})$ such that

ν is dominated by μ and μ is σ-finite, then there exists a nonnegative measurable function f such that $\nu(A) = \int f I_A d\mu$ for all A in $\mathcal{F}$. This f is called a *Radon-Nikodym derivative* (or *a density*) of ν with respect to μ and is denoted by $\frac{d\nu}{d\mu}$.

Theorem 2.3.7: (*Fatou's lemma*). *Let $\{f_n\}_{n\geq 1}$ be a sequence of nonnegative measurable functions on $(\Omega, \mathcal{F}, \mu)$. Then*

$$\liminf_{n\to\infty} \int f_n d\mu \geq \int \liminf_{n\to\infty} f_n d\mu. \tag{3.9}$$

Proof: Let $g_n(\omega) \equiv \inf\{f_j(\omega) : j \geq n\}$. Then $\{g_n\}_{n\geq 1}$ is a sequence of nonnegative, nondecreasing measurable functions on $(\Omega, \mathcal{F}, \mu)$ such that $g_n(\omega) \uparrow g(\omega) \equiv \liminf_{n\to\infty} f_n(\omega)$. By the MCT,

$$\int g_n d\mu \uparrow \int g d\mu.$$

But by monotonicity

$$\int f_n d\mu \geq \int g_n d\mu \text{ for each } n \geq 1,$$

and hence, (3.9) follows. □

Remark 2.3.6: In (3.9), the inequality can be strict. For example, take $f_n = I_{[n,\infty)}$, $n \geq 1$, on the measure space $(\mathbb{R}, \mathcal{B}(\mathbb{R}), m)$ where m is the Lebesgue measure. For another example, consider $f_n = nI_{[0,\frac{1}{n}]}$, $n \geq 1$, on the finite measure space $([0,1], \mathcal{B}([0,1]), m)$.

Definition 2.3.4: (*The integral of a measurable function*). Let f be a real valued measurable function on a measure space $(\Omega, \mathcal{F}, \mu)$. Let $f^+ = fI_{\{f\geq 0\}}$ and $f^- = -fI_{\{f<0\}}$. The integral of f with respect to μ, denoted by $\int f d\mu$, is defined as

$$\int f d\mu = \int f^+ d\mu - \int f^- d\mu,$$

provided that at least one of the integrals on the right side is finite.

Remark 2.3.7: Note that both f^+ and f^- are nonnegative measurable functions and $f = f^+ - f^-$ and $|f| = f^+ + f^-$. Further, the integrals $\int f^+ d\mu$ and $\int f^- d\mu$ in Definition 2.3.4 are defined via Definition 2.3.3.

Definition 2.3.5: (*Integrable functions*). A measurable function f on a measure space $(\Omega, \mathcal{F}, \mu)$ is said to be *integrable* with respect to μ if $\int |f| d\mu < \infty$.

Since $|f| = f^+ + f^-$, it follows that f is *integrable* iff both f^+ and f^- are integrable, i.e., $\int f^+ d\mu < \infty$ and $\int f^- d\mu < \infty$. In the following, whenever the integral of f or its integrability is discussed, the measurability of f will be assumed to hold.

Remark on notation: The $\int f d\mu$ is also written as

$$\int_\Omega f(\omega)\mu(d\omega) \quad \text{and} \quad \int_\Omega f(\omega)d\mu(\omega).$$

Definition 2.3.6: Let f be a measurable function on a measure space $(\Omega, \mathcal{F}, \mu)$ and $A \in \mathcal{F}$. Then integral of f over A with respect to μ, denoted by $\int_A f d\mu$, is defined as

$$\int_A f d\mu \equiv \int f I_A d\mu, \tag{3.10}$$

provided the right side is well defined.

Definition 2.3.7: (L^p *spaces*). Let $(\Omega, \mathcal{F}, \mu)$ be a measure space and $0 < p \leq \infty$. Then $L^p(\Omega, \mathcal{F}, \mu)$ is defined as

$$\begin{aligned} L^p(\Omega, \mathcal{F}, \mu) &\equiv \{f : |f|^p \text{ is integrable with respect to } \mu\} \\ &= \left\{f : \int |f|^p d\mu < \infty\right\} \quad \text{for} \quad 0 < p < \infty, \end{aligned}$$

and

$$L^\infty(\Omega, \mathcal{F}, \mu) \equiv \Big\{f : \mu(\{|f| > K\}) = 0 \quad \text{for some} \quad K \in (0, \infty)\Big\}.$$

The following is an extension of Proposition 2.3.1 to integrable functions.

Proposition 2.3.8: *Let f, $g \in L^1(\Omega, \mathcal{F}, \mu)$. Then*

(i) $\int (\alpha f + \beta g) d\mu = \alpha \int f d\mu + \beta \int g d\mu$ *for any* $\alpha, \beta \in \mathbb{R}$.

(ii) $f \geq g$ *a.e.* $(\mu) \Rightarrow \int f d\mu \geq \int g d\mu$.

(iii) $f = g$ *a.e.* $(\mu) \Rightarrow \int f d\mu = \int g d\mu$.

Proof: It is easy to verify (Problem 2.32) that if $h = h_1 - h_2$ where h_1 and h_2 are nonnegative functions in $L^1(\Omega, \mathcal{F}, \mu)$, then h is also in $L^1(\Omega, \mathcal{F}, \mu)$ and

$$\int h d\mu = \int h_1 d\mu - \int h_2 d\mu. \tag{3.11}$$

Note that $h \equiv \alpha f + \beta g$ can be written as

$$\begin{aligned} & (\alpha^+ - \alpha^-)(f^+ - f^-) + (\beta^+ - \beta^-)(g^+ - g^-) \\ = \; & (\alpha^+ f^+ + \alpha^- f^- + \beta^+ g^+ + \beta^- g^-) \\ & - (\alpha^+ f^- + \alpha^- f^+ + \beta^+ g^- + \beta^- g^+) \\ = \; & h_1 - h_2, \quad \text{say}. \end{aligned}$$

Since $f,\ g \in L^1(\Omega, \mathcal{F}, \mu)$, it follows that h_1 and $h_2 \in L^1(\Omega, \mathcal{F}, \mu)$. Further, they are nonnegative and by (3.11), $h \in L^1(\Omega, \mathcal{F}, \mu)$ and

$$\int h d\mu = \int h_1 d\mu - \int h_2 d\mu.$$

Now apply Proposition 2.3.1 to each of the terms on the right side and regroup the terms to get

$$\int h d\mu = \alpha \int f d\mu + \beta \int g d\mu.$$

Proofs of (ii) and (iii) are left as an exercise. □

Remark 2.3.8: By Proposition 2.3.8, if f and $g \in L^1(\Omega, \mathcal{F}, \mu)$, then so does $\alpha f + \beta g$ for all $\alpha, \beta \in \mathbb{R}$. Thus, $L^1(\Omega, \mathcal{F}, \mu)$ is a vector space over $\mathbb{R}$. Further, if one sets

$$\|f\|_1 \equiv \int |f| d\mu,$$

(and identifies a function f with its equivalence class under the relation $f \sim g$ iff $f = g$ a.e. (μ)), then $\|\cdot\|_1$ defines a norm on $L^1(\Omega, \mathcal{F}, \mu)$ and makes it a *normed linear space* (cf. Chapter 3). A similar remark also holds for $L^p(\Omega, \mathcal{F}, \mu)$ for $1 < p \leq \infty$.

Next note that by Proposition 2.3.8, if $f = 0$ a.e. (μ), then $\int f d\mu = 0$. However, the converse is not true. But if f is nonnegative a.e. (μ), then the converse is true as shown below.

Proposition 2.3.9: *Let f be a measurable function on $(\Omega, \mathcal{F}, \mu)$ and let f be nonnegative a.e. (μ). Then*

$$\int f d\mu = 0 \quad \textit{iff} \quad f = 0 \quad a.e. \quad (\mu).$$

Proof: It is enough to prove the "only if" part. Let $D = \{\omega : f(\omega) > 0\}$ and $D_n = \{\omega : f(\omega) > \frac{1}{n}\}$, $n \geq 1$. Then $D = \cup_{n \geq 1} D_n$. Since $f \geq f I_{D_n}$ a.e. (μ),

$$0 = \int f d\mu \geq \int f I_{D_n} d\mu \geq \frac{1}{n} \mu(D_n) \Rightarrow \mu(D_n) = 0 \quad \text{for each} \quad n \geq 1.$$

Also $D_n \uparrow D$ and so by m.c.f.b.,

$$\mu(D) = \lim_{n \to \infty} \mu(D_n) = 0\,.$$

Hence, Proposition 2.3.9 follows. □

A dual to the above proposition is the next one.

Proposition 2.3.10: *Let $f \in L^1(\Omega, \mathcal{F}, \mu)$. Then, $|f| < \infty$ a.e. (μ).*

Proof: Let $C_n = \{\omega : |f(\omega)| > n\}$, $n \geq 1$ and let $C = \{\omega : |f(\omega)| = \infty\}$. Then $C_n \downarrow C$ and

$$\int |f| d\mu \geq \int |f| I_{C_n} d\mu \geq n\mu(C_n) \Rightarrow \mu(C_n) \leq \frac{\int |f| d\mu}{n}.$$

Since $\int |f| d\mu < \infty$, $\lim_{n\to\infty} \mu(C_n) = 0$. Hence, by m.c.f.a., $\mu(C) = \lim_{n\to\infty} \mu(C_n) = 0$. □

The next result is a useful convergence theorem for integrals.

Theorem 2.3.11: *(The extended dominated convergence theorem or EDCT). Let $(\Omega, \mathcal{F}, \mu)$ be a measure space and let $f_n, g_n : \Omega \to \mathbb{R}$ be $\langle \mathcal{F}, \mathbb{R} \rangle$-measurable functions such that $|f_n| \leq g_n$ a.e. (μ) for all $n \geq 1$. Suppose that*

(i) $g_n \to g$ a.e. (μ) and $f_n \to f$ a.e. (μ);

(ii) $g_n, g \in L^1(\Omega, \mathcal{F}, \mu)$ and $\int |g_n| d\mu \to \int |g| d\mu$ as $n \to \infty$. Then, $f \in L^1(\Omega, \mathcal{F}, \mu)$,

$$\lim_{n\to\infty} \int f_n d\mu = \int f d\mu \quad \textit{and} \quad \lim_{n\to\infty} \int |f_n - f| d\mu = 0. \tag{3.12}$$

Two important special cases of Theorem 2.3.11 will be stated next. When $g_n = g$ for all $n \geq 1$, one has the standard version of the dominated convergence theorem.

Corollary 2.3.12: *(Lebesgue's dominated convergence theorem or DCT). If $|f_n| \leq g$ a.e. (μ) for all $n \geq 1$, $\int g d\mu < \infty$ and $f_n \to f$ a.e. (μ), then $f \in L^1(\Omega, \mathcal{F}, \mu)$,*

$$\lim_{n\to\infty} \int f_n d\mu = \int f d\mu \quad \textit{and} \quad \lim_{n\to\infty} \int |f_n - f| d\mu = 0. \tag{3.13}$$

Corollary 2.3.13: *(The bounded convergence theorem or BCT). Let $\mu(\Omega) < \infty$. If there exist a $0 < k < \infty$ such that $|f_n| \leq k$ a.e. (μ) and $f_n \to f$ a.e. (μ), then*

$$\lim_{n\to\infty} \int f_n d\mu = \int f d\mu \quad \textit{and} \quad \lim_{n\to\infty} \int |f_n - f| d\mu = 0. \tag{3.14}$$

Proof: Take $g(\omega) \equiv k$ for all $\omega \in \Omega$ in the previous corollary. □

Proof of Theorem 2.3.11: By Fatou's lemma,

$$\int |f| d\mu \leq \liminf_{n\to\infty} \int |f_n| d\mu \leq \liminf_{n\to\infty} \int |g_n| d\mu = \int |g| d\mu < \infty.$$

Hence, f is integrable. For proving the second part, let $h_n = f_n + g_n$ and $\gamma_n = g_n - f_n, n \geq 1$. Then, $\{h_n\}_{n\geq 1}$ and $\{\gamma_n\}_{n\geq 1}$ are sequences of nonnegative integrable functions. By Fatou's lemma and (ii),

$$\begin{aligned}\int (f+g)d\mu &= \int \liminf_{n\to\infty} h_n d\mu \\ &\leq \liminf_{n\to\infty} \int h_n d\mu \\ &= \liminf_{n\to\infty} \left[\int g_n d\mu + \int f_n d\mu\right] \\ &= \int g d\mu + \liminf_{n\to\infty} \int f_n d\mu.\end{aligned}$$

Similarly,

$$\int (g-f)d\mu \leq \int g d\mu - \limsup_{n\to\infty} \int f_n d\mu.$$

By Proposition 2.3.8, $\int (g \pm f)d\mu = \int g d\mu \pm \int f d\mu$. Hence,

$$\int f d\mu \leq \liminf_{n\to\infty} \int f_n d\mu$$

and

$$\limsup_{n\to\infty} \int f_n d\mu \leq \int f d\mu$$

yielding that $\lim_{n\to\infty} \int f_n d\mu = \int f d\mu$. For the last part, apply the above argument to f_n and g_n replaced by $\tilde{f}_n \equiv |f - f_n|$ and $\tilde{g}_n \equiv g_n + |f|$, respectively. □

Theorem 2.3.14: (*An approximation theorem*). *Let μ_F be a Lebesgue-Stieltjes measure on $(\mathbb{R}, \mathcal{B}(\mathbb{R}))$. Let $f \in L^p(\mathbb{R}, \mathcal{B}(\mathbb{R}), \mu_F)$, $0 < p < \infty$. Then, for any $\delta > 0$, there exist a step function h and a continuous function g with compact support (i.e., g vanishes outside a bounded interval) such that*

$$\int |f - h|^p d\mu < \delta, \tag{3.15}$$

$$\int |f - g|^p d\mu < \delta, \tag{3.16}$$

where a step function h is a function of the form $h = \sum_{i=1}^k c_i I_{A_i}$ with $k < \infty$, $c_1, c_2, \ldots, c_k \in \mathbb{R}$ and $A_1, A_2, \ldots, A_k$ being bounded disjoint intervals.

Proof: Let $f_n(\cdot) = f(\cdot) I_{B_n}(\cdot)$ where $B_n = \{x : |x| \leq n, |f(x)| \leq n\}$. By the DCT, for every $\epsilon > 0$, there exists an N_ϵ such that for all $n \geq N_\epsilon$,

$$\int |f - f_n|^p d\mu_F < \epsilon. \tag{3.17}$$

Since $|f_{N_\epsilon}(\cdot)| \leq N_\epsilon$ on $[-N_\epsilon, N_\epsilon]$, for any $\eta > 0$, there exists a simple function $\tilde{f}$ such that

$$\sup\{|f_{N_\epsilon}(x) - \tilde{f}(x)| : x \in \mathbb{R}\} < \eta. \quad \text{(cf. (3.7))} \tag{3.18}$$

Next, using Problem 1.32 (b), one can show that for any $\eta > 0$, there exists a step function h (depending on η) such that

$$\int |\tilde{f}_{N_\epsilon} - h|^p d\mu_F < \eta. \tag{3.19}$$

Since $f - h = f - f_{N_\epsilon} + f_{N_\epsilon} - \tilde{f} + \tilde{f} - h$,

$$|f - h|^p \leq C_p\Big(|f - f_{N_\epsilon}|^p + |f_{N_\epsilon} - \tilde{f}|^p + |\tilde{f} - h|^p\Big),$$

where C_p is a constant depending only on p.

This in turn yields, from (3.17)–(3.19),

$$\int |f - h|^p d\mu_F \leq C_p(\epsilon + (\mu_F\{x : |x| \leq N_\epsilon\})\eta^p + \eta). \tag{3.20}$$

Given $\delta > 0$, choose $\epsilon > 0$ first and then $\eta > 0$ such that the right side of (3.20) above is less than δ.

Next, given any step function h and $\eta > 0$ there exists a continuous function g with compact support such that $\mu_F\{x : h(x) \neq g(x)\} < \eta$ (cf. Problem 1.32 (g)). Now (3.16) follows from (3.15). □

Remark 2.3.9: The approximation (3.16) remains valid if g is restricted to the class of all infinitely differentiable functions with compact support. Further it remains valid for $0 < p < \infty$ for Lebesgue-Stieltjes measures on any Euclidean space.

Remark 2.3.10: The above approximation theorem fails for $p = \infty$. For example, consider the function $f(x) \equiv 1$ in $L^\infty(m)$.

2.4 Riemann and Lebesgue integrals

Let f be a real valued bounded function on a bounded interval $[a, b]$. Recall the definition of the *Riemann integral.* Let $P = \{x_0, x_1, \ldots, x_n\}$ be a finite *partition* of $[a, b]$, i.e., $x_0 = a < x_1 < x_2 < x_{n-1} < x_n = b$ and $\Delta = \Delta(P) \equiv \max\{(x_{i+1} - x_i) : 0 \leq i \leq n-1\}$ be the *diameter* of P. Let $M_i = \sup\{f(x) : x_i \leq x \leq x_{i+1}\}$ and $m_i = \inf\{f(x) : x_i \leq x \leq x_{i+1}\}$, $i = 0, 1, \ldots, n-1$.

Definition 2.4.1: The *upper-* and *lower-Riemann sums* of f w.r.t. the partition P are, respectively, defined as

$$U(f, P) \equiv \sum_{i=0}^{n-1} M_i \cdot (x_{i+1} - x_i) \tag{4.1}$$

and

$$L(f,P) \equiv \sum_{i=0}^{n-1} m_i \cdot (x_{i+1} - x_i). \tag{4.2}$$

It is easy to verify that if $Q = \{y_0, y_1, \dots, y_k\}$ is another partition satisfying $P \subset Q$, then $U(f,P) \geq U(f,Q) \geq L(f,Q) \geq L(f,P)$. Let $\mathcal{P}$ denote the collection of all finite partitions of $[a,b]$.

Definition 2.4.2: The *upper-Riemann integral* $\overline{\int} f$ is defined as

$$\overline{\int} f = \inf_{P \in \mathcal{P}} U(f,P) \tag{4.3}$$

and the *lower-Riemann integral* $\underline{\int} f$, by

$$\underline{\int} f = \sup_{P \in \mathcal{P}} L(f,P). \tag{4.4}$$

It can be shown (cf. Problem 2.23) that if $\{P_n\}_{n\geq 1}$ is any sequence of partitions such that $\Delta(P_n) \to 0$ as $n \to \infty$ and $P_n \subset P_{n+1}$ for each $n \geq 1$, then $U(P_n, f) \downarrow \overline{\int} f$ and $L(P_n, f) \uparrow \underline{\int} f$.

Definition 2.4.3: f is said to be *Riemann integrable* if

$$\overline{\int} f = \underline{\int} f. \tag{4.5}$$

The common value is denoted by $\oint_{[a,b]} f$.

Fix a sequence $\{P_n\}_{n\geq 1}$ of partitions such that $P_n \subset P_{n+1}$ for all $n \geq 1$ and $\Delta(P_n) \to 0$ as $n \to \infty$. Let $P_n = \{x_{n0} = a < x_{n1} < x_{n2} \dots < x_{nk_n} = b\}$. For $i = 0, 1, \dots, k_n - 1$, let

$$\begin{aligned} \phi_n(x) &\equiv \sup\{f(t) : x_i \leq t \leq x_{i+1}\}, \quad x \in [x_i, x_{i+1}) \\ \psi_n(x) &\equiv \inf\{f(t) : x_i \leq t \leq x_{i+1}\}, \quad x \in [x_i, x_{i+1}) \end{aligned}$$

and let $\phi_n(b) = \psi_n(b) = 0$. Then, ϕ_n and ψ_n are step functions on $[a,b]$ and hence, are Borel measurable. Further, since f is bounded, so are ϕ_n and ψ_n and hence, are integrable on $[a,b]$ w.r.t. the Lebesgue measure m. The Lebesgue integrals of ϕ_n and ψ_n are given by $\int_{[a,b]} \phi_n dm = U(P_n, f)$ and $\int_{[a,b]} \psi_n dm = L(P_n, f)$.

It can be shown (Problem 2.24) that for all $x \notin \cup_{n\geq 1} P_n$, as $n \to \infty$,

$$\phi_n(x) \to \phi(x) \equiv \lim_{\delta \downarrow 0} \sup\{f(y) : |y - x| < \delta\} \tag{4.6}$$

and

$$\psi_n(x) \to \psi(x) \equiv \lim_{\delta \downarrow 0} \inf\{f(y) : |y - x| < \delta\} \tag{4.7}$$

Thus, ϕ and ψ, being limits of Borel measurable functions (except possibly on a countable set), are Borel measurable. By the BCT (Corollary 2.3.13),

$$\overline{\int} f = \lim_{n\to\infty} \int \phi_n dm = \int \phi dm$$

and

$$\underline{\int} f = \lim_{n\to\infty} \int \psi_n dm = \int \psi dm.$$

Thus, f is Riemann integrable on $[a,b]$, iff $\int \phi dm = \int \psi dm$, iff $\int(\phi - \psi)dm = 0$. Since $\phi(x) \geq f(x) \geq \psi(x)$ for all x, this holds iff $\phi = f = \psi$ a.e $[m]$. It can be shown that f is continuous at x_0 iff $\phi(x_0) = f(x_0) = \psi(x_0)$ (Problem 2.8). Summarizing the above discussion, one gets the following theorem.

Theorem 2.4.1: *Let f be a bounded function on a bounded interval $[a,b]$. Then f is Riemann integrable on $[a,b]$ iff f is continuous a.e. (m) on $[a,b]$. In this case, f is Lebesgue integrable on $[a,b]$ and the Lebesgue integral $\int_{[a,b]} f dm$ equals the Riemann integral $\oint_{[a,b]} f$, i.e., the two integrals coincide.*

It should be noted that Lebesgue integrability need not imply Riemann integrability. For example, consider $f(x) \equiv I_{\mathbb{Q}_1}(x)$ where $\mathbb{Q}_1$ is the set of rationals in $[0,1]$ (Problem 2.25).

The functions ϕ and ψ defined in (4.6) and (4.7) above are called, respectively, the *upper* and the *lower envelopes* of the function f. They are *semicontinuous* in the sense that for each $\alpha \in \mathbb{R}$, the sets $\{x : \phi(x) < \alpha\}$ and $\{x : \psi(x) > \alpha\}$ are open (cf. Problem 2.8).

Remark 2.4.1: The key difference in the definitions of Riemann and Lebesgue integrals is that in the former the *domain of f* is partitioned while in the latter the *range of f* is partitioned.

2.5 More on convergence

Let $\{f_n\}_{n\geq 1}$ and f be measurable functions from a measure space $(\Omega, \mathcal{F}, \mu)$ to $\bar{\mathbb{R}}$, the set of extended real numbers. There are several notions of convergence of $\{f_n\}_{n\geq 1}$ to f. The following two have been discussed earlier.

Definition 2.5.1: $\{f_n\}_{n\geq 1}$ *converges to f pointwise* if

$$\lim_{n\to\infty} f_n(\omega) = f(\omega) \quad \text{for all} \quad \omega \text{ in } \Omega.$$

Definition 2.5.2: $\{f_n\}_{n\geq 1}$ *converges to f almost everywhere* (μ), denoted by $f_n \to f$ a.e. (μ), if there exists a set B in $\mathcal{F}$ such that $\mu(B) = 0$ and

$$\lim_{n\to\infty} f_n(\omega) = f(\omega) \quad \text{for all} \quad \omega \in B^c. \tag{5.1}$$

Now consider some more notions of convergence.

Definition 2.5.3: $\{f_n\}_{n\geq 1}$ *converges to* f *in measure* (w.r.t. μ), denoted by $f_n \longrightarrow^m f$, if for each $\epsilon > 0$,

$$\lim_{n\to\infty} \mu(\{|f_n - f| > \epsilon\}) = 0. \tag{5.2}$$

Definition 2.5.4: Let $0 < p < \infty$. Then, $\{f_n\}_{n\geq 1}$ *converges to* f *in* $L^p(\mu)$, denoted by $f_n \longrightarrow^{L^p} f$, if $\int |f_n|^p d\mu < \infty$ for all $n \geq 1$, $\int |f|^p d\mu < \infty$ and

$$\lim_{n\to\infty} \int |f_n - f|^p d\mu = 0. \tag{5.3}$$

Clearly, (5.3) is equivalent to $\|f_n - f\|_p \to 0$ as $n \to \infty$, where for any $\mathcal{F}$-measurable function g and any $0 < p < \infty$,

$$\|g\|_p = \Big(\int |g|^p d\mu\Big)^{\min\{\frac{1}{p},1\}}. \tag{5.4}$$

For $p = 1$, this is also called *convergence in absolute deviation* and for $p = 2$, *convergence in mean square.*

Definition 2.5.5: $\{f_n\}_{n\geq 1}$ *converges to* f *uniformly* (over Ω) if

$$\lim_{n\to\infty} \sup\{|f_n(\omega) - f(\omega)| : \omega \in \Omega\} = 0. \tag{5.5}$$

Definition 2.5.6: $\{f_n\}_{n\geq 1}$ *converges to* f *in* $L^\infty(\mu)$ if

$$\lim_{n\to\infty} \|f_n - f\|_\infty = 0, \tag{5.6}$$

where for any $\mathcal{F}$-measurable function g on $(\Omega, \mathcal{F}, \mu)$,

$$\|g\|_\infty = \inf\big\{K : K \in (0,\infty), \mu(\{|g| > K\}) = 0\big\}. \tag{5.7}$$

Definition 2.5.7: $\{f_n\}_{n\geq 1}$ *converges to* f *nearly uniformly* (μ) if for every $\epsilon > 0$, there exists a set $A \in \mathcal{F}$ such that $\mu(A) < \epsilon$ and on A^c, $f_n \to f$ uniformly, i.e., $\sup\{|f_n(\omega) - f(\omega)| : \omega \in A^c\} \to 0$ as $n \to \infty$.

The notion of convergence in Definition 2.5.7 is also called almost uniform convergence in some books (cf. Royden (1988)). The sequence $f_n \equiv nI_{[0,1/n]}$ on $(\Omega = [0,1], \mathcal{B}([0,1]), m)$ converges to $f \equiv 0$ nearly uniformly but not in $L^\infty(m)$.

When μ is a probability measure, there is another useful notion of convergence, known as convergence in distribution, that is defined in terms of the induced measures $\{\mu f_n^{-1}\}_{n\geq 1}$ and μf^{-1}. This notion of convergence will be treated in detail in Chapter 9.

Next, the connections between some of these notions of convergence are explored.

Theorem 2.5.1: *Suppose that $\mu(\Omega) < \infty$. Then, $f_n \to f$ a.e. (μ) implies $f_n \longrightarrow^m f$.*

The proof is left as an exercise (Problem 2.26). The hypothesis that '$\mu(\Omega) < \infty$' in Theorem 2.5.1 cannot be dispensed with as seen by taking $f_n = I_{[n,\infty)}$ on $\mathbb{R}$ with Lebesgue measure. Also, $f_n \longrightarrow^m f$ does not imply $f_n \to f$ a.e. (μ) (Problem 2.46), but the following holds.

Theorem 2.5.2: *Let $f_n \longrightarrow^m f$. Then, there exists a subsequence $\{n_k\}_{k\geq 1}$ such that $f_{n_k} \to f$ a.e. (μ).*

Proof: Since $f_n \longrightarrow^m f$, for each integer $k \geq 1$, there exists an n_k such that for all $n \geq n_k$

$$\mu\Big(\{|f_n - f| > 2^{-k}\}\Big) < 2^{-k}. \tag{5.8}$$

W.l.o.g., that assume $n_{k+1} > n_k$ for all $k \geq 1$. Let $A_k = \{|f_{n_k} - f| > 2^{-k}\}$. By Corollary 2.3.5,

$$\int \Big(\sum_{k=1}^{\infty} I_{A_k}\Big) d\mu = \sum_{k=1}^{\infty} \int I_{A_k} d\mu = \sum_{k=1}^{\infty} \mu(A_k),$$

which, by (5.8), is finite. Hence, by Proposition 2.3.10, $\sum_{k=1}^{\infty} I_{A_k} < \infty$ a.e. (μ). Now observe that $\sum_{k=1}^{\infty} I_{A_k}(\omega) < \infty \Rightarrow |f_{n_k}(\omega) - f(\omega)| \leq 2^{-k}$ for all k large $\Rightarrow \lim_{k\to\infty} f_{n_k}(\omega) = f(\omega)$. Thus, $f_{n_k} \to f$ a.e. (μ). □

Remark 2.5.1: From the above result it follows that the extended dominated convergence theorem (Theorem 2.3.11) remains valid if convergence a.e. of $\{f_n\}_{n\geq 1}$ and of $\{g_n\}_{n\geq 1}$ are replaced by convergence in measure for both (Problem 2.37).

Theorem 2.5.3: *Let $\{f_n\}_{n\geq 1}$, f be measurable functions on a measure space $(\Omega, \mathcal{F}, \mu)$. Let $f_n \longrightarrow^{L^p} f$ for some $0 < p < \infty$. Then $f_n \longrightarrow^m f$.*

Proof: For each $\epsilon > 0$, let $A_n = \{|f_n - f| \geq \epsilon\}$, $n \geq 1$. Then

$$\int |f_n - f|^p d\mu \geq \int_{A_n} |f_n - f|^p d\mu \geq \epsilon^p \mu(A_n).$$

Since $f_n \to f$ in L^p, $\int |f_n - f|^p d\mu \to 0$ and hence, $\mu(A_n) \to 0$. □

It turns out that $f_n \longrightarrow^m f$ need not imply $f_n \longrightarrow^{L^p} f$, even if $\{f_n : n \geq 1\} \cup \{f\}$ is contained in $L^p(\Omega, \mathcal{F}, \mu)$. For example, let $f_n = nI_{[0,\frac{1}{n}]}$ and $f \equiv 0$ on the Lebesgue space $([0,1], \mathcal{B}([0,1]), m)$, where m is the Lebesgue measure. Then $f_n \longrightarrow^m f$ but $\int |f_n - f| \equiv 1$ for all $n \geq 1$. However, under some additional conditions, convergence in measure does imply convergence in L^p. Here are two results in this direction.

Theorem 2.5.4: *(Scheffe's theorem). Let $\{f_n\}_{n\geq 1}, f$ be a collection of nonnegative measurable functions on a measure space $(\Omega, \mathcal{F}, \mu)$. Let $f_n \to f$ a.e. (μ), $\int f_n d\mu \to \int f d\mu$ and $\int f d\mu < \infty$. Then*

$$\lim_{n\to\infty} \int |f_n - f| d\mu = 0.$$

Proof: Let $g_n = f - f_n$, $n \geq 1$. Since $f_n \to f$ a.e. (μ), both g_n^+ and g_n^- go to zero a.e. (μ). Further, $0 \leq g_n^+ \leq f$ and by hypothesis $\int f d\mu < \infty$. Thus, by the DCT, it follows that

$$\int g_n^+ d\mu \to 0.$$

Next, note that by hypothesis, $\int g_n d\mu \to 0$. Thus, $\int g_n^- d\mu = \int g_n^+ d\mu - \int g_n d\mu \to 0$ and hence, $\int |g_n| d\mu = \int g_n^+ d\mu + \int g_n^- d\mu \to 0$. □

Corollary 2.5.5: *Let $\{f_n\}_{n\geq 1}$, f be probability density functions on a measure space $(\Omega, \mathcal{F}, \mu)$. That is, for all $n \geq 1$, $\int f_n d\mu = \int f d\mu = 1$ and $f_n, f \geq 0$ a.e. (μ). If $f_n \to f$ a.e. (μ), then*

$$\lim_{n\to\infty} \int |f_n - f| d\mu = 0.$$

Remark 2.5.2: The above theorem and the corollary remain valid if the convergence of f_n to f a.e. (μ) is replaced by $f \longrightarrow^m f$.

Corollary 2.5.6: *Let $\{p_{nk}\}_{k\geq 1}$, $n = 1, 2, \ldots$ and $\{p_k\}_{k\geq 1}$ be sequences of nonnegative numbers satisfying $\sum_{k=1}^{\infty} p_{nk} = 1 = \sum_{k=1}^{\infty} p_k$. Let $p_{nk} \to p_k$ as $n \to \infty$ for each $k \geq 1$. Then $\sum_{k=1}^{\infty} |p_{nk} - p_k| \to 0$.*

Proof: Apply Corollary 2.5.5 with $\mu =$ the counting measure on $(\mathbb{N}, \mathcal{P}(\mathbb{N}))$. □

A more general result in this direction that does not require f_n, f to be nonnegative, involves the concept of *uniform integrability.* Let $\{f_\lambda : \lambda \in \Lambda\}$ be a collection of functions in $L^1(\Omega, \mathcal{F}, \mu)$. Then for each $\lambda \in \Lambda$, by the DCT and the integrability of f_λ,

$$a_\lambda(t) \equiv \int_{\{|f_\lambda|>t\}} |f_\lambda| d\mu \to 0 \quad \text{as} \quad t \to \infty. \tag{5.9}$$

The notion of uniform integrability requires that the integrals $a_\lambda(t)$ go to zero uniformly in $\lambda \in \Lambda$ as $t \to \infty$.

Definition 2.5.8: The collection of functions $\{f_\lambda : \lambda \in \Lambda\}$ in $L^1(\Omega, \mathcal{F}, \mu)$ is *uniformly integrable* (or *UI*, in short) if

$$\sup_{\lambda \in \Lambda} a_\lambda(t) \to 0 \text{ as } t \to \infty. \tag{5.10}$$

The following proposition summarizes some of the main properties of UI families of functions.

Proposition 2.5.7: *Let $\{f_\lambda : \lambda \in \Lambda\}$ be a collection of μ-integrable functions on $(\Omega, \mathcal{F}, \mu)$.*

(i) If Λ is finite, then $\{f_\lambda : \lambda \in \Lambda\}$ is UI.

(ii) If $K \equiv \sup\{\int |f_\lambda|^{1+\epsilon} d\mu : \lambda \in \Lambda\} < \infty$ for some $\epsilon > 0$, then $\{f_\lambda : \lambda \in \Lambda\}$ is UI.

(iii) If $|f_\lambda| \leq g$ a.e. (μ) and $\int g d\mu < \infty$, then $\{f_\lambda : \lambda \in \Lambda\}$ is UI.

(iv) If $\{f_\lambda : \lambda \in \Lambda\}$ and $\{g_\gamma : \gamma \in \Gamma\}$ are UI, then so is $\{f_\lambda + g_\gamma : \lambda \in \Lambda, \gamma \in \Gamma\}$.

(v) If $\{f_\lambda : \lambda \in \Lambda\}$ is UI and $\mu(\Omega) < \infty$, then

$$\sup_{\lambda \in \Lambda} \int |f_\lambda| d\mu < \infty. \tag{5.11}$$

Proof: By hypothesis, $a_\lambda(t) \equiv \int_{\{|f_\lambda|>t\}} |f_\lambda| d\mu \to 0$ as $t \to \infty$ for each λ. If Λ is finite this implies that $\sup\{a_\lambda(t) : \lambda \in \Lambda\} \to 0$ as $t \to \infty$. This proves (i).

To prove (ii), note that since $1 < |f_\lambda|/t$ on the set $\{|f_\lambda| > t\}$,

$$\sup_{\lambda \in \Lambda} a_\lambda(t) \leq \sup_{\lambda \in \Lambda} \int_{|f_\lambda|>t} |f_\lambda| \Big[|f_\lambda|/t\Big]^\epsilon d\mu \leq K t^{-\epsilon} \to 0 \quad \text{as} \quad t \to \infty.$$

For (iii), note that for each $t \in \mathbb{R}$, the function $h_t(x) \equiv x I_{(t,\infty)}(x)$, $x \in \mathbb{R}$ is nondecreasing. Hence, by the integrability of g,

$$\sup_{\lambda \in \Lambda} a_\lambda(t) = \sup_{\lambda \in \Lambda} \int h_t(|f_\lambda|) d\mu \leq \int h_t(g) d\mu = \int_{\{g>t\}} g d\mu \to 0 \text{ as } t \to \infty.$$

To prove (iv), for $t > 0$, let $a(t) = \sup_{\lambda \in \Lambda} \int h_t(|f_\lambda|) d\mu$ and $b(t) = \sup_{\gamma \in \Gamma} \int h_t(|g_\gamma|) d\mu$. Then, for any $\lambda \in \Lambda$, and $\gamma \in \Gamma$,

$$\begin{aligned} & \int_{\{|f_\lambda + g_\gamma|>t\}} |f_\lambda + g_\gamma| d\mu = \int h_t\big(|f_\lambda + g_\gamma|\big) d\mu \\ \leq \ & \int h_t\big(2\max\{|f_\lambda|, |g_\gamma|\}\big) d\mu \end{aligned}$$

$$\begin{aligned}
&= \int h_t(2|f_\lambda|)I(|f_\lambda| \geq |g_\gamma|)d\mu + \int h_t(2|g_\gamma|)I(|f_\lambda| < |g_\gamma|)d\mu \\
&\leq 2\int_{\{|f_\lambda|>t/2\}} |f_\lambda|d\mu + 2\int_{\{|g_\gamma|>t/2\}} |g_\gamma|d\mu \\
&\leq 2[a(t/2) + b(t/2)].
\end{aligned}$$

By hypothesis, both $a(t)$ and $b(t) \to 0$ as $t \to \infty$, thus proving (iv).

Next consider (v). Since $\{f_\lambda\} : \lambda \in \Lambda\}$ is UI, there exists a $T > 0$ such that

$$\sup_{\lambda\in\Lambda} \int h_T(|f_\lambda|)d\mu \leq 1.$$

Hence,

$$\begin{aligned}
\sup_{\lambda\in\Lambda} \int |f_\lambda|d\mu &= \sup_{\lambda\in\Lambda} \left\{ \int_{\{|f_\lambda|\leq T\}} |f_\lambda|d\mu + \int h_T(|f_\lambda|)d\mu \right\} \\
&\leq T\mu(\Omega) + 1 < \infty.
\end{aligned}$$

This completes the proof of the proposition. □

Remark 2.5.3: In the above proposition, part (ii) can be improved as follows: *Let* $\phi : \mathbb{R}_+ \to \mathbb{R}_+$ *be nondecreasing and* $\frac{\phi(x)}{x} \uparrow \infty$ *as* $x \uparrow \infty$. *If* $\sup_{\lambda\in\Lambda} \int \phi(|f_\lambda|)d\mu < \infty$, *then* $\{f_\lambda : \lambda \in \Lambda\}$ *is UI* (Problem 2.27). A converse to this result is true. That is, if $\{f_\lambda : \lambda \in \Lambda\}$ is UI then there exists such a function ϕ. Some examples of such ϕ's are $\phi(x) = x^k,\ \ k > 1,\ \ \phi(x) = x(\log x)^\beta I(x > 1),\ \beta > 0$, and $\phi(x) = \exp(\beta x),\ \beta > 0$.

In part (v) of Proposition 2.5.7, (5.11) does not imply UI. For example, consider the sequence of functions $f_n = nI_{[0,\frac{1}{n}]}, n = 1, 2, \ldots$ on $[0, 1]$. On the other hand, (5.11) with an additional condition becomes necessary and sufficient for UI.

Proposition 2.5.8: *Let* $f \in L^1(\Omega, \mathcal{F}, \mu)$. *Then for every* $\epsilon > 0$, *there exists a* $\delta > 0$ *such that* $\mu(A) < \delta \Rightarrow \int_A |f|d\mu < \epsilon$.

Proof: Fix $\epsilon > 0$. By the DCT, there exists a $t > 0$ such that $\int_{\{|f|>t\}} |f|d\mu < \epsilon/2$. Hence, for any $A \in \mathcal{F}$ with $\mu(A) \leq \delta \equiv \frac{\epsilon}{2t}$,

$$\begin{aligned}
\int_A |f|d\mu &\leq \int_{A\cap\{|f|\leq t\}} |f|d\mu + \int_{\{|f|>t\}} |f|d\mu \\
&\leq t\mu(A) + \int_{\{|f|>t\}} |f|d\mu \\
&\leq \epsilon,
\end{aligned}$$

proving the claim. □

The above proposition shows that for every $f \in L^1(\Omega, \mathcal{F}, \mu)$, the measure (cf. Corollary 2.3.6)

$$\nu_{|f|}(A) \equiv \int_A |f| d\mu \tag{5.12}$$

on $(\Omega, \mathcal{F})$ satisfies the condition that $\nu_{|f|}(A)$ is small if $\mu(A)$ is small, i.e., for every $\epsilon > 0$, there exists a $\delta > 0$ such that $\mu(A) < \delta \Rightarrow \nu_{|f|}(A) < \epsilon$. This property is referred to as the *absolute continuity* of the measure ν_f w.r.t. μ.

Definition 2.5.9: Given a family $\{f_\lambda : \lambda \in \Lambda\} \subset L^1(\Omega, \mathcal{F}, \mu)$, the measures $\{\nu_{|f_\lambda|} : \lambda \in \Lambda\}$ as defined in (5.12) above are *uniformly absolutely continuous* w.r.t. μ (or *u.a.c. (μ)*, in short) if for every $\epsilon > 0$, there exists a $\delta > 0$ such that

$$\mu(A) < \delta \Rightarrow \sup \left\{ \nu_{|f_\lambda|}(A) : \lambda \in \Lambda \right\} < \epsilon.$$

Theorem 2.5.9: *Let $\{f_\lambda : \lambda \in \Lambda\} \subset L^1(\Omega, \mathcal{F}, \mu)$ and $\mu(\Omega) < \infty$. Then, $\{f_\lambda : \lambda \in \Lambda\}$ is UI iff $\sup_{\lambda \in \Lambda} \int |f_\lambda| d\mu < \infty$ and $\left\{\nu_{|f_\lambda|}(\cdot) : \lambda \in \Lambda\right\}$ is u.a.c. (μ).*

Proof: Let $\left\{f_\lambda : \lambda \in \Lambda\right\}$ be UI. Then, since $\mu(\Omega) < \infty$, L_1 boundedness of $\{f_\lambda : \lambda \in \Lambda\}$ follows from Proposition 2.5.7 (v). To establish u.a.c. (μ), fix $\epsilon > 0$. By UI, there exists an N such that

$$\sup_{\lambda \in \Lambda} \int_{\{|f_\lambda| > N\}} |f_\lambda| d\mu < \epsilon/2.$$

Let $\delta = \frac{\epsilon}{2N}$ and let $A \in \mathcal{F}$ be such that $\mu(A) < \delta$. Then, as in the proof of Proposition 2.5.8 above,

$$\sup_{\lambda \in \Lambda} \int_A |f_\lambda| d\mu \leq N\mu(A) + \epsilon/2 < \epsilon.$$

Conversely, suppose $\left\{f_\lambda : \lambda \in \Lambda\right\}$ is L^1 bounded and u.a.c. (μ). Then, for every $\epsilon > 0$, there exists a $\delta_\epsilon > 0$ such that

$$\sup_{\lambda \in \Lambda} \int_A |f_\lambda| d\mu < \epsilon \text{ if } \mu(A) < \delta_\epsilon. \tag{5.13}$$

Also, for any nonnegative f in $L_1(\Omega, \mathcal{F}, \mu)$ and $t > 0$, $\int f d\mu \geq \int_{\{f > t\}} f d\mu \geq t\mu(\{f \geq t\})$ which implies that

$$\mu(\{f \geq t\}) \leq \frac{\int f d\mu}{t}.$$

(This is known as *Markov's inequality* – see Chapter 3). Hence, it follows that

$$\sup_{\lambda \in \Lambda} \mu(\{|f_\lambda| \geq t\}) \leq \Big[\sup_{\lambda \in \Lambda} \int |f_\lambda| d\mu\Big] / t. \tag{5.14}$$

Now given $\epsilon > 0$, choose T_ϵ such that $\left[\sup_{\lambda\in\Lambda}\int|f_\lambda|d\mu\right]/T_\epsilon < \delta_\epsilon$ where δ_ϵ is as in (5.13). Then, by (5.14), it follows that

$$\sup_{\lambda\in\Lambda}\int_{\{|f_\lambda|\geq T_\epsilon\}}|f_\lambda|d\mu < \epsilon,$$

i.e., $\{f_\lambda : \lambda \in \Lambda\}$ is UI. □

Theorem 2.5.10: *Let $(\Omega, \mathcal{F}, \mu)$ be a measure space with $\mu(\Omega) < \infty$, and let $\{f_n : n \geq 1\} \subset L^1(\Omega, \mathcal{F}, \mu)$ be such that $f_n \to f$ a.e. (μ) and f is $\langle\mathcal{F}, \mathcal{B}(\mathbb{R})\rangle$-measurable. If $\{f_n : n \geq 1\}$ is UI, then f is integrable and*

$$\lim_{n\to\infty}\int|f_n - f|d\mu = 0.$$

Remark 2.5.4: In view of Proposition 2.5.7 (iii), Theorem 2.5.10 yields convergence of $\int f d\mu$ to $\int f d\mu$ under *weaker* conditions than the DCT, provided $\mu(\Omega) < \infty$. However, even under the restriction $\mu(\Omega) < \infty$, UI of $\{f_n\}_{n\geq1}$ is a sufficient, but *not* a necessary condition for convergence of $\int f_n d\mu$ to $\int f d\mu$ (Problem 2.28). In the special case where f_n's are nonnegative (and $\mu(\Omega) < \infty$), $\int f_n d\mu \to \int f d\mu < \infty$ if and only if $\{f_n\}_{n\geq1}$ are UI (Problem 2.29). On the other hand, when $\mu(\Omega) = +\infty$, UI is no longer sufficient to guarantee the convergence of $\int f_n d\mu$ to $\int f d\mu$ (Problem 2.30). Thus, the notion of UI is useful mainly for *finite* measures and, in particular, probability measures.

Proof: By Proposition 2.5.7 and Fatou's lemma,

$$\int|f|d\mu \leq \liminf_{n\to\infty}\int|f_n|d\mu \leq \sup_{n\geq1}\int|f_n|d\mu < \infty$$

and hence f is integrable.

Next for $n \in \mathbb{N}, t \in (0, \infty)$, define the functions $g_n = |f_n - f|$, $\underline{g}_{n,t} = g_n I(|g_n| \leq t)$ and $\bar{g}_{n,t} = g_n I(|g_n| > t)$. Since $\{f_n\}_{n\geq1}$ is UI and f is integrable, by Proposition 2.5.7 (iv), $\{g_n\}_{n\geq1}$ is UI. Hence, for any $\epsilon > 0$, there exists $t_\epsilon > 0$ such that

$$\sup_{n\geq1}\int\bar{g}_{n,t}d\mu < \epsilon \quad \text{for all} \quad t \geq t_\epsilon. \tag{5.15}$$

Next note that for any $t > 0$, since $f_n \to f$ a.e. (μ), $\underline{g}_{n,t} \to 0$ a.e. (μ), and $|\underline{g}_{n,t}| \leq t$ and $\int(t)d\mu = t\mu(\Omega) < \infty$. Hence, by the DCT,

$$\lim_{n\to\infty}\int\underline{g}_{n,t}d\mu = 0 \quad \text{for all} \quad t > 0. \tag{5.16}$$

By (5.15) and (5.16) with $t = t_\epsilon$, we get

$$0 \leq \limsup_{n\to\infty} \int |f_n - f| d\mu \leq \sup_{n\geq 1} \int \bar{g}_{n,t} d\mu + \lim_{n\to\infty} \int \underline{g}_{n,t} d\mu \leq \epsilon$$

for all $\epsilon > 0$. Hence, $\int |f_n - f| d\mu \to 0$ as $n \to \infty$. □

The next result concerns connections between the notions of almost everywhere convergence and almost uniform convergence.

Theorem 2.5.11: *(Egorov's theorem). Let $f_n \to f$ a.e. (μ) and $\mu(\Omega) < \infty$. Then $f_n \to f$ nearly uniformly (μ) as in Definition 2.5.7.*

Proof: For $j, n, r \in \mathbb{N}$, define the sets

$$\begin{aligned} A_{jr} &= \{\omega : |f_j(\omega) - f(\omega)| \geq r^{-1}\} \\ B_{nr} &= \bigcup_{j\geq n} A_{jr}, \quad C_r = \bigcap_{n\geq 1} B_{nr} \\ D &= \bigcup_{r\geq 1} C_r\,. \end{aligned}$$

It is easy to verify that D is the set of points where f_n does not convergence to f. That is,

$$D = \{\omega : f_n(\omega) \not\to f(\omega)\}.$$

By hypothesis $\mu(D) = 0$. This implies $\mu(C_r) = 0$ for all $r \geq 1$. Since $B_{nr} \downarrow C_r$ as $n \to \infty$ and $\mu(\Omega) < \infty$, by m.c.f.a., $\mu(C_r) = 0 \Rightarrow \lim_{n\to\infty} \mu(B_{nr}) = 0$. So for all $r \geq 1$, $\epsilon > 0$, there exists $k_r \in \mathbb{N}$ such that $\mu(B_{k_r,r}) < \frac{\epsilon}{2^r}$. Let $A = \cup_{r\geq 1} B_{k_r,r}$. Then $\mu(A) < \varepsilon$ and $A^c = \cap_{r\geq 1} B^c_{k_r,r}$. Also, for each $r \in \mathbb{N}$, $A^c \subset B^c_{k_r,r}$. So for any $n \geq 1$,

$$\begin{aligned} & \sup\{|f_n(\omega) - f(\omega)| : \omega \in A^c\} \\ \leq\ & \sup\{|f_n(\omega) - f(\omega)| : \omega \in B^c_{k_r,r}\} \quad \text{for all} \quad r \in \mathbb{N} \\ \leq\ & \frac{1}{r} \quad \text{if} \quad n \geq k_r. \end{aligned}$$

That is, $\sup\{|f_n(\omega) - f(\omega)| : \omega \in A^c\} \to 0$ as $n \to \infty$. □

Theorem 2.5.12: *(Lusin's theorem). Let $F : \mathbb{R} \to \mathbb{R}$ be a nondecreasing function and let $\mu = \mu^*_F$ be the corresponding Lebesgue-Stieltjes measure on $(\mathbb{R}, \mathcal{M}_{\mu^*_F})$. Let $f : \mathbb{R} \to \mathbb{R}$ be $\langle \mathcal{M}_{\mu^*_F}, \mathcal{B}(\mathbb{R})\rangle$-measurable. Let $\mu(\{x : |f(x)| = \infty\}) = 0$. Then for every $\epsilon > 0$, there exists a continuous function $g : \mathbb{R} \to \mathbb{R}$ such that $\mu(\{x : f(x) \neq g(x)\}) < \epsilon$.*

Proof: Fix $-\infty < a < b < \infty$. Since $\mu([a, b]) < \infty$ and $\mu(\{x : |f(x)| = \infty\}) = 0$, for each $\epsilon > 0$, there exists $K \in (0, \infty)$ such that

$$\mu(\{x : a \leq x \leq b, |f(x)| > K\}) < \frac{\epsilon}{2}.$$

Let $f_K(x) = f(x)I_{[a,b]}(x)I_{\{|f|\le K\}}(x)$. Then $f_K : \mathbb{R} \to \mathbb{R}$ is bounded, $\langle \mathcal{M}_{\mu_F^*}, \mathcal{B}(\mathbb{R})\rangle$-measurable and zero for $|x| > K$. Consider now the following claim: *For every $\epsilon > 0$, there exists a continuous function $g : [a,b] \to \mathbb{R}$ such that*

$$\mu\bigl(\{x : f_K(x) \neq g(x), a \le x \le b\}\bigr) < \frac{\epsilon}{2}. \tag{5.17}$$

Clearly this implies that

$$\mu\bigl(\{x : a \le x \le b, f(x) \neq g(x)\}\bigr) < \epsilon. \tag{5.18}$$

Fix $\delta > 0$. Now for each $n \in \mathbb{Z}$, take $[a,b] = [n, n+1]$, $\epsilon = \frac{\delta}{2^{|n|+2}}$, apply (5.18) and call the resulting continuous function g_n. Let $\tilde{g}$ be a continuous function from $\mathbb{R} \to \mathbb{R}$ such that $\mu\bigl(\{x : n \le x \le n+1, \tilde{g}(x) \neq g_n(x)\}\bigr) < \frac{\delta}{2^{|n|+2}}$. This can be done by setting $\tilde{g}(x) = g_n(x)$ for $n \le x \le (n+1) - \delta_n$ and linear on $\bigl[(n+1) - \delta_n, n+1\bigr]$ for some $0 < \delta_n < \frac{\delta}{2^{|n|+3}}$. Then

$$\begin{aligned}
\mu\bigl(\{x : f(x) \neq \tilde{g}(x)\}\bigr) &\le \sum_{n=-\infty}^{\infty} \mu\bigl(\{x : n \le x \le n+1,\ f(x) \neq \tilde{g}(x)\}\bigr) \\
&\le \sum_{n=-\infty}^{\infty} \mu\bigl(\{x : n \le x \le n+1, f(x) \neq g_n(x)\}\bigr) \\
&\quad + \sum_{n=-\infty}^{\infty} \mu\bigl(\{x : n \le x \le n+1,\ g_n(x) \neq \tilde{g}(x)\}\bigr) \\
&< 2\sum_{n=-\infty}^{\infty} \frac{\delta}{2^{|n|+2}} \le 2\delta.
\end{aligned}$$

So it suffices to establish (5.17). Since $f_K : \mathbb{R} \to \mathbb{R}$ is bounded and $\langle \mathcal{M}_{\mu_F^*}, \mathcal{B}(\mathbb{R})\rangle$-measurable, for each $\epsilon > 0$, there exists a simple function $s(x) \equiv \sum_{i=1}^{k} c_i I_{A_i}(x)$, with $A_i \subset [a,b]$, $A_i \in \mathcal{M}_{\mu_F^*}$, $\{A_i : 1 \le i \le k\}$ are disjoint, $\mu(A_i) < \infty$, and $c_i \in \mathbb{R}$ for $i = 1, \ldots, k$, such that $|f(x) - s(x)| < \frac{\epsilon}{4}$ for all $a \le x \le b$. By Theorem 1.3.4, for each A_i and $\eta > 0$, there exist a finite number of open disjoint intervals $I_{ij} = (a_{ij}, b_{ij}), j = 1, \ldots n_i$ such that

$$\mu\Bigl(A_i \,\triangle\, \bigcup_{j=1}^{n_i} I_{ij}\Bigr) < \frac{\eta}{2k}.$$

Now as in Problem 1.32 (g), there exists a continuous function g_{ij} such that

$$\mu\Bigl(g_{ij}^{-1}\{1\} \,\triangle\, I_{ij}\Bigr) < \frac{\eta}{kn_i},\ j = 1, 2, \ldots, n_i,\ i = 1, 2, \ldots, k.$$

Let

$$g_i \equiv \sum_{j=1}^{n_i} g_{ij},\ 1 \le i \le k.$$

Then $\mu(A_i \triangle g_i^{-1}\{1\}) < \frac{\eta}{k}$. Let $g = \sum_{i=1}^k c_i g_i$. Then $\mu(\{s \neq g\}) < \eta$. Hence for every $\epsilon > 0$, $\eta > 0$, there is a continuous function $g_{\epsilon,\eta} : [a,b] \to \mathbb{R}$ such that

$$\mu\big(\{x : a \leq x \leq b, |f_K(x) - g_{\epsilon,\eta}(x)| > \epsilon\}\big) < \eta.$$

Now for each $n \geq 1$, let

$$h_n(\cdot) \equiv g_{\frac{1}{2^n},\frac{1}{2^n}}(\cdot) \quad \text{and} \quad A_n = \big\{x : a \leq x \leq b, |f_K(x) - h_n(x)| > \frac{1}{2^n}\big\}.$$

Then, $\mu(A_n) < \frac{1}{2^n}$ and hence

$$\sum_{n=1}^{\infty} \mu(A_n) < \infty.$$

By the MCT, this implies that $\int_{[a,b]} \Big(\sum_{n=1}^{\infty} I_{A_n}\Big) d\mu < \infty$ and hence

$$\sum_{n=1}^{\infty} I_{A_n} < \infty \quad \text{a.e.} \quad \mu.$$

Thus $h_n \to f_K$ a.e. μ on $[a,b]$. By Egorov's theorem for any $\epsilon > 0$, there is a set $A_\epsilon \in \mathcal{B}([a,b])$ such that

$$\mu(A_\epsilon^c) < \epsilon/2 \quad \text{and} \quad h_n \to f_K \quad \text{uniformly on} \quad A_\epsilon.$$

By the inner regularity (Corollary 1.3.5) of μ, there is a compact set $D \subset A_\epsilon$ such that

$$\mu(A_\epsilon \backslash D) < \epsilon/2.$$

Since $h_n \to f_K$ uniformly on A_ϵ, f_K is continuous on A_ϵ and hence on D. It can be shown that there exists a continuous function $g : [a,b] \to \mathbb{R}$ such that $g = f_K$ on D (Problem 2.8 (e)). A more general result extending a continuous function defined on a closed set to the whole space is known as Tietze's extension theorem (see Munkres (1975)). Thus $\mu(\{x : a \leq x \leq b, f_K(x) \neq g(x)\}) < \epsilon$. This completes the proof of (5.17) and hence that of the proposition. $\square$

Remark 2.5.5: (*Littlewood's principles*). As pointed out in Section 1.3, Theorems 1.3.4, 2.5.11, and 2.5.12 constitute J. E. Littlewood's three principles: *every $\mathcal{M}_{\mu_F^*}$ measurable set is nearly a finite union of intervals, every a.e. convergent sequence is nearly uniformly convergent and every $\mathcal{M}_{\mu_F^*}$-measurable function is nearly continuous.*

2.6 Problems

2.1 Let Ω_i, $i = 1,2$ be two nonempty sets and $T : \Omega_1 \to \Omega_2$ be a map. Then for any collection $\{A_\alpha : \alpha \in I\}$ of subsets of Ω_2, show that

$$\begin{aligned} T^{-1}\Big(\bigcup_{\alpha\in I} A_\alpha\Big) &= \bigcup_{\alpha\in I} T^{-1}(A_\alpha) \\ \text{and } T^{-1}\Big(\bigcap_{\alpha\in I} A_\alpha\Big) &= \bigcap_{\alpha\in I} T^{-1}(A_\alpha). \end{aligned}$$

Further, $\big(T^{-1}(A)\big)^c = T^{-1}(A^c)$ for all $A \subset \Omega_2$. (These are known as *de Morgan's* laws.)

2.2 Let $\{A_i\}_{i\geq 1}$ be a collection of disjoint sets in a measurable space $(\Omega, \mathcal{F})$.

(a) Let $\{g_i\}_{i\geq 1}$ be a collection of $\langle\mathcal{F}, \mathcal{B}(\mathbb{R})\rangle$-measurable functions from Ω to $\mathbb{R}$. Show that $\sum_{i=1}^\infty I_{A_i} g_i$ converges on $\mathbb{R}$ and is $\langle\mathcal{F}, \mathcal{B}(\mathbb{R})\rangle$-measurable.

(b) Let $\mathcal{G} \equiv \sigma\langle\{A_i : i \geq 1\}\rangle$. Show that $h : \Omega \to \mathbb{R}$ is $\langle\mathcal{G}, \mathcal{B}(\mathbb{R})\rangle$-measurable iff $g(\cdot)$ is constant on each A_i.

2.3 Let $f, g : \Omega \to \mathbb{R}$ be $\langle\mathcal{F}, \mathcal{B}(\mathbb{R})\rangle$-measurable. Set

$$h(\omega) = \frac{f(\omega)}{g(\omega)} I(g(\omega) \neq 0), \quad \omega \in \Omega.$$

Verify that h is $\langle\mathcal{F}, \mathcal{B}(\mathbb{R})\rangle$-measurable.

2.4 Let $g : \Omega \to \bar{\mathbb{R}}$ be such that for every $r \in \mathbb{R}$, $g^{-1}((-\infty, r]) \in \mathcal{F}$. Show that g is $\langle\mathcal{F}, \mathcal{B}(\bar{\mathbb{R}})\rangle$-measurable.

2.5 Prove Proposition 2.1.6.

(**Hint:** Show that

$$\sigma\langle\{f_\lambda : \lambda \in \Lambda\}\rangle = \bigcup_{L\in\mathcal{C}} \sigma\langle\{f_\lambda : \lambda \in L\}\rangle$$

where $\mathcal{C}$ is the collection of all countable subsets of Λ.)

2.6 Let X_i, $i = 1,2,3$ be random variables on a probability space $(\Omega, \mathcal{F}, P)$. Consider the random equation (in $t \in \mathbb{R}$):

$$X_1(\omega)t^2 + X_2(\omega)t + X_3(\omega) = 0. \tag{6.1}$$

(a) Show that $A \equiv \{\omega \in \Omega :$ Equation (6.1) has two distinct real roots$\} \in \mathcal{F}$.

(b) Let $T_1(\omega)$ and $T_2(\omega)$ denote the two roots of (6.1) on A. Let

$$f_i(w) = \begin{cases} T_i(w) & \text{on } A \\ 0 & \text{on } A^c \end{cases},$$

$i = 1, 2$. Show that (f_1, f_2) is $\langle \mathcal{F}, \mathcal{B}(\mathbb{R}^2) \rangle$-measurable.

2.7 Let $M \equiv ((X_{ij})), 1 \le i, j \le k$, be a (random) matrix of random variables X_{ij} defined on a probability space $(\Omega, \mathcal{F}, P)$.

(a) Show that $Y_1 \equiv det(M)$ (the determinant of M) and $Y_2 \equiv tr(M)$(the trace of M) are both $\langle \mathcal{F}, \mathcal{B}(\mathbb{R}) \rangle$-measurable.

(b) Show also that $Y_3 \equiv$ the largest eigenvalue of $M'M$ is $\langle \mathcal{F}, \mathcal{B}(\mathbb{R}) \rangle$-measurable, where M' is the transpose of M.

(**Hint:** Use the result that $Y_3 = \sup_{x \ne 0} \frac{(x'M'Mx)}{(x'x)}$.)

2.8 Let $f : \mathbb{R} \to \mathbb{R}$. Let $\bar{f}(x) = \inf_{\delta > 0} \sup_{|y-x|<\delta} f(y)$ and $\underline{f}(x) = \sup_{\delta > 0} \inf_{|y-x|<\delta} f(y)$, $x \in \mathbb{R}$.

(a) Show that for any $t \in \mathbb{R}$,

$$\{x : \bar{f}(x) < t\}$$

is open and hence, $\bar{f}$ is $\langle \mathcal{B}(\mathbb{R}), \mathcal{B}(\mathbb{R}) \rangle$-measurable.

(b) Show that for any $t > 0$,

$$\{x : \bar{f}(x) - \underline{f}(x) < t\} \equiv \bigcup_{r \in \mathbb{Q}} \{x : \bar{f}(x) < t + r, \underline{f}(x) > r\}$$

and hence is open.

(c) Show that f is continuous at some x_0 in $\mathbb{R}$ iff $\bar{f}(x_0) = \underline{f}(x_0)$.

(d) Show that the set $C_f \equiv \{x : f(\cdot) \text{ is continuous at } x\}$ is a G_δ set, i.e., an intersection of a countable number of open sets, and hence, C_f is a Borel set.

(e) Let D be a closed set in $\mathbb{R}$. Let $f : D \to \mathbb{R}$ be continuous on D. Show that there exists a $g : \mathbb{R} \to \mathbb{R}$ continuous such that $g = f$ on D.

(**Hint:** Note that D^c is open in $\mathbb{R}$ and hence it can be expressed as a countable union of disjoint open intervals $\{I_j = (a_j, b_j) : 1 \le j \le k \le \infty\}$. Note that $a_j, b_j \in D$ for all j except for possibly the j's for which $a_j = -\infty$ or $b_j = +\infty$. Let

$$g(x) \equiv \begin{cases} f(x) & \text{if } x \in D \\ f(a_j) + \frac{(x-a_j)}{(b_j-a_j)}(f(b_j) - f(a_j)) & \text{if } x \in (a_j, b_j), \\ & \quad a_j, b_j \in D \\ f(b_j) & \text{if } a_j = -\infty, \\ & \quad x \in (a_j, b_j) \\ f(a_j) & \text{if } b_j = \infty, \\ & \quad x \in (a_j, b_j). \end{cases}$$

Now verify that g has the required properties.)

2.9 Prove Proposition 2.2.1 using Problem 2.1.

2.10 (a) Show that for any $x \in \mathbb{R}$ and any random variable X with cdf $F_X(\cdot)$, $P(X < x) = F_X(x-) \equiv \lim_{y \uparrow x} F_X(y)$.

(b) Show that $F_c(\cdot)$ in (2.5) is continuous.

2.11 Let $F : \mathbb{R} \to \mathbb{R}$ be nondecreasing.

(a) Show that for $x \in \mathbb{R}$, $F(x-) \equiv \lim_{y \uparrow x} F(y)$ and $F(x+) = \lim_{y \downarrow x} F(y)$ exist and satisfy $F(x-) \leq F(x) \leq F(x+)$.

(b) Let $D \equiv \{x : F(x+) - F(x-) > 0\}$. Show that D is at most countable.

(**Hint:** Show that

$$D = \bigcup_{n \geq 1} \bigcup_{r \geq 1} D_{n,r},$$

where $D_{n,r} = \{x : |x| \leq n, F(x+) - F(x-) > \frac{1}{r}\}$ and that each $D_{n,r}$ is finite.)

2.12 Suppose that (i) and (iii) of Proposition 2.2.3 hold. Show that for any a_1 in $\mathbb{R}$ and $-\infty < a_2 \leq b_2 < \infty$, $F(a_2, a_1) \leq F(b_2, a_1)$ and $F(a_1, a_2) \leq F(a_1, b_2)$, (i.e., F is monotone coordinatewise).

2.13 Let $F : \mathbb{R}^k \to \mathbb{R}$ be such that:

(a) for $x_1 = (x_{11}, x_{12}, \ldots, x_{1k})$ and $x_2 = (x_{21}, x_{22}, \ldots, x_{2k})$ with $x_{1i} \leq x_{2i}$ for $i = 1, 2, \ldots k$,

$$\Delta F(x_1, x_2) \equiv \sum_{a \in A} (-1)^{s(a)} F(a) \geq 0,$$

where $A \equiv \{a = (a_1, a_2, \ldots, a_k) : a_i \in \{x_{1i}, x_{2i}\}, i = 1, 2, \ldots, k\}$ and for a in A, $s(a) = |\{i : a_i = x_{1i}, i = 1, 2, \ldots, k\}|$ is the number of indices i for which $a_i = x_{1i}$.

(b) For each $i = 1, 2, \ldots, k$, $\lim\limits_{x_{i1} \downarrow -\infty} F(x_i) = 0$.

Let $\mathcal{C}_k$ be the semialgebra of sets of the form $\{A : A = A_1 \times \ldots \times A_k, A_i \in \mathcal{C}$ for all $1 \leq i \leq k\}$ where $\mathcal{C}$ is the semialgebra in $\mathbb{R}$ defined in (3.7). Set $\mu_F(A) \equiv \Delta F(x_1, x_2)$ if $A = (x_{11}, x_{21}] \times (x_{12}, x_{22}] \times \ldots \times (x_{1k}, x_{2k}]$ is bounded and set $\mu_F(A) = \lim_{n\to\infty} \mu(A \cap J_n)$, where $J_n = (-n, n]^k$.

(i) Show that F is coordinatewise monotone, i.e., if $x = (x_1, \ldots, x_k)$, $y = (y_1, \ldots, y_k)$ and $x_i \leq y_i$ for every $i = 1, \ldots, k$, then

$$F(y) \geq F(x).$$

(ii) Show that $\mathcal{C}$ is a semialgebra and μ_F is a measure on $\mathcal{C}$ by using the Heine-Borel theorem as in Problem 1.22 and 1.23.

2.14 Let $m(\cdot)$ denote the Lebesgue measure on $(\mathbb{R}, \mathcal{B}(\mathbb{R}))$. Let $T : \mathbb{R} \to \mathbb{R}$ be the map $T(x) = x^2$. Evaluate the induced measure $mT^{-1}(A)$ of the set A, where

(a) $A = [0, t], t > 0$.

(b) $A = (-\infty, 0)$.

(c) $A = \{1, 2, 3, \ldots\}$.

(d) $A = \bigcup_{i=1}^{\infty} (i^2, (i + \frac{1}{i^2})^2)$.

(e) $A = \bigcup_{i=1}^{\infty} (i^2, (i + \frac{1}{i})^2)$.

2.15 Consider the probability space $((0, 1), \mathcal{B}((0, 1)), m)$, where $m(\cdot)$ is the Lebesgue measure.

(a) Let Y_1 be the random variable $Y_1(x) \equiv \sin 2\pi x$ for $x \in (0, 1)$. Find the cdf of Y_1.

(b) Let Y_2 be the random variable $Y_2(x) \equiv \log x$ for $x \in (0, 1)$. Find the cdf of Y_2.

(c) Let $F : \mathbb{R} \to \mathbb{R}$ be a cdf . For $0 < x < 1$, let

$$\begin{aligned} F_1^{-1}(x) &= \inf\{y : y \in \mathbb{R}, F(y) \geq x\} \\ F_2^{-1}(x) &= \sup\{y : y \in \mathbb{R}, F(y) \leq x\}. \end{aligned}$$

Let Z_i be the random variable defined by

$$Z_i = F_i^{-1}(x) \quad 0 < x < 1, \quad i = 1, 2.$$

(i) Find the cdf of $Z_i, i = 1, 2$.

(**Hint:** Verify using the right continuity of F that for any $0 < x < 1$, $t \in \mathbb{R}$, $F(t) \geq x \Leftrightarrow F_1^{-1}(x) \leq t$.)

(ii) Show also that $F_1^{-1}(\cdot)$ is left continuous and $F_2^{-1}(\cdot)$ is right continuous.

2.16 (a) Let $(\Omega, \mathcal{F}_1, \mu)$ be a σ-finite measure space. Let $T : \Omega \to \mathbb{R}$ be $\langle \mathcal{F}, \mathcal{B}(\mathbb{R}) \rangle$-measurable. Show, by an example, that the induced measure μT^{-1} need not be σ-finite.

(b) Let $(\Omega_i, \mathcal{F}_i)$ be measurable spaces for $i = 1, 2$ and let $T : \Omega_1 \to \Omega_2$ be $\langle \mathcal{F}_1, \mathcal{F}_2 \rangle$-measurable. Show that any measure μ on $(\Omega_1, \mathcal{F}_1)$ is σ-finite if μT^{-1} is σ-finite on $(\Omega_2, \mathcal{F}_2)$.

2.17 Let $(\Omega, \mathcal{F}, \mu)$ be a measure space and let $f : \Omega \to [0, \infty]$ be such that it admits two representations

$$f = \sum_{i=1}^{k} c_i I_{A_i} \quad \text{and} \quad f = \sum_{j=1}^{\ell} d_j I_{B_j},$$

where $c_i, d_j \in [0, \infty]$,and A_i and $B_j \in \mathcal{F}$ for all i, j. Show that

$$\sum_{i=1}^{k} c_i \mu(A_i) = \sum_{j=1}^{\ell} d_j \mu(B_j).$$

(**Hint:** Express A_i and B_j as finite unions of a common collection of disjoint sets in $\mathcal{F}$.)

2.18 (a) Prove Proposition 2.3.1.

(b) In the proof of Proposition 2.3.2, verify that $D_n \uparrow D$.

(c) Verify Remark 2.3.1.

(**Hint:** Let

$$\begin{aligned} A_n &= \{\omega : f_{n-1}(\omega) \geq f_n(\omega),\ g_{n+1}(\omega) \geq g_n(\omega)\} \\ A &= \Big(\bigcap_{n\geq 1} A_n\Big) \bigcap \Big\{\omega : \lim_{n\to\infty} g_n(\omega) = g(\omega), \\ &\qquad\qquad \lim_{n\to\infty} f_n(\omega) = f(\omega)\Big\} \end{aligned}$$

and $\tilde{f}_n = f_n I_A$, $\tilde{g}_n = g_n I_A$. Verify that $\mu(A^c) = 0$ and apply Proposition 2.3.2 to $\{\tilde{f}_n\}_{n\geq 1}$ and $\{\tilde{g}_n\}_{n\geq 1}$.)

2.19 (a) Verify that $f_n(\cdot)$ defined in (3.7) satisfies $f_n(\omega) \uparrow f(\omega)$ for all ω in Ω.

(b) Verify that the sequence $\{h_n\}_{n\geq 1}$ of Remark 2.3.3 satisfies $\lim_{n\to\infty} h_n = f$ a.e. (μ), and $\lim_{n\to\infty} \int h_n d\mu = M$.

2.20 Apply Corollary 2.3.5 to show that for any collection $\{a_{ij} : i, j \in \mathbb{N}\}$ of nonnegative numbers,

$$\sum_{i=1}^{\infty} \Bigg(\sum_{j=1}^{\infty} a_{ij} \Bigg) = \sum_{j=1}^{\infty} \Bigg(\sum_{i=1}^{\infty} a_{ij} \Bigg).$$

2.21 Let $g : \mathbb{R} \to \mathbb{R}$.

(a) Recall that $\lim_{t\to\infty} g(t) = L$ for some L in $\mathbb{R}$ if for every $\epsilon > 0$, there exists a $T_\epsilon < \infty$ such that $t \geq T_\epsilon \Rightarrow |g(t) - L| < \epsilon$. Show that $\lim_{t\to\infty} g(t) = L$ for some L in $\mathbb{R}$ iff $\lim_{n\to\infty} g(t_n) = L$ for every sequence $\{t_n\}_{n\geq 1}$ with $\lim_{n\to\infty} t_n = \infty$.

(b) Formulate and prove a similar result when $\lim_{t\to a} g(t) = L$ for some $a, L \in \mathbb{R}$.

2.22 Let $\{f_t : t \in \mathbb{R}\} \subset L^1(\Omega, \mathcal{F}, \mu)$.

(a) (*The continuous version of the MCT*). Suppose that $f_t \uparrow f$ as $t \uparrow \infty$ a.e. (μ) and for each t, $f_t \geq 0$ a.e. (μ). Show that

$$\int f_t d\mu \uparrow \int f d\mu.$$

(b) (*The continuous version of the DCT*). Suppose there exists a nonnegative $g \in L^1(\Omega, \mathcal{F}, \mu)$ such that for each t, $|f_t| \leq g$ a.e. (μ) and as $t \to \infty$, $f_t \to f$ a.e. (μ). Then $f \in L^1(\Omega, \mathcal{F}, \mu)$ and $\int |f_t - f| d\mu \to 0$ and hence, $\int f_t d\mu \to \int f d\mu$, as $t \to \infty$.

2.23 Let $f : [a, b] \to \mathbb{R}$ be bounded where $-\infty < a < b < \infty$. Let $\{P_n\}_{n\geq 1}$ be a sequence of partitions such that $\Delta(P_n) \to 0$. Show that as $n \to \infty$,

$$U(P_n, f) \to \overline{\int} f \quad \text{and} \quad L(P_n, f) \to \underline{\int} f,$$

where $\overline{\int} f$ and $\underline{\int} f$ are as defined in (4.3) and (4.4), respectively.

(**Hint:** Given $\epsilon > 0$, fix a partition $P = \{x_0 = a < x_1 < \ldots < x_k = b\}$ such that $\overline{\int} f < U(P, f) < \overline{\int} + \epsilon$. Let $\delta = \min_{0\leq i\leq k-1}(x_{i+1} - x_i)$. Choose n large such that the diameter $\Delta(P_n) < \delta$. Verify that

$$U(P_n, f) < U(P, f) + kB\Delta(P_n)$$

where $B = \sup\{|f(x)| : a \leq x \leq b\}$ and conclude that $\overline{\lim}_n U(P_n, f) \leq \overline{\int} f + \epsilon$.)

2.24 Establish (4.6) and (4.7).

(**Hint:** Show that for every x and any $\epsilon > 0$, $\phi_n(x) \leq \phi(x) + \epsilon$ for all n large and that for $x \notin \cup_{n\geq 1} P_n$, $\phi_n(x) \geq \phi(x)$ for all n.)

2.25 If $f(x) = I_{\mathbb{Q}_1}(x)$ where $\mathbb{Q}_1 = \mathbb{Q} \cap [0, 1]$, $\mathbb{Q}$ being the set of all rationals, then show that for any partition P, $U(P, f) = 1$ and $L(P, f) = 0$.

2.26 Establish Theorem 2.5.1.

(**Hint:** Verify that

$$\begin{aligned} D &\equiv \{\omega : f_j(\omega) \not\to f(\omega)\} \\ &= \bigcup_{r\geq 1} \bigcap_{n\geq 1} \bigcup_{j\geq n} A_{jr}, \end{aligned}$$

where $A_{jr} = \{|f_j - f| > \frac{1}{r}\}$. Show that since $\mu(D) = 0$ and $\mu(\Omega) < \infty$, $\mu(D_{rn}) \to 0$ as $n \to \infty$ for each $r \in \mathbb{N}$, where $D_{rn} = \cup_{j\geq n} A_{jr}$.)

2.27 Let $\phi : \mathbb{R}_+ \to \mathbb{R}_+$ be nondecreasing and $\frac{\phi(x)}{x} \uparrow \infty$ as $x \uparrow \infty$. Also, let $\{f_\lambda : \lambda \in \Lambda\}$ be a subset of $L^1(\Omega, \mathcal{F}, \mu)$. Show that if $\sup_{\lambda\in\Lambda} \int \phi(|f_\lambda|)d\mu < \infty$, then $\{f_\lambda : \lambda \in \Lambda\}$ is UI.

2.28 Let μ be the Lebesgue measure on $([-1, 1], \mathcal{B}([-1, 1]))$. For $n \geq 1$, define $f_n(x) = nI_{(0,n^{-1})}(x) - nI_{(-n^{-1},0)}(x)$ and $f(x) \equiv 0$ for $x \in [-1, 1]$. Show that $f_n \to f$ a.e. (μ) and $\int f_n d\mu \to \int f d\mu$ but $\{f_n\}_{n\geq 1}$ is not UI.

2.29 Let $\{f_n : n \geq 1\} \cup \{f\} \subset L^1(\Omega, \mathcal{F}, \mu)$.

(a) Show that $\int |f_n - f|(d\mu) \to 0$ iff $f_n \longrightarrow^m f$ and $\int |f_n| d\mu \to \int |f| d\mu$.

(b) Show further that if $\mu(\Omega) < \infty$ then the above two are equivalent to $f_n \longrightarrow^m f$ and $\{f_n\}$ UI.

2.30 For $n \geq 1$, let $f_n(x) = n^{-1/2} I_{(0,n)}(x)$, $x \in \mathbb{R}$, and let $f(x) = 0$, $x \in \mathbb{R}$. Let m denote the Lebesgue measure on $(\mathbb{R}, \mathcal{B}(\mathbb{R}))$. Show that $f_n \to f$ a.e. (m) and $\{f_n\}_{n\geq 1}$ is UI, but $\int f_n dm \not\to \int f dm$.

2.31 (*Computing integrals w.r.t. the Lebesgue measure*). Let $f \in L^1(\mathbb{R}, \mathcal{M}_m, m)$ where $(\mathbb{R}, \mathcal{M}_m, m)$ is the real line with Lebesgue σ-algebra, and Lebesgue measure, i.e., $m = \mu_F^*$ where $F(x) \equiv x$. The definition of $\int f dm$ as $\int f^+ dm - \int f^- dm$ involves computing $\int f^+ dm$ and $\int f^- dm$ which in turn is given in terms of approximating by integrals of simple nonnegative functions. This is not a very practical procedure. For f that happens to be continuous a.e. and bounded on finite intervals, one can compute the Riemann integral of f over finite intervals and pass to the limit. Justify the following steps:

(a) Let f be continuous a.e. and bounded on finite intervals and $f \in L^1(\mathbb{R}, \mathcal{M}_m, m)$. Show that for $-\infty < a < b < \infty$, $f \in L^1([a, b], \mathcal{M}_m, m)$ and

$$\int_{[a,b]} f dm = \oint_{[a,b]} f(x) dx,$$

where the right side denotes the Riemann integral of f on $[a, b]$.

(b) If, in addition, $f \in L^1(\mathbb{R}, \mathcal{M}_m, m)$, then

$$\int_{\mathbb{R}} f dm = \lim_{\substack{a \to -\infty \\ b \to +\infty}} \int_{[a,b]} f dm.$$

(c) If f is continuous a.e. and $\in L^1(\mathbb{R}, \mathcal{M}_m, m)$, then

$$\int_{\mathbb{R}} f dm = \lim_{\substack{a \to -\infty \\ b \to +\infty \\ c \to \infty}} \int_{[a,b]} \phi_c(f) dm$$

where $\phi_c(f) = f(x)I(|f(x)| \leq c) + cI(f(x) > c) - cI(f(x) < -c)$.

(d) Apply the above procedure to compute $\int_{\mathbb{R}} f dm$ for

(i) $f(x) = \frac{1}{1+x^2}$,

(ii) $f(x) = e^{-x^2/2}$,

(iii) $f(x) = e^{+x}e^{-x^2/2}$.

2.32 (a) Let $a \in \mathbb{R}$. Show that if a_1, a_2 are nonnegative such that $a = a_1 - a_2$ then $a_1 \geq a^+$, $a_2 \geq a^-$ and $a_1 - a^+ = a_2 - a^-$.

(b) Let $f = f_1 - f_2$ where f_1, f_2 are nonnegative and are in $L^1(\Omega, \mathcal{F}, \mu)$. Show that $f \in L^1(\Omega, \mathcal{F}, \mu)$ and $\int f d\mu = \int f_1 d\mu - \int f_2 d\mu$.

2.33 Let $(\Omega, \mathcal{F}, \mu)$ be a measure space. Let $f : \Omega \times (a, b) \to \mathbb{R}$ be such that for each $a < t < b$, $f(\cdot, t) \in L^1(\Omega, \mathcal{F}, \mu)$.

(a) Suppose for each $a < t < b$,

(i) $\lim_{h \to 0} f(\omega, t+h) = f(\omega, t)$ a.e. (μ).

(ii) $\sup_{|h| \leq 1} |f(\omega, t+h)| \leq g_1(\omega, t)$, where $g_1(\cdot, t) \in L^1(\Omega, \mathcal{F}, \mu)$.

Show that $\phi(t) \equiv \int_\Omega f(\omega, t) d\mu$ is continuous on (a, b).

(b) Suppose for each $a < t < b$.

(i) $\lim_{h \to 0} \dfrac{f(\omega, t+h) - f(\omega, t)}{h} = g_2(\omega, t)$ exists a.e. (μ),

(ii) $\sup_{0 \leq |h| \leq 1} \left| \dfrac{f(\omega, t+h) - f(\omega, t)}{h} \right| \leq G(\omega, t)$ a.e. (μ),

(iii) $G(\omega, t) \in L^1(\Omega, \mathcal{F}, \mu)$.

Show that

$$\phi(t) \equiv \int_\Omega f(\omega, t) d\mu$$

is differentiable on (a, b).

(**Hint:** Use the continuous version of DCT (cf. Problem 2.22).)

2.34 Let $A \equiv ((a_{ij}))$ be an infinite matrix of real numbers. Suppose that for each j, $\lim_{i\to\infty} a_{ij} = a_j$ exists in $\mathbb{R}$ and $\sup_i |a_{ij}| \le b_j$, where $\sum_{j=1}^{\infty} b_j < \infty$.

(a) Show by an application of the DCT that

$$\lim_{i\to\infty} \sum_{j=1}^{\infty} |a_{ij} - a_j| = 0.$$

(b) Show the same directly, i.e., without using the DCT.

2.35 Using the above problem or otherwise, show that for any sequence $\{x_n\}_{n\ge1}$ with $\lim_{n\to\infty} x_n = x$ in $\mathbb{R}$,

$$\lim_{n\to\infty} \left(1 + \frac{x_n}{n}\right)^n = \sum_{j=0}^{\infty} \frac{x^j}{j!} \equiv \exp(x).$$

2.36 (a) Let $\{f_n\}_{n\ge1} \subset L^1(\Omega, \mathcal{F}, \mu)$ such that $f_n \to 0$ in $L^1(\mu)$. Show that $\{f_n\}_{n\ge1}$ is UI.

(b) Let $\{f_n\}_{n\ge1} \subset L^p(\Omega, \mathcal{F}, \mu)$, $0 < p < \infty$, such that $\mu(\Omega) < \infty$, $\{|f_n|^p\}_{n\ge1}$ is UI and $f_n \longrightarrow^m f$. Show that $f \in L^p(\mu)$ and $f_n \to f$ in $L^p(\mu)$.

2.37 (a) Show that for a sequence of real numbers $\{x_n\}_{n\ge1}$, $\lim_{n\to\infty} x_n = x \in \bar{\mathbb{R}}$ holds iff every subsequence $\{x_{n_j}\}_{j\ge1}$ of $\{x_n\}_{n\ge1}$ has a further subsequence $\{x_{n_{jk}}\}_{k\ge1}$ such that $\lim_{k\to\infty} x_{n_{jk}} = x$.

(b) Use (a) and Theorem 2.5.2 to show that the extended DCT (Theorem 2.3.11) is valid if the a.e. convergence of $\{f_n\}_{n\ge1}$ and $\{g_n\}_{n\ge1}$ is replaced by convergence in measure.

2.38 Let $(\mathbb{R}, \mathcal{M}_{\mu_F^*}, \mu_F^*)$ be a Lebesgue-Stieltjes measure space generated by a $F : \mathbb{R} \to \mathbb{R}$ nondecreasing. Let $f : \mathbb{R} \to \bar{\mathbb{R}}$ be $\mathcal{M}_{\mu_F^*}$-measurable such that $|f| < \infty$ a.e. (μ_F^*). Show that for every $k \in \mathbb{N}$ and for every $\epsilon, \eta \in (0, \infty)$, there exists a continuous function $g : \mathbb{R} \to \mathbb{R}$ such that $g(x) = 0$ for $|x| > k$ and $\mu_F^*(\{x : |x| \le k, |f(x) - g(x)| > \eta\} < \epsilon$.

(**Hint:** Complete the following:
STEP I: For all $\epsilon > 0$, there exists $M_{k,\epsilon} \in (0, \infty)$ such that $\mu_F^*(\{x : |x| \le k, |f(x)| > M_{k,\epsilon}\}) < \epsilon$.
STEP II: For $\eta > 0$, there exists a simple function $s(\cdot)$ such that $\mu_F^*(\{x : |x| \le k, |f(x)| \le M_{k,\epsilon}, |f(x) - s(x)| > \eta\}) = 0$.
STEP III: For $\delta > 0$, there exists a continuous function $g(\cdot)$ such that $g \equiv 0$ for $|x| > k$ and $\mu_F^*(\{x : |x| \le k, s(x) \ne g(x)\}) < \delta$.)

2.39 Recall from Corollary 2.3.6 that for $g \in L^1(\Omega, \mathcal{F}, \mu)$ and nonnegative $\nu_g(A) \equiv \int_A g d\mu$ is a measure. Next for any $\mathcal{F}$-measurable function h, show that $h \in L^1(\nu_g)$ iff $h \cdot g \in L^1(\mu)$ and $\int h d\nu_g = \int hg d\mu$.

(**Hint:** Verify first for h simple and nonnegative, next for h nonnegative, and finally for any h.)

2.40 Prove the BCT, Corollary 2.3.13, using Egorov's theorem (Theorem 2.5.11).

2.41 Deduce the DCT from the BCT with the notation as in Corollary 2.3.12.

(**Hint:** Apply the BCT to the measure space $(\Omega, \mathcal{F}, \nu_g)$ and functions $h_n = \frac{f_n}{g} I(g > 0)$, $h = \frac{f}{g} I(g > 0)$ where ν_g is as in Problem 2.39.)

2.42 (*Change of variables formula*). Let $(\Omega_i, \mathcal{F}_i)$, $i = 1, 2$ be two measurable spaces. Let $f : \Omega_1 \to \Omega_2$ be $\langle \mathcal{F}_1, \mathcal{F}_2 \rangle$-measurable, $h : \Omega_2 \to \mathbb{R}$ be $\langle \mathcal{F}_2, \mathcal{B}(\mathbb{R}) \rangle$-measurable, and μ_1 be a measure on $(\Omega_1, \mathcal{F}_1)$. Show that $g \equiv h \circ f$, i.e., $g(\omega) \equiv h(f(\omega))$ for $\omega \in \Omega_1$ is in $L^1(\mu)$ iff $h(\cdot) \in L^1(\Omega_2, \mathcal{F}_2, \mu_2)$ where $\mu_2 = \mu_1 f^{-1}$ iff $I(\cdot) \in L^1\big(\mathbb{R}, \mathcal{B}(\mathbb{R}), \mu_3 \equiv \mu_2 h^{-1}\big)$ where $I(\cdot)$ is the identity function in $\mathbb{R}$, i.e., $I(x) \equiv x$ for all $x \in \mathbb{R}$ and also that

$$\int_{\Omega_1} g d\mu_1 = \int_{\Omega_2} h d\mu_2 = \int_{\mathbb{R}} x d\mu_3.$$

2.43 Let $\phi(x) \equiv \frac{1}{\sqrt{2\pi}} e^{-x^2/2}$ be the standard $N(0, 1)$ p.d.f. on $\mathbb{R}$. Let $\{(\mu_n, \sigma_n)\}_{n \geq 1}$ be a sequence in $\mathbb{R} \times \mathbb{R}_+$. Suppose $\mu_n \to \mu$ and $\sigma_n \to \sigma$ as $n \to \infty$. Let $f_n(x) = \frac{1}{\sigma_n} \phi\big(\frac{x - \mu_n}{\sigma_n}\big)$, $f(x) = \frac{1}{\sigma} \phi\big(\frac{x - \mu}{\sigma}\big)$ and $\nu_n(A) = \int_A f_n dm$, $\nu(A) = \int_A f dm$ for any Borel set A in $\mathbb{R}$, where $m(\cdot)$ is the Lebesgue measure on $\mathbb{R}$. Using Scheffe's theorem, verify that, as $n \to \infty$, $\nu_n(\cdot) \to \nu(\cdot)$ uniformly on $\mathcal{B}(\mathbb{R})$ and that for any $h : \mathbb{R} \to \mathbb{R}$, bounded and Borel measurable,

$$\int h d\nu_n \to \int h d\nu.$$

2.44 Let $f_n(x) = c_n \big(1 - \frac{x}{n}\big)^n I_{[0,n]}(x)$, $x \in \mathbb{R}$, $n \geq 1$.

(a) Find c_n such that $\int f_n dm = 1$.

(b) Show that $\lim_{n \to \infty} f_n(x) \equiv f(x)$ exists for all x in $\mathbb{R}$ and that f is a p.d.f.

(c) For $A \in \mathcal{B}(\mathbb{R})$, let

$$\nu_n(A) \equiv \int_A f_n dm \quad \text{and} \quad \nu(A) = \int_A f dm.$$

Show that $\nu_n \to \nu$ uniformly on $\mathcal{B}(\mathbb{R})$.

2.45 Let $(\Omega, \mathcal{F}, \mu)$ be a measure space and $f : \Omega \to \mathbb{R}$ be $\mathcal{F}$-measurable. Suppose that $\int_\Omega |f| d\mu < \infty$ and D is a closed set in $\mathbb{R}$ such that for all $B \in \mathcal{F}$ with $\mu(B) > 0$,

$$\frac{1}{\mu(B)} \int_B f d\mu \in D.$$

Show that $f(\omega) \in D$ a.e. μ.

(**Hint:** Show that for $x \notin D$, there exists $r > 0$ such that $\mu\{\omega : |f(\omega) - x| < r\} = 0$.)

2.46 Find a sequence of nonnegative continuous functions $\{f_n\}_{n\geq 1}$ on [0,1] such that $\int_{[0,1]} f_n dm \to 0$ but $\{f_n(x)\}_{n\geq 1}$ does not converge for any x.

(**Hint:** Let for $m \geq 1$, $1 \leq k \leq m$, $g_{m,k} = I_{\left[\frac{k-1}{m}, \frac{k}{m}\right]}$ and $\{f_n\}_{n\geq 1}$ be a reordering of $\{g_{n,k} : 1 \leq k \leq n, n \geq 1\}$.)

2.47 Let $\{f_n\}_{n\geq 1}$ be a sequence of continuous functions from [0,1] to [0,1] such that $f_n(x) \to 0$ as $n \to \infty$ for all $0 \leq x \leq 1$. Show that $\int_0^1 f_n(x)dx \to 0$ as $n \to \infty$ (where the integral is the Riemann integral) by two methods: one by using BCT and one without using BCT. Show also that if μ is a finite measure on $([0, 1], \mathcal{B}([0, 1]))$, then $\int_{[0,1]} f_n d\mu \to 0$ as $n \to \infty$.

2.48 (*Invariance of Lebesgue measure under translation and reflection.*) Let $m(\cdot)$ be Lebesgue measure on $(\mathbb{R}, \mathcal{B}(\mathbb{R}))$.

(a) For any $E \in \mathcal{B}(\mathbb{R})$ and $c \in \mathbb{R}$, define $-E \equiv \{x : -x \in E\}$ and $E + c \equiv \{y : y = x + c, x \in E\}$. Show that

$$\begin{aligned} m(-E) &= m(E) \quad \text{and} \\ m(E+c) &= m(E) \end{aligned}$$

for all $E \in \mathcal{B}(\mathbb{R})$ and $c \in \mathbb{R}$.

(b) For any $f \in L^1(\mathbb{R}, \mathcal{B}(\mathbb{R}), m)$ and $c \in \mathbb{R}$, let $\tilde{f}(x) \equiv f(-x)$ and $f_c(x) \equiv f(x + c)$. Show that

$$\int \tilde{f} dm = \int f_c dm = \int f dm.$$

3
L^p-Spaces

3.1 Inequalities

This section contains a number of useful inequalities.

Theorem 3.1.1: (*Markov's inequality*). *Let f be a nonnegative measurable function on a measure space $(\Omega, \mathcal{F}, \mu)$. Then for any $0 < t < \infty$,*

$$\mu(\{f \geq t\}) \leq \frac{\int f d\mu}{t}. \tag{1.1}$$

Proof: Since f is nonnegative, $\int f d\mu \geq \int_{(\{f \geq t\})} f d\mu \geq t\mu(\{f \geq t\})$. □

Corollary 3.1.2: *Let X be a random variable on a probability space $(\Omega, \mathcal{F}, P)$. Then, for $r > 0$, $t > 0$,*

$$P(|X| \geq t) \leq \frac{E|X|^r}{t^r}.$$

Proof: Since $\{|X| \geq t\} = \{|X|^r \geq t^r\}$ for all $t > 0$, $r > 0$, this follows from (1.1). □

Corollary 3.1.3: (*Chebychev's inequality*). *Let X be a random variable with $EX^2 < \infty$, $E(X) = \mu$ and $Var(X) = \sigma^2$. Then for any $0 < k < \infty$,*

$$P(|X - \mu| \geq k\sigma) \leq \frac{1}{k^2}.$$

Proof: Follows from Corollary 3.1.2 with X replaced by $X - \mu$ and with $r = 2$, $t = k\sigma$. □

Corollary 3.1.4: *Let $\phi : \mathbb{R}_+ \to \mathbb{R}_+$ be nondecreasing. Then for any random variable X and $0 < t < \infty$,*

$$P(|X| \geq t) \leq \frac{E\phi(|X|)}{\phi(t)}.$$

Proof: Use (1.1) and the fact that $|X| \geq t \Rightarrow \phi(|X|) \geq \phi(t)$. □

Corollary 3.1.5: (*Cramer's inequality*). *For any random variable X and $t > 0$,*

$$P(X \geq t) \leq \inf_{\theta > 0} \frac{E(e^{\theta X})}{e^{\theta t}}.$$

Proof: For $t > 0$, $\theta > 0$, $P(X \geq t) = P\big(e^{\theta X} \geq e^{\theta t}\big) \leq \frac{E(e^{\theta X})}{e^{\theta t}}$, by (1.1). □

Definition 3.1.1: A function $\phi : (a, b) \to \mathbb{R}$ is called *convex* if for all $0 \leq \lambda \leq 1$, $a < x \leq y < b$,

$$\phi(\lambda x + (1 - \lambda)y) \leq \lambda\phi(x) + (1 - \lambda)\phi(y). \tag{1.2}$$

Geometrically, this means that for the graph of $y = \phi(x)$ on (a, b), for each fixed $t \in (0, \infty)$, the chord over the interval $(x, x + t)$ turns in the counterclockwise direction as x increases.

More precisely, the following result holds.

Proposition 3.1.6: *Let $\phi : (a, b) \to \mathbb{R}$. Then ϕ is convex iff for all $a < x_1 < x_2 < x_3 < b$,*

$$\frac{\phi(x_2) - \phi(x_1)}{x_2 - x_1} \leq \frac{\phi(x_3) - \phi(x_2)}{x_3 - x_2}, \tag{1.3}$$

which is equivalent to

$$\frac{\phi(x_2) - \phi(x_1)}{x_2 - x_1} \leq \frac{\phi(x_3) - \phi(x_1)}{(x_3 - x_1)} \leq \frac{\phi(x_3) - \phi(x_2)}{x_3 - x_2}. \tag{1.4}$$

Proof: Let ϕ be convex and $a < x_1 < x_2 < x_3 < b$. Then one can write $x_2 = \lambda x_1 + (1 - \lambda)x_3$ with $\lambda = \frac{(x_3 - x_2)}{(x_3 - x_1)}$. So by (1.2),

$$\begin{aligned}
\phi(x_2) &= \phi(\lambda x_1 + (1 - \lambda)x_3) \\
&\leq \lambda\phi(x_1) + (1 - \lambda)\phi(x_3) \\
&= \frac{(x_3 - x_2)}{(x_3 - x_1)}\phi(x_1) + \frac{(x_2 - x_1)}{(x_3 - x_1)}\phi(x_3)
\end{aligned}$$

which is equivalent to (1.3). Also, since

$$\frac{\phi(x_3)-\phi(x_1)}{(x_3-x_1)} = \lambda\frac{\phi(x_3)-\phi(x_2)}{(x_3-x_2)} + (1-\lambda)\frac{\phi(x_2)-\phi(x_1)}{(x_2-x_1)},$$

(1.4) follows from (1.3).

Conversely, suppose (1.4) holds for all $a < x_1 < x_2 < x_3 < b$. Then given $a < x < y < b$ and $0 < \lambda < 1$, set $x_1 = x$, $x_2 = \lambda x + (1-\lambda)y$, $x_3 = y$ and apply (1.4) to verify (1.2). □

The following properties of a convex function are direct consequences of (1.3). The proof is left as an exercise (Problem 3.1).

Proposition 3.1.7: *Let $\phi : (a,b) \to \mathbb{R}$ be convex. Then,*

(i) For each $x \in (a,b)$,

$$\phi'_+(x) \equiv \lim_{y \downarrow x} \frac{\phi(y)-\phi(x)}{(y-x)}, \quad \phi'_-(x) \equiv \lim_{y \uparrow x} \frac{\phi(y)-\phi(x)}{(y-x)}$$

exist and are finite.

(ii) Further, $\phi'_-(\cdot) \le \phi'_+(\cdot)$ and both are nondecreasing on (a,b).

(iii) $\phi'(\cdot)$ exists except on the countable set of discontinuity points of ϕ'_+ and ϕ'_-.

(iv) For any $a < c < d < b$, ϕ is Lipschitz on $[c,d]$, i.e., there exists a constant $K < \infty$ such that

$$|\phi(x)-\phi(y)| \le K|x-y| \quad \textit{for all} \quad c \le x,\ y \le d.$$

(v) For any $a < c$, $x < b$,

$$\phi(x)-\phi(c) \ge \phi'_+(c)(x-c) \quad \text{and} \quad \phi(x)-\phi(c) \ge \phi'_-(c)(x-c). \quad (1.5)$$

By the mean value theorem, a sufficient condition for (1.3) and hence, for the convexity of ϕ is that ϕ be differentiable on (a,b) and ϕ' be nondecreasing. A further sufficient condition for this is that ϕ be twice differentiable on (a,b) and ϕ'' be nonnegative. This is stated as

Proposition 3.1.8: *Let ϕ be twice differentiable on (a,b) and ϕ'' be nonnegative on (a,b). Then ϕ is convex on (a,b).*

Example 3.1.1: The following functions are convex in the given intervals:

(a) $\phi(x) = |x|^p$, $p \ge 1$, $(-\infty,\infty)$.

(b) $\phi(x) = e^x$, $(-\infty,\infty)$.

(c) $\phi(x) = -\log x, \quad (0, \infty)$.

(d) $\phi(x) = x \log x, \quad (0, \infty)$.

Remark 3.1.1: By Proposition 3.1.7 (iii), the convexity of ϕ implies that ϕ' exists except on a countable set. For example, the function $\phi(x) = |x|$ is convex on $\mathbb{R}$; it is differentiable at all $x \neq 0$. Similarly, it is easy to construct a piecewise linear convex function ϕ with a countable number of points where ϕ is not differentiable.

The following is an important inequality for convex functions.

Theorem 3.1.9: (*Jensen's inequality*). *Let f be a measurable function on a probability space $(\Omega, \mathcal{F}, P)$ with $P(f \in (a, b)) = 1$ for some interval (a, b), $-\infty \leq a < b \leq \infty$ and let $\phi : (a, b) \to \mathbb{R}$ be convex. Then*

$$\phi\Big(\int f dP\Big) \leq \int \phi(f) dP, \tag{1.6}$$

provided $\int |f| dP < \infty$ and $\int |\phi(f)| dP < \infty$.

Remark 3.1.2: In terms of random variables, this says that for any random variable X on a probability space $(\Omega, \mathcal{F}, P)$ with $P(X \in (a, b)) = 1$ and for any function ϕ that is convex on (a, b),

$$\phi(EX) \leq E\phi(X), \tag{1.7}$$

provided $E|X| < \infty$ and $E|\phi(X)| < \infty$, where for any Borel measurable function h, $Eh(X) \equiv \int h(X) dP$.

Proof of Theorem 3.1.9: Let $c = \int f dP$. Applying (1.5), one gets

$$Y(\omega) \equiv \phi(f(\omega)) - \phi(c) - \phi_+'(c)(f(\omega) - c) \geq 0 \quad \text{a.e. } (P), \tag{1.8}$$

which, when integrated, yields $\int Y(\omega) P(d\omega) \geq 0$. Since $\int (f(\omega) - c) P(d\omega) = 0$, (1.6) follows. $\square$

Remark 3.1.3: Suppose that equality holds in (1.6). Then, it follows that $\int Y(\omega) P(d\omega) = 0$. By (1.8), this implies

$$\phi(f(\omega)) - \phi(c) = \phi_+'(c)(f(\omega) - c) \quad \text{a.e. } (P).$$

Thus, if ϕ is a strictly convex function (i.e., strict inequality holds in (1.2) for all x, $y \in (a, b)$ and $0 < \lambda < 1$), then equality holds in (1.6) iff $f(\omega) = c$ a.e. (P).

The following are easy consequences of Jensen's inequality (Problem 3.3).

Proposition 3.1.10: *Let $k \geq 1$ be an integer.*

(i) *Let* $a_1, a_2, \ldots, a_k$ *be real and* $p_1, p_2, \ldots, p_k$ *be positive numbers such that* $\sum_{i=1}^k p_i = 1$. *Then*

$$\sum_{i=1}^k p_i \exp(a_i) \geq \exp\Big(\sum_{i=1}^k p_i a_i\Big). \tag{1.9}$$

(ii) *Let* $b_1, b_2, \ldots, b_k$ *be nonnegative numbers and* $p_1, p_2, \ldots, p_k$ *be as in (i). Then*

$$\sum_{i=1}^k p_i b_i \geq \prod_{i=1}^k b_i^{p_i}, \tag{1.10}$$

and in particular,

$$\frac{1}{k}\sum_{i=1}^k b_i \geq \Big(\prod_{i=1}^k b_i\Big)^{\frac{1}{k}}, \tag{1.11}$$

i.e., the arithmetic mean of $b_1, \ldots, b_k$ *is greater than or equal to the geometric mean of the* b_i*'s. Further, equality holds in (1.10) iff* $b_1 = b_2 = \ldots = b_k$.

(iii) *For any* a, b *real and* $1 \leq p < \infty$,

$$|a+b|^p \leq 2^{p-1}(|a|^p + |b|^p). \tag{1.12}$$

Inequality (1.10) is useful in establishing the following:

Theorem 3.1.11: (*Hölder's inequality*). *Let* $(\Omega, \mathcal{F}, \mu)$ *be a measure space. Let* $1 < p < \infty$, $f \in L^p(\Omega, \mathcal{F}, \mu)$ *and* $g \in L^q(\Omega, \mathcal{F}, \mu)$ *where* $q = \frac{p}{(p-1)}$. *Then*

$$\int |fg| d\mu \ \leq \ \Big(\int |f|^p d\mu\Big)^{\frac{1}{p}} \Big(\int |g|^q d\mu\Big)^{\frac{1}{q}}, \tag{1.13}$$

$$\text{i.e.,} \quad \|fg\|_1 \ \leq \ \|f\|_p \|g\|_q.$$

If $\|fg\|_1 \neq 0$, *then equality holds in (1.13) iff* $|f|^p = c|g|^q$ *a.e.* (μ) *for some constant* $c \in (0, \infty)$.

Proof: W.l.o.g. assume that $\int |f|^p d\mu > 0$ and $\int |g|^q d\mu > 0$. Fix $\omega \in \Omega$. Let $p_1 = \frac{1}{p}$, $p_2 = \frac{1}{q}$, $b_1 = c_1|f(\omega)|^p$, and $b_2 = c_2|g(\omega)|^q$, where $c_1 = (\int |f|^p d\mu)^{-1}$ and $c_2 = (\int |g|^q d\mu)^{-1}$. Then applying (1.10) with $k = 2$ yields

$$\frac{c_1}{p}|f(\omega)|^p + \frac{c_2}{q}|g(\omega)|^q \geq c_1^{\frac{1}{p}} c_2^{\frac{1}{q}} |f(\omega)g(\omega)|. \tag{1.14}$$

Integrating w.r.t. μ yields

$$1 \geq c_1^{\frac{1}{p}} c_2^{\frac{1}{q}} \int |f(\omega)g(\omega)| d\mu(\omega)$$

which is equivalent to (1.13).

Next, equality in (1.13) implies equality in (1.14) a.e. (μ). Since $1 < p < \infty$, by the last part of Proposition 3.1.10 (ii), this implies that $b_1 = b_2$ a.e. (μ), i.e., $|f(\omega)|^p = c_2 c_1^{-1} |g(\omega)|^q$ a.e. (μ). □

Remark 3.1.4: (*Hölder's inequality for* $p = 1, q = \infty$). Let $f \in L^1(\Omega, \mathcal{F}, \mu)$ and $g \in L^\infty(\Omega, \mathcal{F}, \mu)$. Then $|fg| \le |f| \, \|g\|_\infty$ a.e. (μ) and hence,

$$\|fg\|_1 \equiv \int |fg| d\mu \le \|f\|_1 \|g\|_\infty.$$

If equality holds in the above inequality, then $|f|(\|g\|_\infty - |g|) = 0$ a.e. (μ) and hence, $|g| = \|g\|_\infty$ on the set $\{|f| > 0\}$ a.e. (μ).

The next two corollaries follow directly from Theorem 3.1.11. The proof is left as an exercise (Problem 3.9).

Corollary 3.1.12: (*Cauchy-Schwarz inequality*). *Let* f, $g \in L^2(\Omega, \mathcal{F}, \mu)$. *Then*

$$\begin{aligned} \int |fg| d\mu &\le \Big(\int |f|^2 d\mu\Big)^{\frac{1}{2}} \Big(\int |g|^2 d\mu\Big)^{\frac{1}{2}}, \qquad (1.15) \\ \textit{i.e.,} \quad \|fg\|_1 &\le \|f\|_2 \|g\|_2. \end{aligned}$$

Corollary 3.1.13: *Let* $k \in \mathbb{N}$. *Let* $a_1, a_2, \ldots, a_k$, $b_1, b_2, \ldots, b_k$ *be real numbers and* $c_1, c_2, \ldots, c_k$ *be positive real numbers.*

(i) Then, for any $1 < p < \infty$,

$$\sum_{i=1}^k |a_i b_i| c_i \le \Big(\sum_{i=1}^k |a_i|^p c_i\Big)^{\frac{1}{p}} \Big(\sum_{i=1}^k |b_i|^q c_i\Big)^{\frac{1}{q}}, \qquad (1.16)$$

where $q = \frac{p-1}{p}$.

(ii)

$$\sum_{i=1}^k |a_i b_i| c_i \le \Big(\sum_{i=1}^k |a_i|^2 c_i\Big)^{\frac{1}{2}} \Big(\sum_{i=1}^k |b_i|^2 c_i\Big)^{\frac{1}{2}}. \qquad (1.17)$$

Next, as an application of Hölder's inequality, one gets

Theorem 3.1.14: (*Minkowski's inequality*). *Let* $1 < p < \infty$ *and* $f, g \in L^p(\Omega, \mathcal{F}, \mu)$. *Then*

$$\begin{aligned} \Big(\int |f+g|^p d\mu\Big)^{\frac{1}{p}} &\le \Big(\int |f|^p d\mu\Big)^{\frac{1}{p}} + \Big(\int |g|^p d\mu\Big)^{\frac{1}{p}}, \\ \textit{i.e.,} \quad \|f+g\|_p &\le \|f\|_p + \|g\|_q. \qquad (1.18) \end{aligned}$$

Proof: Let $h_1 = |f+g|$, $h_2 = |f+g|^{p-1}$. Then by (1.12),

$$|f+g|^p \leq 2^{p-1}(|f|^p + |g|^p),$$

implying that $h_1 \in L^p(\Omega, \mathcal{F}, \mu)$ and $h_2 \in L^q(\Omega, \mathcal{F}, \mu)$, where $q = \frac{p}{(p-1)}$. Since $|f+g|^p = h_1 h_2 \leq |f|h_2 + |g|h_2$, by Hölder's inequality,

$$\int |f+g|^p d\mu \leq \Big(\int |f|^p d\mu\Big)^{\frac{1}{p}} \Big(\int h_2^q\Big)^{\frac{1}{q}} + \Big(\int |g|^p d\mu\Big)^{\frac{1}{p}} \Big(\int h_2^q\Big)^{\frac{1}{q}}. \quad (1.19)$$

But $\int h_2^q = \int |f+g|^p$ and so (1.19) yields (1.18). □

Remark 3.1.5: Inequality (1.18) holds for both $p = 1$ and $p = \infty$.

3.2 L^p Spaces

3.2.1 Basic properties

Let $(\Omega, \mathcal{F}, \mu)$ be a measure space. Recall the definition of $L^p(\Omega, \mathcal{F}, \mu)$, $0 < p \leq \infty$, from Section 2.5 as the set of all measurable functions f on $(\Omega, \mathcal{F}, \mu)$ such that $\|f\|_p < \infty$, where for $0 < p < \infty$,

$$\|f\|_p = \Big(\int |f|^p d\mu\Big)^{\min\{\frac{1}{p}, 1\}}$$

and for $p = \infty$,

$$\|f\|_\infty \equiv \sup\{k : \mu(\{|f| > k\}) > 0\}$$

(called the *essential supremum* of f). In this section and elsewhere, $L^p(\mu)$ denotes $L^p(\Omega, \mathcal{F}, \mu)$. The following proposition shows that $L^p(\mu)$ is a vector space over $\mathbb{R}$.

Proposition 3.2.1: *For* $0 < p \leq \infty$,

$$f, g \in L^p(\mu), \quad a, b \in \mathbb{R} \quad \Rightarrow \quad af + bg \in L^p(\mu). \quad (2.1)$$

Proof:

CASE 1: $0 < p \leq 1$. For any two positive numbers x, y,

$$\Big(\frac{x}{x+y}\Big)^p + \Big(\frac{y}{x+y}\Big)^p \geq \frac{x}{x+y} + \frac{y}{x+y} = 1.$$

Hence, for all $x, y \in (0, \infty)$

$$(x+y)^p \leq x^p + y^p. \quad (2.2)$$

It is easy to check that (2.2) continues to hold if $x, y \in [0, \infty)$. This yields $|af + bg|^p \leq |a|^p|f|^p + |b|^p|g|^p$, which, in turn, yields (2.1) by integration.

CASE 2: $1 < p < \infty$. By (1.12),

$$|af + bg|^p \leq 2^{p-1}(|af|^p + |bg|^p).$$

Integrating both sides of the above inequality yields (2.1).

CASE 3: $p = \infty$. By definition, there exist constants $K_1 < \infty$ and $K_2 < \infty$ such that

$$\mu(\{|f| > K_1\}) = 0 = \mu(\{|g| > K_2\}).$$

This implies that $\mu(\{|af + bg| > K\}) = 0$ for any $K > |a|K_1 + |b|K_2$. Hence, $af + bg \in L^\infty(\mu)$ and

$$\|af + bg\|_\infty \leq |a|\,\|f\|_\infty + |b|\,\|g\|_\infty\,.$$

□

Recall that a set $\mathbb{S}$ with a function $d : \mathbb{S} \times \mathbb{S} \to [0, \infty]$ is called a *metric space* if for all x, y, $z \in \mathbb{S}$,

(i) $d(x, y) = d(y, x)$ *(symmetry)*

(ii) $d(x, y) \leq d(x, z) + d(y, z)$ *(triangle inequality)*

(iii) $d(x, y) = 0$ iff $x = y$

and the function $d(\cdot, \cdot)$ is called a *metric* on $\mathbb{S}$. Some examples are:

(a) $\mathbb{S} = \mathbb{R}^k$ with $d(x, y) = \left(\sum_{i=1}^k |x_i - y_i|^2\right)^{\frac{1}{2}}$;

(b) $\mathbb{S} = C[0, 1]$, the space of all continuous functions on $[0, 1]$ with $d(f, g) = \sup\{|f(x) - g(x)| : 0 \leq x \leq 1\}$;

(c) $\mathbb{S} =$ a nonempty set, and $d(x, y) = 1$ if $x \neq y$ and 0 if $x = y$. (This $d(\cdot, \cdot)$ is called the *discrete metric* on $\mathbb{S}$.)

The L^p-norm $\|\cdot\|_p$ can be used to introduce a distance notion in $L^p(\mu)$ for $0 < p \leq \infty$.

Definition 3.2.1: For f, $g \in L^p(\mu)$, $0 < p \leq \infty$, let

$$d_p(f, g) \equiv \|f - g\|_p. \tag{2.3}$$

Note that, for any f, g, $h \in L^p(\mu)$ and $1 \leq p \leq \infty$,

(i) $d_p(f, g) = d_p(g, f) \geq 0$ *(nonnegativity and symmetry)*, and

(ii) $d_p(f, h) \leq d_p(f, g) + d_p(g, h)$ *(triangle inequality)*, which follows by Minkowski's inequality (Theorem 3.1.14) for $1 \le p < \infty$, and by Proposition 3.2.1 for $p = \infty$. However, $d_p(f, g) = 0$ implies only that $f = g$ a.e. (μ). Thus, $d_p(\cdot, \cdot)$ of (2.3) satisfies conditions (i) and (ii) of being a metric and it satisfies condition (iii) as well, provided any two functions f and g that agree a.e. (μ) are regarded as the same element of $L^p(\mu)$. This leads to the following:

Definition 3.2.2: For f, g in $L^p(\mu)$, f is called *equivalent* to g and is written as $f \sim g$, if $f = g$ a.e. (μ).

It is easy to verify that the relation $\sim$ of Definition 3.2.2 is an *equivalence relation*, i.e., it satisfies

(i) $f \sim f$ for all f in $L^p(\mu)$ *(reflexive)*

(ii) $f \sim g \Rightarrow g \sim f$ *(symmetry)*

(iii) $f \sim g, g \sim h \Rightarrow f \sim h$ *(transitive)*.

This equivalence relation $\sim$ divides $L^p(\mu)$ into disjoint equivalence classes such that in each class all elements are equivalent. The notion of distance between these classes may be defined as follows:

$$d_p([f], [g]) \equiv d_p(f, g)$$

where $[f]$ and $[g]$ denote, respectively, the equivalence classes of functions containing f and g. It can be verified that this is a metric on the set of equivalence classes. In what follows, the equivalence class $[f]$ is identified with the element f. With this identification, $(L^p(\mu), d_p(\cdot, \cdot))$ becomes a metric space for $1 \le p \le \infty$.

Remark 3.2.1: For $0 < p < 1$, if one defines

$$d_p(f, g) \equiv \int |f - g|^p d\mu, \tag{2.4}$$

then $(L^p(\mu), d_p)$ becomes a metric space (with the same identification as above of functions with their equivalence classes). The triangle inequality follows from (2.2).

Recall that a metric space $(\mathbb{S}, d)$ is called *complete* if every *Cauchy sequence* in $(\mathbb{S}, d)$ converges to an element in $\mathbb{S}$, i.e., if $\{x_n\}_{n \ge 1}$ is a sequence in $\mathbb{S}$ such that for every $\epsilon > 0$, there exists a N_ϵ such that $n, m \ge N_\epsilon \Rightarrow d(x_n, x_m) \le \epsilon$, then there exists an element x in $(\mathbb{S}, d)$ such that $\lim_{n \to \infty} d(x_n, x) = 0$. The next step is to establish the *completeness of* $L^p(\mu)$.

Theorem 3.2.2: *For $0 < p \le \infty$, $(L^p(\mu)$, $d_p(\cdot, \cdot))$ is complete, where d_p is as in (2.3).*

Proof: Let $\{f_n\}_{n\geq 1}$ be a Cauchy sequence in $L^p(\mu)$ for $0 < p < \infty$. The main steps in the proof are as follows:

(I) there exists a subsequence $\{n_k\}_{k\geq 1}$ such that $\{f_{n_k}\}_{k\geq 1}$ converges a.e. (μ) to a limit function f;

(II) $\lim_{k\to\infty} d_p(f_{n_k}, f) = 0$;

(III) $\lim_{n\to\infty} d_p(f_n, f) = 0$.

STEP (I): Let $\{\epsilon_k\}_{k\geq 1}$ and $\{\delta_k\}_{k\geq 1}$ be sequences of positive numbers decreasing to zero. Since $\{f_n\}_{n\geq 1}$ is Cauchy, for each $k \geq 1$, there exists an integer n_k such that

$$\int |f_n - f_m|^p d\mu \leq \epsilon_k \quad \text{for all} \quad n, m \geq n_k. \tag{2.5}$$

W.l.o.g., let $n_{k+1} > n_k$ for each $k \geq 1$. Then, by Markov's inequality (Theorem 3.1.1),

$$\mu(\{|f_{n_{k+1}} - f_{n_k}| \geq \delta_k\}) \leq \delta_k^{-p} \int |f_{n_{k+1}} - f_{n_k}|^p d\mu \leq \delta_k^{-p} \epsilon_k. \tag{2.6}$$

Let $A_k = \{|f_{n_{k+1}} - f_{n_k}| \geq \delta_k\}$, $k = 1, 2, \ldots$ and $A = \limsup_{k\to\infty} A_k \equiv \bigcap_{j=1}^{\infty} \bigcup_{k\geq j} A_k$. If $\{\epsilon_k\}_{k\geq 1}$ and $\{\delta_k\}_{k\geq 1}$ satisfy

$$\sum_{k=1}^{\infty} \delta_k^{-p} \epsilon_k < \infty, \tag{2.7}$$

then by (2.6), $\sum_{k=1}^{\infty} \mu(A_k) < \infty$ and hence, as in the proof of Theorem 2.5.2, $\mu(A) = 0$.

Note that for ω in A^c, $|f_{n_{k+1}}(\omega) - f_{n_k}(\omega)| < \delta_k$ for all k large. Thus, if $\sum_{k=1}^{\infty} \delta_k < \infty$, then for ω in A^c, $\{f_{n_k}(\omega)\}_{k\geq 1}$ is a Cauchy sequence in $\mathbb{R}$ and hence, it converges to some $f(\omega)$ in $\mathbb{R}$. Setting $f(\omega) = 0$ for $\omega \in A$, one gets

$$\lim_{k\to\infty} f_{n_k} = f \quad \text{a.e. } (\mu).$$

A choice of $\{\epsilon_k\}_{k\geq 1}$ and $\{\delta_k\}_{k\geq 1}$ such that $\sum_{k=1}^{\infty} \delta_k < \infty$ and (2.7) holds is given by $\epsilon_k = 2^{-(p+1)k}$ and $\delta_k = 2^{-k}$. This completes Step (I).

STEP (II): By Fatou's lemma, part (I), and (2.5), for any $k \geq 1$ fixed,

$$\epsilon_k \geq \liminf_{j\to\infty} \int |f_{n_k} - f_{n_{k+j}}|^p d\mu \geq \int |f_{n_k} - f|^p d\mu \, .$$

Since $f_{n_k} \in L^p(\mu)$, this shows that $f \in L^p(\mu)$. Now, on letting $k \to \infty$, (II) follows.

STEP (III): By triangle inequality, for any $k \geq 1$ fixed,

$$d_p(f_n, f) \leq d_p(f_n, f_{n_k}) + d_p(f_{n_k}, f).$$

By (2.5) and (II), for $n \geq n_k$, the right side above is $\leq 2\tilde{\epsilon}_k$, where $\tilde{\epsilon}_k = \epsilon_k$ if $0 < p < 1$ and $\tilde{\epsilon}_k = \epsilon_k^{1/p}$ if $1 \leq p < \infty$. Now letting $k \to \infty$, (III) follows.

The proof of Theorem 3.2.2 is complete for $0 < p < \infty$. The case $p = \infty$ is left as an exercise (Problem 3.14). □

3.2.2 Dual Spaces

Let $1 \leq p < \infty$. Let $g \in L^q(\mu)$, where $q = \frac{p}{(p-1)}$ if $1 < p < \infty$ and $q = \infty$ if $p = 1$. Let

$$T_g(f) = \int fg d\mu, \quad f \in L^p(\mu). \tag{2.8}$$

By Hölders inequality, $\int |fg| d\mu < \infty$ and so $T_g(\cdot)$ is well defined. Clearly T_g is *linear*, i.e.,

$$T_g(\alpha_1 f_1 + \alpha_2 f_2) = \alpha_1 T_g(f_1) + \alpha_2 T_g(f_2) \tag{2.9}$$

for all $\alpha_1, \alpha_2 \in \mathbb{R}$ and $f_1, f_2 \in L^p(\mu)$.

Definition 3.2.3:

(a) A function $T : L^p(\mu) \to \mathbb{R}$ that satisfies (2.9) is called a *linear functional.*

(b) A linear functional T on $L^p(\mu)$ is called *bounded* if there is a constant $c \in (0, \infty)$ such that

$$|T(f)| \leq c\|f\|_p \quad \text{for all} \quad f \in L^p(\mu).$$

(c) The norm of a bounded linear functional T on $L^p(\mu)$ is defined as

$$\|T\| = \sup\big\{|Tf| : f \in L^p(\mu), \|f\|_p = 1\big\}.$$

By Hölder's inequality (cf. Theorem 3.1.11 and Remark 3.1.4),

$$|T_g(f)| \leq \|g\|_q \|f\|_p \quad \text{for all} \quad f \in L^p(\mu),$$

and hence, T_g is a bounded linear functional on $L^p(\mu)$. This implies that if $d_p(f_n, f) \to 0$, then

$$|T_g(f_n) - T_g(f)| \leq \|g\|_q d_p(f_n, f) \to 0,$$

i.e., T_g is *continuous* on the metric space $(L^p(\mu), d_p)$.

Definition 3.2.4: The set of all continuous linear functionals on $L^p(\mu)$ is called the *dual space* of $L^p(\mu)$ and is denoted by $(L^p(\mu))^*$.

In the next section, it will be shown that continuity of a linear functional on $L^p(\mu)$ implies boundedness. A natural question is whether every continuous linear functional T on $L^p(\mu)$ coincides with T_g for some g in $L^q(\mu)$. The answer is "yes" for $1 \leq p < \infty$, as shown by the following result.

Theorem 3.2.3: (*Riesz representation theorem*). *Let $1 \leq p < \infty$. Let $T : L^p(\mu) \to \mathbb{R}$ be linear and continuous. Then, there exists a g in $L^q(\mu)$ such that $T = T_g$, i.e.,*

$$T(f) = T_g(f) \equiv \int fgd\mu \quad \textit{for all} \quad f \in L^p(\mu), \tag{2.10}$$

where $q = \frac{p}{p-1}$ for $1 < p < \infty$ and $q = \infty$ if $p = 1$.

Remark 3.2.2: Such a representation is not valid for $p = \infty$. That is, there exists continuous linear functionals T on $L^\infty(\mu)$ for which there is no $g \in L^1(\mu)$ such that $T(f) = \int fgd\mu$ for all $f \in L^\infty(\mu)$, provided μ is not concentrated on a finite set $\{\omega_1, \omega_2, \ldots, \omega_k\} \subset \Omega$.

For a proof of Theorem 3.2.3 and the above remark, see Royden (1988) or Rudin (1987).

Next consider the mapping from $(L^p(\mu))^*$ and $L^q(\mu)$ defined by

$$\phi(T_g) = g,$$

where T_g is as defined in (2.10). Then, ϕ is linear, i.e.,

$$\phi(\alpha_1 T_1 + \alpha_2 T_2) = \alpha_1 \phi(T_1) + \alpha_2 \phi(T_2)$$

for all $\alpha_1, \alpha_2 \in \mathbb{R}$ and $T_1, T_2 \in (L^p(\mu))^*$. Further,

$$||\phi(T)||_q = ||T|| \quad \text{for all} \quad T \in (L^p(\mu))^*.$$

Thus, ϕ preserves the vector space structure of $(L^p(\mu))^*$ and the norm. For this reason, it is called an *isometry* between $(L^p(\mu))^*$ and $L^q(\mu)$.

3.3 Banach and Hilbert spaces

3.3.1 Banach spaces

If $(\Omega, \mathcal{F}, \mu)$ is a measure space it was seen in the previous section that the space $L^p(\Omega, \mathcal{F}, \mu)$ of equivalence classes of functions f with $\int_\Omega |f|^p d\mu < \infty$ is a vector space over $\mathbb{R}$ for all $1 \leq p < \infty$ and for $p \geq 1$, $|| \cdot ||_p \equiv (\int |f|^p d\mu)^{1/p}$ satisfies

(i) $||f + g|| \leq ||f|| + ||g||$,

(ii) $||\alpha f|| = |a| \, ||f||$ for every $\alpha \in \mathbb{R}$,

(iii) $||f|| = 0$ iff $f = 0$ a.e. (μ).

The Euclidean spaces $\mathbb{R}^k$ for any $k \in \mathbb{N}$ is also a vector space. Note that for $p \geq 1$, setting $||x||_p \equiv (\sum_{i=1}^k |x_i|^p)^{1/p}$ if $x = (x_1, x_2, \ldots, x_k)$, $(\mathbb{R}^k, ||x||_p)$ may be identified with a special case of $L^p(\Omega, \mathcal{F}, \mu)$, where $\Omega \equiv \{1, 2, \ldots, k\}$, $\mathcal{F} = \mathcal{P}(\Omega)$ and μ is the counting measure.

Generalizing the above examples leads to the notion of a *normed vector space* (also called *normed linear space*). Recall that a *vector space V over* $\mathbb{R}$ is a nonempty set with a binary operation $+$, a function from $V \times V$ to V (called *addition*), and scalar multiplication by the real numbers, i.e., a function from $\mathbb{R} \times V \to V$, $\big((\alpha, v) \to \alpha v\big)$ satisfying

(i) $v_1, v_2 \in V \Rightarrow v_1 + v_2 = v_2 + v_1 \in V$.

(ii) $v_1, v_2, v_3 \in V \Rightarrow (v_1 + v_2) + v_3 = v_1 + (v_2 + v_3)$.

(iii) There exists an element θ, called the *zero vector*, in V such that $v + \theta = v$ for all v in V.

(iv) $\alpha \in \mathbb{R}$, $v \in V \Rightarrow \alpha v \in V$.

(v) $\alpha \in \mathbb{R}$, $v_1, v_2 \in V \Rightarrow \alpha(v_1 + v_2) = \alpha v_1 + \alpha v_2$.

(vi) $\alpha_1, \alpha_2 \in \mathbb{R}$, $v \in V \Rightarrow (\alpha_1 + \alpha_2)v = \alpha_1 v + \alpha_2 v$ and $\alpha_1(\alpha_2 v) = (\alpha_1 \alpha_2)v$.

(vii) $v \in V \Rightarrow 0v = \theta$ and $1v = v$.

Note that from conditions (vi) and (vii) above, it follows that for any v in V, $v + (-1)v = 0 \cdot v = \theta$. Thus for any v in V, $(-1)v$ is the additive inverse and is denoted by $-v$. Conditions (i), (ii), and (iii) are called respectively *commutativity*, *associativity* and the *existence of an additive identity*. Thus V under the operation $+$ is an Abelian (i.e., commutative) group.

Definition 3.3.1: A function f from V to $\mathbb{R}_+$ denoted by $f(v) \equiv ||v||$ is called a *norm* if

(a) $v_1, v_2 \in V \Rightarrow ||v_1 + v_2|| \leq ||v_1|| + ||v_2||$ *(triangle inequality)*

(b) $\alpha \in \mathbb{R}$, $v \in V \Rightarrow ||\alpha v|| = |\alpha| \, ||v||$ *(scalar homogeneity)*

(c) $||v|| = 0$ iff $v = \theta$.

A vector space V with a norm $|| \cdot ||$ defined on it is called a *normed vector space* or *normed linear space* and is denoted as $(V, || \cdot ||)$. Let $d(v_1, v_2) \equiv ||v_1 - v_2||$, $v_1, v_2 \in V$. Then from the definition of $|| \cdot ||$, it follows that d is a metric on V, i.e., (V, d) is a metric space. Recall that a metric space $(\mathbb{S}, d)$

is called *complete* if every Cauchy sequence $\{x_n\}_{n\geq 1}$ in $\mathbb{S}$ converges to an element x in $\mathbb{S}$.

Definition 3.3.2: A *Banach space* is a complete normed linear space $(V, ||\cdot||)$.

It was shown by S. Banach of Poland that all $L^p(\Omega, \mathcal{B}, \mu)$ spaces are Banach spaces, provided $p \geq 1$ and in particular, all Euclidean spaces are Banach spaces. An example of a different kind is the space $C[0,1]$ of all real valued continuous functions on $[0,1]$ with the usual operation of pointwise addition and scalar multiplication, i.e., $(f+g)(x) = f(x) + g(x)$ and $(\alpha f)(x) = \alpha \cdot f(x)$ for all $\alpha \in \mathbb{R}$, $0 \leq x \leq 1$, f, $g \in C[0,1]$ where the norm (called the *supnorm*) is defined by $||f|| = \sup\{|f(x)| = 0 \leq x \leq 1\}$. The verification of the fact that $C[0,1]$ with the supnorm is a Banach space is left as an exercise (Problem 3.22). The space $\mathcal{P}$ of all polynomials on $[0,1]$ is also a normed linear space under the above norm but $(\mathcal{P}, ||\cdot||)$ is *not complete* (Problem 3.23). However for each $n \in \mathbb{N}$, the space $\mathcal{P}_n$ of all polynomials on $[0,1]$ of degree $\leq n$ is a Banach space under the supnorm (Problem 3.26).

Definition 3.3.3: Let V be a vector space. A subset $W \subset V$ is called a *subspace* of V if $v_1, v_2 \in W, \alpha_1, \alpha_2 \in \mathbb{R} \Rightarrow \alpha_1 v_1 + \alpha_2 v_2 \in W$. If $(V, ||\cdot||)$ is a normed vector space and W is a subspace of V, then $(W, ||\cdot||)$ is also a normed vector space. If W is closed in $(V, ||\cdot||)$, then W is called a *closed subspace* of V.

Remark 3.3.1: If $(V, ||\cdot||)$ is a Banach space and W is a closed subspace of V, then $(W, ||\cdot||)$ is also a Banach space.

3.3.2 Linear transformations

Let $(V_i, ||\cdot||_i)$, $i = 1, 2$ be two normed linear spaces over $\mathbb{R}$.

Definition 3.3.4: A function T from V_1 to V_2 is called a *linear transformation* or *linear operator* if α_1, $\alpha_2 \in \mathbb{R}$, x, $y \in V_1 \Rightarrow T(\alpha_1 x + \alpha_2 y) = \alpha_1 T(x) + \alpha_2 T(y)$.

Definition 3.3.5: A linear operator T from $(V_1, ||\cdot||_1)$ to $(V_2, ||\cdot||_2)$ is called *bounded* if $||T|| \equiv \sup\{||Tx||_2 : ||x||_1 < 1\} < \infty$, i.e., the image of the *unit ball* in $(V_1, ||\cdot||_1)$ is contained in a ball of finite radius centered at the zero in V_2.

Here is a summary of some important results on this topic. By linearity of T, $T(\frac{x}{||x||}) = \frac{1}{||x||}T(x)$ for any $x \neq 0$. It follows that T is *bounded* iff there exists $k < \infty$ such that for any $x \in V_1$, $||Tx||_2 \leq k||x||_1$. Clearly, then k can be taken to be $||T||$. Also by linearity, if T is bounded, then $||Tx_1 - Tx_2|| = ||T(x_1 - x_2)|| \leq ||T||\,||x_1 - x_2||$ and so the map T is *continuous* (indeed, uniformly so). It turns out that if a linear operator T is continuous at some x_0 in V_1, then T is continuous on all of V_1 and is bounded (Problem 3.28 (a)).

Now let $B(V_1, V_2)$ be the space of all bounded linear operators from $(V_1, ||\cdot||_1)$ to $(V_2, ||\cdot||_2)$. For $T_1, T_2 \in B(V_1, V_2)$, α_1, α_2 in $\mathbb{R}$, let $(\alpha_1 T_1 + \alpha_2 T_2)$ be defined by $(\alpha_1 T_1 + \alpha_2 T_2)(x) \equiv \alpha_1 T_1(x) + \alpha_2 T_1(x)$ for all x in V_1. Then it can be verified that $(\alpha_1 T_1 + \alpha_2 T_2)$ also belongs to $B(V_1, V_2)$ and

$$||T|| \equiv \sup\{||Tx||_2 : ||x||_1 \leq 1\} \tag{3.1}$$

is a norm on $B(V_1, V_2)$. Thus $(B(V_1, V_2), ||\cdot||)$ is also a normed linear space. If $(V_2, ||\cdot||_2)$ is complete, then it can be shown that $(B(V_1, V_2), ||\cdot||)$ is also a Banach space (Problem 3.28 (b)). In particular, if $(V_2, ||\cdot||_2)$ is the real line, the space $(B(V_1, \mathbb{R}), ||\cdot||)$ is a Banach space.

3.3.3 Dual spaces

Definition 3.3.6: The space of all bounded linear functions from $(V_1, ||\cdot||)$ to $\mathbb{R}$ (also called *bounded linear functionals*), denoted by V_1^*, is called the *dual space* of V_1.

Thus, for any normed linear space $(V_1, ||\cdot||_1)$ (that need not be complete), the dual space $(V_1^*, ||\cdot||)$ is always a Banach space, where $||T|| \equiv \sup\{|Tx| : ||x||_1 < 1\}$ for $T \in V_1^*$. If $(V_1, ||\cdot||_1) = L^p(\Omega, \mathcal{F}, \mu)$ for some measure space $(\Omega, \mathcal{F}, \mu)$ and $1 \leq p < \infty$, by the *Riesz representation theorem* (see Theorem 3.2.3), the dual space may be identified with $L^q(\Omega, \mathcal{F}, \mu)$ where q is the conjugate of p, i.e., $\frac{1}{p} + \frac{1}{q} = 1$. However, as pointed out earlier in Section 3.2, this is not true for $p = \infty$. That is, the dual of $L^\infty(\Omega, \mathcal{F}, \mu)$ is not $L^1(\Omega, \mathcal{F}, \mu)$ unless $(\Omega, \mathcal{F}, \mu)$ is a measure space where Ω is a finite set $\{w_1, w_2, \ldots, w_k\}$ and $\mathcal{F} = \mathcal{P}(\Omega)$. An example for the $p = \infty$ case can be constructed for the space ℓ_∞ of all bounded sequences of real numbers (cf. Royden (1988)).

The representation of the dual space of the Banach space $C[0,1]$ with supnorm is in terms of finite signed measures (cf. Section 4.2).

Theorem 3.3.1: (*Riesz*). *Let $T : C[0,1] \to \mathbb{R}$ be linear and bounded. Then there exists two finite measures μ_1 and μ_2 on $[0,1]$ such that for any $f \in C[0,1]$*

$$T(f) = \int f d\mu_1 - \int f d\mu_2.$$

For a proof see Royden (1988) or Rudin (1987) (see also Problem 3.27).

3.3.4 Hilbert space

A vector space V over $\mathbb{R}$ is called a *real innerproduct* space if there exists a function $f : V \times V \to \mathbb{R}$, denoted by $f(x,y) \equiv \langle x,y\rangle$ (and called the *innerproduct*) that satisfies

(i) $\langle x,y\rangle = \langle y,x\rangle$ for all $x, y \in V$,

(ii) (*linearity*) $\langle \alpha_1 x_1 + \alpha_2 x_2, y\rangle = \alpha_1\langle x,y\rangle + \alpha_2\langle x_2,y\rangle$ for all $\alpha_1,\ \alpha_2 \in \mathbb{R}$, $x_1,\ x_2,\ y \in V$,

(iii) $\langle x,x\rangle \geq 0$ for all $x \in V$ and $\langle x,x\rangle = 0$ iff $x = \theta$, the zero vector of V.

Using the fact that the quadratic function $\varphi(t) = \langle x+ty, x+ty\rangle = \langle x,x\rangle + 2t\langle x,y\rangle + t^2\langle y,y\rangle$ is nonnegative for all $t \in \mathbb{R}$, one gets the *Cauchy-Schwarz* inequality

$$|\langle x,y\rangle| \leq \sqrt{\langle x,x\rangle\langle y,y\rangle} \quad \text{for all} \quad x,y \in V.$$

Now setting $||x|| = \sqrt{\langle x,x\rangle}$ and using the Cauchy-Schwarz inequality, one verifies that $||x||$ is a norm on V and thus $(V, ||\cdot||)$ is a normed linear space. Further, the function $\langle x,y\rangle$ from $V \times V$ to $\mathbb{R}$ is continuous (Problem 3.29) under the norm $||(x_1,x_2)|| = ||x_1|| + ||x_2||$, $(x_1,x_2) \in V \times V$.

Definition 3.3.7: Let $(V, \langle\cdot,\cdot\rangle)$ be a real innerproduct space. It is called a *Hilbert space* if $(V, ||\cdot||)$ is a Banach space, i.e., if it is complete.

It was seen in Section 3.2 that for any measure space $(\Omega, \mathcal{F}, \mu)$, the space $L^2(\Omega, \mathcal{F}, \mu)$ of all equivalence classes of functions $f : \Omega \to \mathbb{R}$ satisfying $\int |f|^2 d\mu < \infty$ is a complete innerproduct space with the innerproduct $\langle f,g\rangle = \int fg d\mu$ and hence, a Hilbert space. It turns out that every Hilbert space H is an $L^2(\Omega, \mathcal{F}, \mu)$ for some $(\Omega, \mathcal{F}, \mu)$. (The axiom of choice or its equivalent, the Hausdorff's maximality principle, is required for a proof of this. See Rudin (1987).) This is in contrast to the Banach space case where every $L^p(\Omega, \mathcal{F}, \mu)$ with $p \geq 1$ is a Banach space but not conversely, i.e., every Banach space need not be an $L^p(\Omega, \mathcal{F}, \mu)$.

Next for each x in a Hilbert space H, let $T_x : H \to \mathbb{R}$ be defined by $T_x(y) = \langle x,y\rangle$. By the defining properties of $\langle x,y\rangle$ and the Cauchy-Schwarz inequality, it is easy to verify that T_x is a bounded linear function on H, i.e.,

$$T_x(\alpha_1 y_1 + \alpha_2 y_2) = \alpha_1 T_x(y_1) + \alpha_2 T_x(y_2) \quad \text{for all} \quad \alpha_1, \alpha_2 \in \mathbb{R}, y_1, y_2 \in H \tag{3.2}$$

and

$$|T_x(y)| \leq ||x||\,||y|| \quad \text{for all} \quad y \in H. \tag{3.3}$$

Thus $T_x \in H^*$, the dual space. It is an important result (see Theorem 3.3.3 below) that every $T \in H^*$ is equal to T_x for some x in H and $||T|| = ||x||$. Thus H^* can be identified with H.

Definition 3.3.8: Let $(V, \langle \cdot, \cdot \rangle)$ be an inner product space. Two vectors x, y in V are said to be *orthogonal* and written as $x \perp y$ if $\langle x, y \rangle = 0$.

A collection $B \subset V$ is called *orthogonal* if $x, y \in B$, $x \neq y \Rightarrow \langle x, y \rangle = 0$. The collection B is called *orthonormal* if it is orthogonal and in addition for all x in B, $\|x\| = 1$. Note that if $x \perp y$, then $\|x - y\|^2 = \langle x - y, x - y \rangle = \langle x, x \rangle + \langle y, y \rangle = \|x\|^2 + \|y\|^2$ and so if B is an orthonormal set, then $x, y \in B \Rightarrow$ either $x = y$ or $\|x - y\| = \sqrt{2}$. Thus, if V is separable under the metric $d(x, y) = \|x - y\|$ (i.e., there exists a countable set $D \subset V$ such that for every x in V and $\epsilon > 0$, there exists a $d \in D$ such that $\|x - d\| < \epsilon$) and if $B \subset V$ is an orthonormal system, then the open ball S_b of radius $\frac{1}{2\sqrt{2}}$ around each $b \in B$ satisfies $\{S_b \cap D : b \in B\}$ are disjoint and nonempty. Thus B is countable.

Now let $(V, \langle \cdot, \cdot \rangle)$ be a separable innerproduct space and $B \subset V$ be an orthonormal system.

Definition 3.3.9: The *Fourier coefficients* of a vector x in V with respect on *orthonormal set* B is the set $\{\langle x, b \rangle : b \in B\}$.

Since V is separable, B is countable. Let $B = \{b_i : i \in \mathbb{N}\}$. For a given $x \in V$, let $c_i = \langle x, b_i \rangle$, $i \geq 1$. Let $x_n \equiv \sum_{i=1}^{n} c_i b_i$, $n \in \mathbb{N}$. The sequence $\{x_n\}_{n \geq 1}$ is called the *partial sum sequence of the Fourier expansion of the vector x w.r.t. the orthonormal set B*.

A natural question is: *when does $\{x_n\}_{n \geq 1}$ converge to x?* By the linearity property in the definition of the innerproduct $\langle \cdot, \cdot \rangle$, it follows that

$$0 \leq \|x - x_n\|^2 = \langle x - x_n, x - x_n \rangle = \langle x, x \rangle - 2\langle x, x_n \rangle + \langle x_n, x_n \rangle$$

and

$$\langle x, x_n \rangle = \sum_{i=1}^{n} c_i \langle x, b_i \rangle = \sum_{i=1}^{n} c_i^2.$$

Since $\{b_i\}_{i \geq 1}$ are orthonormal,

$$\|x_n\|^2 = \langle x_n, x_n \rangle = \sum_{i=1}^{n} c_i^2 = \langle x, x_n \rangle.$$

Thus,

$$0 \leq \|x - x_n\|^2 = \|x\|^2 - \|x_n\|^2 = \|x\|^2 - \sum_{i=1}^{n} c_i^2,$$

leading to

Proposition 3.3.2: *(Bessel's inequality). Let $\{b_i\}_{i \geq 1}$ be orthonormal in an innerproduct space $(V, \langle \cdot, \cdot \rangle)$. Then, for any x in V,*

$$\sum_{i=1}^{\infty} \langle x, b_i \rangle^2 \leq \|x\|^2. \tag{3.4}$$

Now let $(V, \langle\cdot,\cdot\rangle)$ be a *Hilbert space*. Since for $m > n$,

$$\|x_n - x_m\|^2 = \sum_{i=n+1}^{m} \langle x, b_i\rangle^2,$$

it follows from Bessel's inequality that $\{x_n\}_{n\geq 1}$ is a Cauchy sequence and since V is complete, there is a y in V such that $x_n \to y$. This implies (by the continuity of $\langle x, y\rangle$) that $\langle x, b_i\rangle = \lim_{n\to\infty}\langle x_n, b_i\rangle = \langle y, b_i\rangle \Rightarrow \langle x-y, b_i\rangle = 0$ for all $i \geq 1$. Thus, it follows that $\langle x, b_i\rangle = \langle y, b_i\rangle$ for all $i \geq 1$. The last relation implies $y = x$ iff the set $\{b_i\}_{i\geq 1}$ satisfies the property that

$$\langle z, b_i\rangle = 0 \quad \text{for all} \quad i \geq 1 \Rightarrow \|z\| = 0. \tag{3.5}$$

This property is called the *completeness of B*. Thus $B \equiv \{b_i\}_{i\geq 1}$ is a *complete orthonormal set* for a Hilbert space $H \equiv (V, \langle\cdot,\cdot\rangle)$, iff for every vector x,

$$\sum_{i=1}^{\infty} c_i^2 = \|x\|^2, \tag{3.6}$$

where $c_i = \langle x, b_i\rangle$, $i \geq 1$, which in turn holds iff

$$\left\| x - \sum_{i=1}^{n} c_i b_i \right\| \to 0 \quad \text{as} \quad n \to \infty. \tag{3.7}$$

Conversely, if $\{c_i\}_{i\geq 1}$ is a sequence of real numbers such that $\sum_{i=1}^{\infty} c_i^2 < \infty$, then the sequence $\{x_n \equiv \sum_{i=1}^{n} c_i b_i\}_{n\geq 1}$ is Cauchy and hence converges to an x in V. Thus the Hilbert space H can be identified with the space ℓ_2 of all *square summable* sequences $\{\{c_i\}_{i\geq 1} : \sum_{i=1}^{\infty} c_i^2 < \infty\}$, in the sense that the map $\varphi : x \to \{c_i\}_{i\geq 1}$, where $c_i = \langle x, b_i\rangle$, $i \geq 1$, preserves the algebraic structure as well as the innerproduct, i.e., φ is a linear operator from H to ℓ_2 and $\langle \varphi(x), \varphi(y)\rangle = \langle x, y\rangle$ for all $x, y \in H$. Such a φ is called an *isometric isomorphism* between H to ℓ_2. Note also that ℓ_2 is simply $L_2(\Omega, \mathcal{F}, \mu)$ where $\Omega \equiv \mathbb{N}$, $\mathcal{F} = \mathcal{P}(\mathbb{N})$, and μ, the counting measure. It can be shown (using the axiom of choice) that every separable Hilbert space does possess a complete orthonormal system, i.e, an *orthonormal basis.*

Next some examples are given. Here, unless otherwise indicated, H denotes the Hilbert space and B denotes an orthonormal basis of H.

Example 3.3.1:

(a) $H \equiv \ell_2 = \{(x_1, x_2, \ldots) : x_i \in \mathbb{R}, \sum_{i=1}^{\infty} x_i^2 < \infty\}$.
$B \equiv \{e_i : i \geq 1\}$ where $e_i \equiv (0, 0, \ldots, 1, 0, \ldots)$ with 1 in the ith position and 0 elsewhere.

(b) $H \equiv L^2\left([0, 2\pi], \mathcal{B}([0, 2\pi]), \mu\right)$ where $\mu(A) = \frac{1}{2\pi} m(A)$, $m(\cdot)$ being Lebesgue measure.
$B \equiv \{\cos nx : n = 0, 1, 2, \ldots, \} \cup \{\sin nx : n = 1, 2, \ldots\}$. (For a proof, see Chapter 5.)

(c) Let $H \equiv L^2(\mathbb{R}, \mathcal{B}(\mathbb{R}), \mu)$ where μ is a finite measure such that $\int |x|^k d\mu < \infty$ for all $k = 1, 2, \ldots$.
Let $B_1 \equiv \{1, x, x^2, \ldots\}$ and B be the orthonormal set generated by applying the Gram-Schmidt procedure to B_1 (see Problem 3.31). It can be shown that B is a basis for H (Problem 3.39). When μ is the standard normal distribution, the elements of B are called *Hermite* polynomials.

For one more example, i.e. Haar functions, see Problem 3.40.

Theorem 3.3.3: (*Riesz representation*). *Let H be a separable Hilbert space. Then every bounded linear functional T on $H \to \mathbb{R}$ can be represented as $T \equiv T_{x_0}$ for some $x_0 \in V$, where $T_{x_0}(y) \equiv \langle y, x_0 \rangle$.*

Proof: Let $B = \{b_i\}_{i \geq 1}$ be an orthonormal basis for H. Let $c_i \equiv T(b_i)$, $i \geq 1$. Then, for $n \geq 1$,

$$\begin{aligned}
\sum_{i=1}^n c_i^2 &= \sum_{i=1}^n c_i T(b_i) \\
&= T\Big(\sum_{i=1}^n c_i b_i\Big) \quad \text{(by the linearity of } T) \\
\Rightarrow \Big|\sum_{i=1}^n c_i^2\Big| &\leq \|T\| \Big\|\sum_{i=1}^n c_i b_i\Big\| = \|T\| \Big(\sum_{i=1}^n c_i^2\Big)^{1/2} \\
\Rightarrow \sum_{i=1}^n c_i^2 &\leq \|T\|^2 \\
\Rightarrow \sum_{i=1}^\infty c_i^2 &< \infty.
\end{aligned}$$

Thus $\{x_n \equiv \sum_{i=1}^n c_i b_i\}_{n \geq 1}$ is Cauchy in H and hence converges to an x_0 in H. By the continuity of T, for any y, $Ty = \lim_{n \to \infty} Ty_n$, where $y_n \equiv \sum_{i=1}^n \langle y, b_i \rangle b_i$, $n \geq 1$. But

$$\begin{aligned}
Ty_n &= \sum_{i=1}^n \langle y, b_i \rangle c_i = \Big\langle y, \sum_{i=1}^n b_i c_i \Big\rangle \\
&= \langle y, x_n \rangle, \quad \text{by the linearity of } T
\end{aligned}$$

Again by continuity of $\langle y, x \rangle$, it follows that $Ty = \langle y, x_0 \rangle$. $\square$

A sufficient condition for an $L^2(\Omega, \mathcal{F}, \mu)$ to be separable is that there exists an at most countable family $\mathcal{A} \equiv \{A_j\}_{j \geq 1}$ of sets in $\mathcal{F}$ such that $\mathcal{F} = \sigma\langle \mathcal{A} \rangle$ and $\mu(A_j) > 0$ for each j. This holds for any σ-finite measure μ on $(\mathbb{R}^k, \mathcal{B}(\mathbb{R}^k))$ (Problem 3.38).

Remark 3.3.2: Assuming the axiom of choice, the Riesz representation theorem remains valid for any Hilbert space, separable or not (Problem 3.43).

3.4 Problems

3.1 Prove Proposition 3.1.7.

(**Hint:** Use (1.4) repeatedly.)

3.2 Let $(\Omega, \mathcal{F}, \mu)$ be a measure space with $\mu(\Omega) \le 1$ and $f : \Omega \to (a, b) \subset \mathbb{R}$ be in $L^1(\Omega, \mathcal{F}, \mu)$. Let $\phi : (a, b) \to \mathbb{R}$ be convex. Show that if $c \equiv \int f d\mu \in (a, b)$ and $\phi(f) \in L^1(\Omega, \mathcal{F}, \mu)$ and $c\phi'_+(c) \ge 0$, then

$$\mu(\Omega)\,\phi\Big(\int f d\mu\Big) \le \int \phi(f) d\mu.$$

3.3 Prove Proposition 3.1.10.

(**Hint:** Apply Jensen's inequality with $\Omega \equiv \{1, 2, \ldots, k\}$, $\mathcal{F} = \mathcal{P}(\Omega)$, $P(\{i\}) = p_i$, $f(i) = a_i$, $i = 1, 2, \ldots, k$, and $\phi(x) = e^x$ to get (i). Deduce (ii) from (i) and Remark 3.1.3. For (iii), consider $\phi(x) = |x|^p$.)

3.4 Give an example of a convex function ϕ on $(0, 1)$ with a finite number of points where it is not differentiable. Can this be extended to the countable case? Uncountable case?

(**Hint:** Note that $\phi'_+(\cdot)$ and $\phi'_-(\cdot)$ are both monotone and hence have at most a countable number of discontinuity points.)

3.5 Let $\phi : (a, b) \to \mathbb{R}$ be convex.

(a) Using the definition and induction, show that

$$\phi\Big(\sum_{i=1}^{n} p_i x_i\Big) \le \sum_{i=1}^{n} p_i \phi(x_i)$$

for any $n \ge 2$, $x_1, x_2 \ldots, x_n$ in (a, b) and $\{p_1, p_2, \ldots, p_n\}$, a probability distribution.

(b) Use (a) to prove Jensen's inequality for any bounded ϕ.

3.6 Show that a function $\phi : \mathbb{R} \to \mathbb{R}$ is convex iff

$$\phi\Big(\int_{[0,1]} f dm\Big) \le \int_{[0,1]} \phi(f) dm$$

for every bounded Borel measurable function $f : [0, 1] \to \mathbb{R}$, where $m(\cdot)$ is the Lebesgue measure.

3.7 Let ϕ be convex on (a, b) and $\psi : \mathbb{R} \to \mathbb{R}$ be convex and nondecreasing. Show that $\psi \circ \phi$ is convex on (a, b).

3.8 Let X be a nonnegative random variable on some probability space.

(a) Show that $(EX)(E\frac{1}{X}) \geq 1$. What does this say about the correlation between X and $\frac{1}{X}$?

(b) Let $f, g : \mathbb{R}_+ \to \mathbb{R}_+$ be Borel measurable and such that $f(x)g(x) \geq 1$ for all x in $\mathbb{R}_+$. Show that $Ef(X)Eg(X) \geq 1$.

3.9 Prove Corollary 3.1.13 using Hölder's inequality applied to an appropriate measure space.

3.10 Extend Hölder's inequality as follows. Let $1 < p_i < \infty$, and $f_i \in L^{p_i}(\Omega, \mathcal{F}, \mu)$, $i = 1, 2, \ldots, k$. Suppose $\sum_{i=1}^{k} \frac{1}{p_i} = 1$. Show that $\int (\prod_{i=1}^{k} f_i) d\mu \leq \prod_{i=1}^{k} \|f_i\|_{p_i}$.

(**Hint:** Use Proposition 3.1.10 (ii).)

3.11 Verify Minkowski's inequality for $p = 1$ and $p = \infty$.

3.12 (a) Find $(\Omega, \mathcal{F}, \mu)$, $0 < p < 1$, $f, g \in L^p(\Omega, \mathcal{F}, \mu)$ such that

$$\left(\int |f + g|^p d\mu\right)^{1/p} > \left(\int |f|^p d\mu\right)^{1/p} + \left(\int |g|^p d\mu\right)^{1/p}.$$

(b) Prove (1.18) for $0 < p < 1$ with $\|f\|_p = \int |f|^p d\mu$.

3.13 Let $(\Omega, \mathcal{F}, \mu)$ be a measure space. Let $\{A_k\}_{k \geq 1} \subset \mathcal{F}$ and $\sum_{k=1}^{\infty} \mu(A_k) < \infty$. Show that $\mu\Big(\varlimsup_{k\to\infty} A_k\Big) = 0$, where $\varlimsup_{k\to\infty} A_k = \bigcap_{n=1}^{\infty} \bigcup_{j \geq n} A_j = \{\omega : \omega \in A_j \text{ for infinitely many } j \geq 1\}$.

3.14 Establish Theorem 3.2.2 for $p = \infty$.

(**Hint:** For each $k \geq 1$, choose $n_k \uparrow$ such that $\|f_{n_{k+1}} - f_{n_k}\|_\infty < 2^{-k}$. Show that there exists a set A with $\mu(A^c) = 0$ and for ω in A, $|f_{n_{k+1}}(\omega) - f_{n_k}(\omega)| < 2^{-k}$ for all $k \geq 1$ and now proceed as in the proof for the case $0 < p < \infty$.)

3.15 Let $f, g \in L^p(\Omega, \mathcal{F}, \mu)$, $0 < p < 1$. Show that $d(f, g) = \int |f - g|^p d\mu$ is a metric and $(L^p(\Omega, \mathcal{F}, \mu), d)$ is complete.

3.16 Let $(\Omega, \mathcal{F}, \mu)$ be a measure space and $f : \Omega \to \mathbb{R}$ be $\mathcal{F}$-measurable. Let $A_f = \{p : 0 < p < \infty, \int |f|^p d\mu < \infty\}$.

(a) Show that $p_1, p_2 \in A_f$, $p_1 < p_2$ implies $[p_1, p_2] \subset A_f$.

(**Hint:** Use $\int_{|f| \geq 1} |f|^p d\mu \leq \int_{|f| \geq 1} |f|^{p_2} d\mu$ and $\int_{|f| \leq 1} |f|^p d\mu \leq \int_{|f| \leq 1} |f|^{p_1} d\mu$ for any $p_1 < p < p_2$.)

(b) Let $\psi(p) = \log \int |f|^p d\mu$ for $p \in A_f$. By (a), it is known that A_f is connected, i.e., it is an interval. Show that ψ is convex in the open interior of A_f.

(**Hint:** Use Hölder's inequality.)

(c) Give examples to show that A_f could be a closed interval, an open interval, and semi-open intervals.

(d) If $0 < p_1 < p < p_2$, show that

$$\|f\|_p \le \max\{\|f\|_{p_1}, \|f\|_{p_2}\}$$

(**Hint:** Use (b).)

(e) Show that if $\int |f|^r d\mu < \infty$ for some $0 < r < \infty$, then $\|f\|_p \to \|f\|_\infty$ as $p \to \infty$.

(**Hint:** Show first that for any $K > 0$, $\mu(|f| > K) > 0 \Rightarrow \varliminf_{p\to\infty} \|f\|_p \ge K$. If $\|f\|_\infty < \infty$ and $\mu(\Omega) < \infty$, use the fact that $\|f\|_p \le \|f\|_\infty (\mu(\Omega))^{1/p}$ and reduce the general case under the hypothesis that $\int |f|^p d\mu < \infty$ for some p to this case.)

3.17 Let X be a random variable on a probability space $(\Omega, \mathcal{F}, \mu)$. Recall that $Eh(X) = \int h(X) d\mu$ if $h(X) \in L^1(\Omega, \mathcal{F}, \mu)$.

(a) Show that $(E|X|^{p_1}) \le (E|X|^{p_2})^{\frac{p_1}{p_2}}$ for any $0 < p_1 < p_2 < \infty$.

(b) Show that '=' holds in (a) iff $|X|$ is a constant a.e. (μ).

(c) Show that if $E|X| < \infty$, then $E|\log |X|| < \infty$ and $E|X|^r < \infty$ for all $0 < r < 1$, and $\frac{1}{r}\log(E|X|^r) \to E \log |X|$ as $r \to 0$.

3.18 Let X be a nonnegative random variable.

(a) Show that $EX \log X \ge (EX)(E \log X)$.

(b) Show that $\sqrt{1 + (EX)^2} \le E(\sqrt{1 + X^2}) \le 1 + EX$.

3.19 Let $\Omega = \mathbb{N}$, $\mathcal{F} = \mathcal{P}(\mathbb{N})$, and let μ be the counting measure. Denote $L^p(\Omega, \mathcal{F}, \mu)$ for this case by ℓ_p.

(a) Show that ℓ_p is the set of all sequences $\{x_n\}_{n\ge1}$ such that $\sum_{n=1}^\infty |x_n|^p < \infty$.

(b) For the following sequences, find all $p > 0$ such that they belong to ℓ_p:

(i) $x_n \equiv \frac{1}{n}$, $n \ge 1$.

(ii) $x_n = \frac{1}{n(\log(n+1))^2}$, $n \ge 1$.

3.20 For $1 \le p < \infty$, prove the Riesz representation theorem for ℓ_p. That is, show that if T is a bounded linear functional from $\ell_p \to \mathbb{R}$, then there exists a $y = \{y_i\}_{i \ge 1} \in \ell_q$ such that for any $x = \{x_i\}_{i \ge 1}$ in ℓ_p, $T(x) = \sum_{i=1}^{\infty} x_i y_i$.

(**Hint:** Let $y_i = T(e_i)$ where $e_i = \{e_i(j)\}_{j \ge 1}$, $e_i(j) = 1$ if $i = j$, 0 if $i \ne j$. Use the fact $|T(x)| \le \|T\| \, \|x\|_p$ to show that for each $n \in \mathbb{N}$, $(\sum_{i=1}^{n} |y_i|^q) \le \|T\|^q < \infty$.)

3.21 Let $\Omega = \mathbb{R}$, $\mathcal{F} = \mathcal{B}(\mathbb{R})$, $\mu = \mu_F$ where F is a cdf on $\mathbb{R}$. If $f(x) \equiv x^2$, find $A_f - \{p : 0 < p < \infty,\ f \in L^p(\mathbb{R}, \mathcal{B}(\mathbb{R}), \mu_F)\}$ for the following cases:

(a) $F(x) = \Phi(x)$, the $N(0,1)$ cdf, i.e., $\Phi(x) \equiv \frac{1}{\sqrt{2\pi}} \int_{-\infty}^{x} e^{-u^2/2} du$, $x \in \mathbb{R}$.

(b) $F(x) = \frac{1}{\pi} \int_{-\infty}^{x} \frac{1}{1+u^2} du$, $x \in \mathbb{R}$.

3.22 Show that $C[0,1]$ with the supnorm (i.e., with $\|f\| = \sup\{|f(x)| : 0 \le x \le 1\}$) is a Banach space.

(**Hint:** To verify completeness, let $\{f_n\}_{n \ge 1}$ be a Cauchy sequence in $C[0,1]$. Show that for each $0 \le x \le 1$, $\{f_n(x)\}_{n \ge 1}$ is a Cauchy sequence in $\mathbb{R}$. Let $f(x) = \lim_{n \to \infty} f_n(x)$. Now show that $\sup\{|f_n(x) - f(x)| : 0 \le s \le 1\} \le \lim_{m \to \infty} \|f_n - f_m\|$. Conclude that f_n converges to f uniformly on $[0,1]$ and that $f \in C[0,1]$.)

3.23 Show that the space $(\mathcal{P}, \|\cdot\|)$ of all polynomials on $[0,1]$ with the supnorm is a normed linear space that is not complete.

(**Hint:** Let $g(t) = \frac{1}{1-t/2}$ for $0 \le t \le 1$. Find a sequence of polynomials $\{f_n\}_{n \ge 1}$ in $\mathcal{P}$ that converge to g in supnorm.)

3.24 Show that the function $f(v) \equiv \|v\|$ from a normed linear space $(V, \|\cdot\|)$ to $\mathbb{R}_+$ is continuous.

3.25 Let $(V, \|\cdot\|)$ be a normed linear space. Let $S = \{v : \|v\| < 1\}$. Show that S is an open set in V.

3.26 Show that the space $\mathcal{P}_k$ of all polynomials on $[0,1]$ of degree $\le k$ is a Banach space under the supnorm, i.e., under $\|f\| = \sup\{|f(x)|, 0 \le s \le 1\}$.

(**Hint:** Let $p_n(x) = \sum_{j=0}^{k} a_{nj} x^j$ be a sequence of elements in $\mathcal{P}_k$ that converge in supnorm to some $f(\cdot)$. Show that $\{a_{n1}\}_{n \ge 1}$ converges and recursively, $\{a_{ni}\}_{n \ge 1}$ converges for each i.)

3.27 Let μ be a finite measure on [0,1]. Verify that $T_\mu(f) \equiv \int f d\mu$ is a bounded linear functional on $C[0,1]$ and that $\|T_\mu\| = \mu([0,1])$.

3.28 Let $(V_i, \|\cdot\|_i)$, $i = 1, 2$, be two normed linear spaces over $\mathbb{R}$.

(a) Let $T : V_1 \to V_2$ be a linear operator. Show that if for some x_0, $\|Tx - Tx_0\| \to 0$ as $x \to x_0$, then T is continuous on V_1 and hence bounded.

(b) Show that if $(V_2, \|\cdot\|_2)$ is complete, then $B(V_1, V_2) \equiv \{T \mid T : V_1 \to V_2,\ T \text{ linear and bounded}\}$ is complete under the operator norm defined in (3.1).

In the following, H will denote a real Hilbert space.

3.29 (a) Use the Cauchy-Schwarz inequality to show that the function $f(x, y) = \langle x, y\rangle$ is continuous from $H \times H \to \mathbb{R}$.

(b) (*Parallelogram law*). Show that in an innerproduct space $(V, \langle\cdot,\cdot\rangle)$, for any x, $y \in V$

$$\|x + y\|^2 + \|x - y\|^2 = 2(\|x\|^2 + \|y\|^2)$$

where $\|x\|^2 = \langle x, x\rangle$.

3.30 (a) Let $\{Q_n(x)\}_{n\geq 0}$ be defined on $[0, 2\pi]$ by

$$Q_n(x) = c_n\left(\frac{1 + \cos x}{2}\right)^n$$

where c_n is such that

$$\frac{1}{2\pi}\int_0^{2\pi} Q_n(x)dx = 1.$$

Clearly, $Q_n(\cdot) \geq 0$.

(i) Verify that for each $\delta > 0$,

$$\sup\{Q_n(x) : \delta \leq x \leq 2\pi - \delta\} \to 0 \quad \text{as} \quad n \to \infty.$$

(ii) Use this to show that if $f \in C[0, 2\pi]$ and if

$$P_n(t) \equiv \frac{1}{2\pi}\int_0^{2\pi} Q_n(t - s)f(s)ds, \quad n \geq 0, \tag{4.1}$$

then $P_n \to f$ uniformly on $[0, 2\pi]$.

(b) Use this to give a proof of the completeness of the class C of trigonometric functions.

(c) Show that if $f \in L^1([0, 2\pi])$, then $P_n(\cdot)$ converges to f in $L^1([0, 2\pi])$.

(d) Let $\{\mu_n(\cdot)\}_{n\geq 1}$ be a sequence of probability measures on $(\mathbb{R}, \mathcal{B}(\mathbb{R}))$ such that for each $\delta > 0$, $\mu_n(\{x : |x| \geq \delta\}) \to 0$ as $n \to \infty$. Let $f : \mathbb{R} \to \mathbb{R}$ be Borel measurable. Let $f_n(x) \equiv \int f(x-y)\mu_n(dy)$, $n \geq 1$. Assuming that $f_n(\cdot)$ is well defined and Borel measurable, show that

(i) $f(\cdot)$ continuous at x_0 and bounded $\Rightarrow f_n(x_0) \to f(x_0)$.

(ii) $f(\cdot)$ uniformly continuous and bounded on $\mathbb{R} \Rightarrow f_n \to f$ uniformly on $\mathbb{R}$.

(iii) $f \in L^p(\mathbb{R}, \mathcal{B}(\mathbb{R}), m)$, $0 < p < \infty$, $m(\cdot)$ = Lebesgue measure $\Rightarrow \int |f_n - f|^p dm \to 0$.

(iv) Show that (iii) $\Rightarrow$ (c).

3.31 (*Gram-Schmidt procedure*). Let $B \equiv \{b_n : n \in \mathbb{N}\}$ be a collection of nonzero vector in H. Set

$$\begin{aligned} e_1 &= \frac{b_1}{\|b_1\|} \\ \tilde{e}_2 &= b_2 - \langle b_2, e_1\rangle e_1, \\ e_2 &= \frac{\tilde{e}_2}{\|\tilde{e}_2\|} \quad \text{(provided } \|\tilde{e}_2\| \neq 0\text{), and so on.} \end{aligned}$$

If $\|\tilde{e}_n\| = 0$ for some $n \in \mathbb{N}$, then delete b_n. Let $E \equiv \{e_j : 1 \leq j < k\}$, $k \leq \infty$, be the collection of vectors reached this way.

(a) Show that E is an orthonormal system.

(b) Let H_B denote the closed linear subspace generated by B, i.e.,

$$H_B \equiv \Big\{x : x \in H, \text{ there exists } x_n \text{ of the form } \sum_{j=1}^{n} a_j b_j, \; a_j \in \mathbb{R}, \text{ such that } \|x_n - x\| \to 0\Big\}.$$

Show that H_B is a Hilbert space and E is a basis for H_B.

3.32 Let $H = L^2(\mathbb{R}, B(\mathbb{R}), \mu)$, where μ is a probability measure. Let $B \equiv \{1, x, x^2, \ldots\}$. Assume that $\int |x|^k d\mu < \infty$ for all $k \in \mathbb{N}$. Apply the Gram-Schmidt procedure in Problem 3.31 to the set B for the following cases and evaluate e_1, e_2, e_3.

(a) μ = Uniform $[0, 1]$ distribution.

(b) μ = standard normal distribution.

(c) μ = Exponential (1) distribution.

The orthonormal basis E obtained this way is called *Orthogonal Polynomials* w.r.t. the *given measure*. (See Szego (1939).)

3.33 Let $B \subset H$ be an orthonormal system. Show that for any x in H, $\{b : \langle x, b\rangle \neq 0\}$ is at most countable.

(**Hint:** Show first that if $\{y_\alpha : \alpha \in I\}$ is a collection of nonnegative real numbers such that for some $C < \infty$, $\sum_{\alpha \in F} y_\alpha \leq C$ for all $F \subset I$, F finite, then $\{\alpha : y_\alpha > 0\}$ is at most countable and apply this to the Bessel inequality.)

3.34 Let $B \subset H$. Define $B^\perp \equiv \{x : x \in H,\ \langle x, b\rangle = 0,\ \text{for all } b \in B\}$. Show that $B^\perp$ is a closed subspace of H.

3.35 Let $B \subset H$ be a closed subspace of H.

(a) Using the fact that every Hilbert space admits an orthonormal basis, show that every x in H can be uniquely decomposed as

$$x = y + z \tag{4.2}$$

where $y \in B$ and $z \in B^\perp$ and $\|x\|^2 = \|y\|^2 + \|z\|^2$.

(b) Let $P_B : H \to B$ be defined by $P_B x = y$ where x admits the decomposition in (4.2) above. Verify that P_B is a bounded linear operator from H to B and is of norm 1 if B has at least one nonzero vector. (The operator P_B is called the *projection* onto B.)

(c) Verify that $P_B(P_B x) = P_B x$ for all x in H.

3.36 Let H be separable and $\{x_n\}_{n\geq 1} \subset H$ be such that $\{\|x_n\|_{n\geq 1}\}$ is bounded by some $C < \infty$. Show that there exist a subsequence $\{x_{n_j}\}_{j\geq 1} \subset \{x_n\}_{n\geq 1}$ and an x_0 in H, such that for every y in H,

$$\langle x_{n_j}, y\rangle \to \langle x_0, y\rangle.$$

(**Hint:** Fix an orthonormal basis $B \equiv \{b_n\}_{n\geq 1} \subset H$. Let $a_{ni} = \langle x_n, b_i\rangle$, $n \geq 1$, $i \geq 1$. Using $\sum_{i=1}^\infty a_{ni}^2 \leq C$ for all n and the Bolzano-Wierstrass property, show that

(a) there exists $\{n_j\}_{j\geq 1}$ such that $\lim\limits_{j\to\infty} a_{n_j i} = a_i$ exists for all $i \geq 1$, $\sum_{i=1}^\infty a_i^2 < \infty$,

(b) $\lim\limits_{n\to\infty} \sum_{i=1}^n a_i b_i \equiv x_0$ exists in H,

(c) $\langle x_{n_j}, y\rangle \to \langle x_0, y\rangle$ for all y in H.)

3.37 Let $(V, \langle\cdot,\cdot\rangle)$ be an innerproduct space. Verify that $\langle\cdot,\cdot\rangle$ is *bilinear*, i.e., for α_1, α_2, β_1, $\beta_2 \in \mathbb{R}$, x_1, x_2, y_1, y_2 in V,

$$\begin{aligned}\langle \alpha_1 x_1 + \alpha_2 x_2, \beta_1 y_1 + \beta_2 y_2\rangle &= \alpha_1\beta_1\langle x_1, y_1\rangle + \alpha_1\beta_2\langle x_1, y_2\rangle \\ &\quad + \alpha_2\beta_1\langle x_2, y_1\rangle + \alpha_2\beta_2\langle x_2, y_2\rangle.\end{aligned}$$

State and prove an extension to more than two vectors.

3.38 Let $(\Omega, \mathcal{F}, \mu)$ be a measure space. Suppose that there exists an at most countable family $\mathcal{A} \equiv \{A_j\}_{j\geq 1} \subset \mathcal{F}$ such that $\mathcal{F} = \sigma\langle \mathcal{A} \rangle$ and $\mu(A_j) > 0$ for each $j \geq 1$. Then show that for $0 < p < \infty$, $L^p(\Omega, \mathcal{F}, \mu)$ is separable.

(**Hint:** Show first that for any $A \in \mathcal{F}$ with $\mu(A) < \infty$, and $\epsilon > 0$, there exists a countable subcollection $\mathcal{A}_1$ of $\mathcal{A}$ such that $\mu(A \triangle B) < \epsilon$ where $B = \{\cup A_j : A_j \in \mathcal{A}_1\}$. Now consider the class of functions $\{\sum_{i=1}^n c_i I_{A_i},\, n \geq 1,\, A_i \in \mathcal{A},\, c_i \in \mathbb{Q}\}$.)

3.39 Show that B in Example 3.3.1 (c) is a basis for H.

(**Hint:** Using Theorem 2.3.14 prove that the set of all polynomials are dense in H.)

3.40 (*Haar functions*). For x in $\mathbb{R}$ let $h(x) = I_{[0,1/2)}(x) - I_{[1/2,1)}(x)$. Let $h_{00}(x) \equiv I_{[0,1)}(x)$ and for $k \geq 1$, $0 \leq j < 2^{k-1}$, let $h_{kj}(x) \equiv 2^{\frac{k-1}{2}} h(2^{k-1}x - j)$, $0 \leq x < 1$.

(a) Verify that the family $\{h_{kj}(\cdot),\ k \geq 0,\ 0 \leq j < 2^{k-1}\}$ is an orthonormal set in $L^2([0,1], \mathcal{B}([0,1]), m)$, where $m(\cdot)$ is Lebesgue measure.

(b) Verify that this family is complete by completing the following two proofs:

(i) Show that for indicator function f of dyadic interval of the form $\left[\frac{k}{2^n}, \frac{\ell}{2^n}\right)$, $k < \ell$, the following identity holds:

$$\int f^2 dm = \frac{\ell - k}{2^n} = \sum_{k,j} \Big(\int f h_{kj} dm \Big)^2.$$

Now use the fact the linear combinations of such f's is dense in $L^2[0,1]$.

(ii) For each $f \in L^2([0,1], \mathcal{B}([0,1]), m)$ such that f is orthogonal to the Haar functions, $F(t) \equiv \int_{[0,t]} f dm$, $0 \leq t \leq 1$ is continuous and satisfies $F(\frac{j}{2^n}) = 0$ for all $0 \leq j \leq 2^n$, $n = 1, 2, \ldots$ and hence $F \equiv 0$ implying $f = 0$ a.e.

3.41 Let H be a Hilbert space over $\mathbb{R}$ and M be a closed subspace of H. Let $v_0 \in H$. Show that

$$\min\{\|v - v_0\| : v \in M\} = \max\{\langle v_0, u\rangle, u \in M^\perp,\ \|u\| = 1\},$$

where $M^\perp$ is the orthogonal complement of M, i.e., $M^\perp \equiv \{u : \langle v, u\rangle = 0 \text{ for all } v \in M\}$.

(**Hint:** Use Problem 3.35 (a).)

3.42 Let B be an orthonormal set in a Hilbert space H.

(a) (i) Show that for any x in H and any finite set $\{b_i : 1 \leq i \leq k\} \subset B$, $k < \infty$,

$$\sum_{i=1}^{k} \langle x, b_i \rangle^2 \leq \|x\|^2.$$

(ii) Conclude that for all x in H, $B_x \equiv \{b : \langle x, b \rangle \neq 0,\ b \in B\}$ is at most countable.

(b) Show that the following are equivalent:

(i) B is complete, i.e., $x \in H$, $\langle x, b \rangle = 0$ for all $b \in B \Rightarrow x = 0$.

(ii) For all $x \in B$, there exists an at most countable set $B_x \equiv \{b_i : i \geq 1\}$ such that $\|x\|^2 = \sum_{i=1}^{\infty} \langle x, b_i \rangle^2$.

(iii) For all $x \in B$, $\epsilon > 0$, there exists a finite set $\{b_1, b_2, \ldots, b_k\} \subset B$ such that

$$\left\| x - \sum_{i=1}^{k} \langle x, b_i \rangle b_i \right\| < \epsilon.$$

(iv) If $B \subset B^1$, B^1 an orthonormal set in $H \Rightarrow B = B^1$.

3.43 Extend Theorem 3.3.3 to any Hilbert space assuming that the axiom of choice holds.

(**Hint:** Using the axiom of choice or its equivalent, the Hausdorff maximality principle, it can be shown that every Hilbert space H admits an orthonormal basis B (see Rudin (1987)). Now let T be a bounded linear functional from H to $\mathbb{R}$. Let $f(b) \equiv T(b)$ for b in B. Verify that $\sum_{i=1}^{k} |f(b_i)|^2 \leq \|T\|^2$ for all finite collection $\{b_i : 1 \leq i \leq k\} \subset B$. Conclude that $D \equiv \{b : f(b) \neq 0\}$ is countable. Let $x_0 \equiv \sum_{b \in D} f(b) b$. Now use the proof of Theorem 3.3.3 to show that $T(x) \equiv \langle x, x_0 \rangle$ for all x in H.)

3.44 Let $(V, \|\cdot\|)$ be a normed linear space. Let $\{T_n\}_{n \geq 1}$ and T be bounded linear operators from V to V. The sequence $\{T_n\}_{n \geq 1}$ is said to converge

(a) *weakly* to T if for each w in V^*, the dual of V, and each v in V,

$$w(T_n(v)) \to w(T(v)),$$

(b) *strongly* if for each v in V, $\|T_n v - Tv\| \to 0$,

(c) *uniformly* if $\sup\{\|T_n v - Tv\| : \|v\| \leq 1\} \to 0$.

Let $V_p = L^p(\mathbb{R}, \mathcal{B}(\mathbb{R}), \mu_L)$, $1 \le p \le \infty$. Let $\{h_n\}_{n\ge1} \subset \mathbb{R}$ be such that $h_n \neq 0$, $h_n \to 0$, as $n \to \infty$. Let $(T_n f)(\cdot) \equiv f(\cdot + h_n)$, $Tf(\cdot) \equiv f(\cdot)$. Verify that

(i) $\{T_n\}_{n\ge1}$ and T are bounded linear operators on V_p, $1 \le p \le \infty$,

(ii) for $1 \le p < \infty$, $\{T_n\}_{n\ge1}$ converges to T weakly,

(iii) for $1 \le p < \infty$, $\{T_n\}$ converges to T strongly,

(iv) for $1 \le p < \infty$, $\{T_n\}$ does not converge to T uniformly by showing that for all n, $||T_n - T|| = 1$,

(v) for $p = \infty$, show that T_n does not converge weakly to T.

4

Differentiation

4.1 The Lebesgue-Radon-Nikodym theorem

Definition 4.1.1: Let $(\Omega, \mathcal{F})$ be a measurable space and let μ and ν be two measures on $(\Omega, \mathcal{F})$. The measure μ is said to be *dominated by* ν or *absolutely continuous w.r.t.* ν and written as $\mu \ll \nu$ if

$$\nu(A) = 0 \Rightarrow \mu(A) = 0 \quad \text{for all} \quad A \in \mathcal{F}. \tag{1.1}$$

Example 4.1.1: Let m be the Legesgue measure on $(\mathbb{R}, \mathcal{B}(\mathbb{R}))$ and let μ be the standard normal distribution, i.e.,

$$\mu(A) \equiv \int_A \frac{1}{\sqrt{2\pi}} e^{-x^2/2} m(dx), \quad A \in \mathcal{B}(\mathbb{R}).$$

Then $m(A) = 0 \Rightarrow \mu(A) = 0$ and hence $\mu \ll m$.

Example 4.1.2: Let $\mathbb{Z}_+ \equiv \{0, 1, 2, \ldots\}$ denote the set of all nonnegative integers. Let ν be the counting measure on $\Omega = \mathbb{Z}_+$ and μ be the Poisson (λ) distribution for $0 < \lambda < \infty$, i.e.,

$$\nu(A) = \text{number of elements in } A$$

and

$$\mu(A) = \sum_{j \in A} \frac{e^{-\lambda} \lambda^j}{j!}$$

for all $A \in \mathcal{P}(\Omega)$, the power set of Ω. Since $\nu(A) = 0 \Leftrightarrow A = \emptyset \Leftrightarrow \mu(A) = 0$, it follows that $\mu \ll \nu$ and $\nu \ll \mu$.

Example 4.1.3: Let f be a nonnegative measurable function on a measure space $(\Omega, \mathcal{F}, \nu)$. Let

$$\mu(A) \equiv \int_A f d\nu \quad \text{for all} \quad A \in \mathcal{F}. \tag{1.2}$$

Then, μ is a measure on $(\Omega, \mathcal{F})$ and $\nu(A) = 0 \Rightarrow \mu(A) = 0$ for all $A \in \mathcal{F}$ and hence $\mu \ll \nu$.

The Radon-Nikodym theorem is a sort of converse to Example 4.1.3. It says that if μ and ν are σ-finite measures (see Section 1.2) on a measurable space $(\Omega, \mathcal{F})$ and if $\mu \ll \nu$, then there is a nonnegative measurable function f on $(\Omega, \mathcal{F})$ such that (1.2) holds.

Definition 4.1.2: Let $(\Omega, \mathcal{F})$ be a measurable space and let μ and ν be two measures on $(\Omega, \mathcal{F})$. Then, μ is called *singular* w.r.t. ν and written as $\mu \perp \nu$ if there exists a set $B \in \mathcal{F}$ such that

$$\mu(B) = 0 \quad \text{and} \quad \nu(B^c) = 0. \tag{1.3}$$

Note that μ is singular w.r.t. ν implies that ν is singular w.r.t. μ. Thus, the notion of singularity between two measures μ and ν is symmetric but that of absolutely continuity is not. Note also that if μ and ν are mutually singular and B satisfies (1.3), then for all $A \in \mathcal{F}$,

$$\mu(A) = \mu(A \cap B^c) \quad \text{and} \quad \nu(A) = \nu(A \cap B). \tag{1.4}$$

Example 4.1.4: Let μ be the Lebesgue measure on $(\mathbb{R}, \mathcal{B}(\mathbb{R}))$ and ν be defined as $\nu(A) = \#$ elements in $A \cap \mathbb{Z}$ where $\mathbb{Z}$ is the set of integers. Then

$$\nu(\mathbb{Z}^c) = 0 \quad \text{and} \quad \mu(\mathbb{Z}) = 0$$

and hence (1.3) holds with $B = \mathbb{Z}$. Thus μ and ν are *mutually singular.*

Another example is the pair m and μ_c on [0,1] where μ_c is the Lebesgue-Stieltjes measure generated by the Cantor function (cf. Section 4.5) and m is the Lebesgue measure.

Example 4.1.5: Let μ be the Lebesgue measure restricted to $(-\infty, 0]$ and ν be the Exponential(1) distribution. That is, for any $A \in \mathcal{B}(\mathbb{R})$,

$$\begin{aligned} \mu(A) &= \text{the Lebesgue measure of} \quad A \cap (-\infty, 0]; \\ \nu(A) &= \int_{A \cap (0,\infty)} e^{-x} dx. \end{aligned}$$

Then, $\mu((0,\infty)) = 0$ and $\nu((-\infty, 0]) = 0$, and (1.3) holds with $B = (-\infty, 0]$.

Suppose that μ and ν are two finite measures on a measurable space $(\Omega, \mathcal{F})$. H. Lebesgue showed that μ_1 can be decomposed as a sum of two measures, i.e.,

$$\mu = \mu_a + \mu_s$$

where $\mu_a \ll \nu$ and $\mu_s \perp \nu$. The next theorem is the main result of this section and it combines the above decomposition result of Lebesgue with the Radon-Nikodym theorem mentioned earlier.

Theorem 4.1.1: *Let $(\Omega, \mathcal{F})$ be a measurable space and let μ_1 and μ_2 be two σ-finite measures on $(\Omega, \mathcal{F})$.*

(i) (The Lebesgue decomposition theorem). The measure μ_1 can be uniquely decomposed as

$$\mu_1 = \mu_{1a} + \mu_{1s} \tag{1.5}$$

where μ_{1a} and μ_{1s} are σ-finite measures on $(\Omega, \mathcal{F})$ such that $\mu_{1a} \ll \mu_2$ and $\mu_{1s} \perp \mu_2$.

(ii) (The Radon-Nikodym theorem). There exists a nonnegative measurable function h on $(\Omega, \mathcal{F})$ such that

$$\mu_{1a}(A) = \int_A h d\mu_2 \quad \textit{for all} \quad A \in \mathcal{F}. \tag{1.6}$$

Proof: CASE 1: Suppose that μ_1 and μ_2 are finite measures. Let μ be the measure $\mu = \mu_1 + \mu_2$ and let $H = L^2(\mu)$. Define a linear function T on H by

$$T(f) = \int f d\mu_1. \tag{1.7}$$

Then, by the Cauchy-Schwarz inequality applied to the functions f and $g \equiv 1$,

$$\begin{aligned} |T(f)| &\leq \left(\int f^2 d\mu_1\right)^{\frac{1}{2}} \left(\mu_1(\Omega)\right)^{\frac{1}{2}} \\ &\leq \left(\int f^2 d\mu\right)^{\frac{1}{2}} \left(\mu_1(\Omega)\right)^{\frac{1}{2}}. \end{aligned}$$

This shows that T is a bounded linear functional on H with $||T|| \leq M \equiv (\mu_1(\Omega))^{\frac{1}{2}}$. By the Riesz representation theorem (cf. Theorem 3.3.3 and Remark 3.3.2), there exists a $g \in L^2(\mu)$ such that

$$T(f) = \int f g d\mu \tag{1.8}$$

for all $f \in L^2(\mu)$. Let $f = I_A$ for A in $\mathcal{F}$. Then, (1.7) and (1.8) yield

$$\mu_1(A) = T(I_A) = \int_A g d\mu.$$

But $0 \leq \mu_1(A) \leq \mu(A)$ for all $A \in \mathcal{F}$. Hence the function g in $L^2(\mu)$ satisfies

$$0 \leq \int_A g d\mu \leq \mu(A) \quad \text{for all} \quad A \in \mathcal{F}. \tag{1.9}$$

Let $A_1 = \{0 \leq g < 1\}$, $A_2 = \{g = 1\}$, $A_3 = \{g \notin [0, 1]\}$. Then (1.9) implies that $\mu(A_3) = 0$ (see Problem 4.1). Now define the measures $\mu_{1a}(\cdot)$ and $\mu_{1s}(\cdot)$ by

$$\mu_{1a}(A) \equiv \mu_1(A \cap A_1), \quad \mu_{1s}(A) \equiv \mu_1(A \cap A_2), \quad A \in \mathcal{F}. \tag{1.10}$$

Next it will be shown that $\mu_{1a} \ll \mu_2$ and $\mu_{1s} \perp \mu_2$, thus establishing (1.5). By (1.7) and (1.8), for all $f \in H$,

$$\int f d\mu_1 = \int f g d\mu = \int f g d\mu_1 + \int f g d\mu_2$$

$$\Rightarrow \int f(1-g) d\mu_1 = \int f g d\mu_2. \tag{1.11}$$

Setting $f = I_{A_2}$ yields

$$0 = \mu_2(A_2).$$

From (1.10), since $\mu_{1s}(A_2^c) = 0$, it follows that $\mu_{1s} \perp \mu_2$. Now fix $n \geq 1$ and $A \in \mathcal{F}$. Let $f = I_{A \cap A_1}(1 + g + \ldots + g^{n-1})$. Then (1.11) implies that

$$\int_{A \cap A_1} (1 - g^n) d\mu_1 = \int_{A \cap A_1} g(1 + g + \ldots + g^{n-1}) d\mu_2.$$

Now letting $n \to \infty$, and using the MCT on both sides, yields

$$\mu_{1a}(A) = \int_A I_{A_1} \frac{g}{(1-g)} d\mu_2. \tag{1.12}$$

Setting $h \equiv \frac{g}{1-g} I_{A_1}$ completes the proof of (1.5) and (1.6).

CASE 2: Now suppose that μ_1 and μ_2 are σ-finite. Then there exists a countable partition $\{D_n\}_{\geq 1} \subset \mathcal{F}$ of Ω such that $\mu_1(D_n)$ and $\mu_2(D_n)$ are both finite for all $n \geq 1$. Let $\mu_1^{(n)}(\cdot) \equiv \mu_1(\cdot \cap D_n)$ and $\mu_2^{(n)} \equiv \mu_2(\cdot \cap D_n)$. Then applying 'Case 1' to $\mu_1^{(n)}$ and $\mu_2^{(n)}$ for each $n \geq 1$, one gets measures $\mu_{1a}^{(n)}, \mu_{1s}^{(n)}$ and a function h_n such that

$$\mu_1^{(n)}(\cdot) \equiv \mu_{1a}^{(n)}(\cdot) + \mu_{1s}^{(n)}(\cdot) \tag{1.13}$$

where, for A in $\mathcal{F}$, $\mu_{1a}^{(n)}(A) = \int_A h_n d\mu_2^{(n)} = \int_A h_n I_{D_n} d\mu_2$ and $\mu_{1s}^{(n)}(\cdot) \perp \mu_2^{(n)}$. Since $\mu_1(\cdot) = \sum_{n=1}^{\infty} \mu_1^{(n)}(\cdot)$, it follows from (1.13) that

$$\mu_1(\cdot) = \mu_{1a}(\cdot) + \mu_{1s}(\cdot), \tag{1.14}$$

where $\mu_{1a}(A) \equiv \sum_{n=1}^{\infty} \mu_{1a}^{(n)}(A)$ and $\mu_{1s}(\cdot) = \sum_{n=1}^{\infty} \mu_{1s}^{(n)}(\cdot)$. By the MCT,

$$\mu_{1a}(A) = \int_A h d\mu_2, \quad A \in \mathcal{F},$$

where $h \equiv \sum_{n=1}^{\infty} h_n I_{D_n}$.

Clearly, $\mu_{1a} \ll \mu_2$. The verification of the singularity of μ_{1s} and μ_2 is left as an exercise (Problem 4.2).

It remains to prove the uniqueness of the decomposition. Let $\mu_1 = \mu_a + \mu_s$ and $\mu_1 = \mu_a' + \mu_s'$ be two decompositions of μ_1 where μ_a and μ_a' are absolutely continuous w.r.t. μ_2 and μ_s and μ_s' are singular w.r.t. μ_2. By definition, there exist sets B and B' in $\mathcal{F}$ such that

$$\mu_2(B) = 0, \mu_2(B') = 0, \quad \text{and} \quad \mu_s(B^c) = 0, \mu_s'(B'^c) = 0.$$

Let $D = B \cup B'$. Then $\mu_2(D) = 0$ and $\mu_s(D^c) \leq \mu_s(B^c) = 0$. Similarly, $\mu_s'(D^c) \leq \mu_s'(B'^c) = 0$. Also $\mu_2(D) = 0$ implies $\mu_a(D) = 0 = \mu_a'(D)$. Thus for any $A \in \mathcal{F}$,

$$\mu_a(A) = \mu_a(A \cap D^c) \quad \text{and} \quad \mu_a'(A) = \mu_a'(A \cap D^c).$$

Also

$$\begin{aligned} \mu_s(A \cap D^c) &\leq \mu_s(A \cap B^c) = 0 \\ \mu_s'(A \cap D^c) &\leq \mu_s'(A \cap B'^c) = 0. \end{aligned}$$

Thus, $\mu(A \cap D^c) = \mu_a(A \cap D^c) + \mu_s(A \cap D^c) = \mu_a(A \cap D^c) = \mu_a(A)$ and $\mu(A \cap D^c) = \mu_a'(A \cap D^c) + \mu_s'(A \cap D^c) = \mu_a'(A \cap D^c) = \mu_a'(A)$. Hence, $\mu_a(A) = \mu(A \cap D^c) = \mu_a'(A)$ for every $A \in \mathcal{F}$. That is, $\mu_a = \mu_a'$ and hence, $\mu_s = \mu_s'$. □

Remark 4.1.1: In Theorem 4.1.1, the hypothesis of σ finiteness cannot be dropped. For example, let μ be the Lebesgue measure and ν be the counting measure on $[0,1]$. Then $\mu \ll \nu$ but there does not exist a nonnegative $\mathcal{F}$-measurable function h such that $\mu(A) = \int_A h d\nu$. To see this, if possible, suppose that for some $h \in L^1(\nu)$, $\mu(A) = \int_A h d\nu$ for all $A \in \mathcal{F}$. Note that $\mu([0,1]) = 1$ implies that $\int_{[0,1]} h d\nu < \infty$ and hence, that $B \equiv \{x : h(x) > 0\}$ is countable (Problem 4.3). But μ being the Lebesgue measure, $\mu(B) = 0$ and $\mu(B^c) = 1$. Since by definition, $h \equiv 0$ on B^c, this implies $1 = \mu(B^c) = \int_{B^c} h d\nu = 0$, leading to a contradiction. However, if ν is σ-finite and $\mu \ll \nu$ (μ not necessarily σ-finite), then the Radon-Nikodym theorem holds, i.e., *there exists a nonnegative $\mathcal{F}$-measurable function h such that*

$$\mu(A) = \int_A h d\nu \quad \textit{for all} \quad A \in \mathcal{F}.$$

For a proof, see Royden (1988), Chapter 11.

Definition 4.1.3: Let μ and ν be measures on a measurable space $(\Omega, \mathcal{F})$ and let h be a nonnegative measurable function such that

$$\mu(A) = \int_A h d\nu \quad \text{for all} \quad A \in \mathcal{F}.$$

Then h is called the *Radon-Nikodym derivative* of μ w.r.t. ν and is written as

$$\frac{d\mu}{d\nu} = h.$$

If $\mu(\Omega) < \infty$ and there exist two nonnegative $\mathcal{F}$-measurable functions h_1 and h_2 such that

$$\mu(A) = \int_A h_1 d\nu = \int_A h_2 d\nu$$

for all $A \in \mathcal{F}$, then $h_1 = h_2$ a.e. (ν) and thus the Radon-Nikodym derivative $\frac{d\mu}{d\nu}$ is unique up to equivalence a.e. (ν). This also extends to the case when μ is σ-finite.

The following proposition is easy to verify (cf. Problem 4.4).

Proposition 4.1.2: *Let $\nu, \mu, \mu_1, \mu_2, \ldots$ be σ-finite measures on a measurable space $(\Omega, \mathcal{F})$.*

(i) If $\mu_1 \ll \mu_2$ and $\mu_2 \ll \mu_3$, then $\mu_1 \ll \mu_3$ and

$$\frac{d\mu_1}{d\mu_3} = \frac{d\mu_1}{d\mu_2}\frac{d\mu_2}{d\mu_3} \quad a.e.\ (\mu_3).$$

(ii) Suppose that μ_1 and μ_2 are dominated by μ_3. Then for any $\alpha, \beta \geq 0$, $\alpha\mu_1 + \beta\mu_2$ is dominated by μ_3 and

$$\frac{d(\alpha\mu_1 + \beta\mu_2)}{d\mu_3} = \alpha\frac{d\mu_1}{d\mu_3} + \beta\frac{d\mu_2}{d\mu_3} \quad a.e.\ (\mu_3).$$

(iii) If $\mu \ll \nu$ and $\frac{d\mu}{d\nu} > 0$ a.e. (ν), then $\nu \ll \mu$ and

$$\frac{d\nu}{d\mu} = \left(\frac{d\mu}{d\nu}\right)^{-1} \quad a.e.\ (\mu).$$

(iv) Let $\{\mu_n\}_{n\geq 1}$ be a sequence of measures and $\{\alpha_n\}_{n\geq 1}$ be a sequence of positive real numbers, i.e., $\alpha_n > 0$ for all $n \geq 1$. Define $\mu = \sum_{n=1}^{\infty} \alpha_n \mu_n$.

(a) Then, $\mu \ll \nu$ *iff* $\mu_n \ll \nu$ *for each* $n \geq 1$ *and in this case,*

$$\frac{d\mu}{d\nu} = \sum_{n=1}^{\infty} \alpha_n \frac{d\mu_n}{d\nu} \quad a.e.\ (\nu).$$

(b) $\mu \perp \nu$ *iff* $\mu_n \perp \nu$ *for all* $n \geq 1$.

4.2 Signed measures

Let μ_1 and μ_2 be two finite measures on a measurable space $(\Omega, \mathcal{F})$. Let

$$\nu(A) \equiv \mu_1(A) - \mu_2(A), \quad \text{for all} \quad A \in \mathcal{F}. \tag{2.1}$$

Then $\nu : F \to \mathbb{R}$ satisfies the following:

(i) $\nu(\emptyset) = 0$.

(ii) For any $\{A_n\}_{n\geq 1} \subset \mathcal{F}$ with $A_i \cap A_j = \emptyset$ for $i \neq j$, and with $\sum_{i=1}^{\infty} |\nu(A_i)| < \infty$,

$$\nu(A) = \sum_{i=1}^{\infty} \nu(A_i). \tag{2.2}$$

(iii) Let

$$\|\nu\| \equiv \sup\Big\{ \sum_{i=1}^{\infty} |\nu(A_i)| : \{A_n\}_{n\geq 1} \subset \mathcal{F},\ A_i \cap A_j = \emptyset \text{ for } i \neq j,\ \bigcup_{n\geq 1} A_n = \Omega \Big\}. \tag{2.3}$$

Then, $\|\nu\|$ is finite.

Note that (iii) holds because $\|\nu\| \leq \mu_1(\Omega) + \mu_2(\Omega) < \infty$.

Definition 4.2.1: A set function $\nu : \mathcal{F} \to \mathbb{R}$ satisfying (i), (ii) and (iii) above is called a *finite signed measure.*

The above example shows that the difference of two finite measures is a finite signed measure. It will be shown below that every finite signed measure can be expressed as the difference of two finite measures.

Proposition 4.2.1: *Let* ν *be a finite signed measure on* $(\Omega, \mathcal{F})$. *Let*

$$|\nu|(A) \equiv \sup\Big\{ \sum_{n=1}^{\infty} |\nu(A_n)| : \{A_n\}_{n\geq 1} \subset \mathcal{F}, A_i \cap A_j = \emptyset \ \textit{for} \ i \neq j, \bigcup_{n\geq 1} A_n = A \Big\}. \tag{2.4}$$

Then $|\nu|(\cdot)$ is a finite measure on $(\Omega, \mathcal{F})$.

Proof: That $|\nu|(\Omega) < \infty$ follows from part (iii) of the definition. Thus it is enough to verify that $|\nu|$ is countably additive. Let $\{A_n\}_{n\geq 1}$ be a countable family of disjoint sets in $\mathcal{F}$. Let $A = \cup_{n\geq 1} A_n$. By the definition of $|\nu|$, for all $\epsilon > 0$ and $n \in \mathbb{N}$, there exists a countable family $\{A_{nj}\}_{j\geq 1}$ of disjoint sets in $\mathcal{F}$ with $A_n = \cup_{j\geq 1} A_{nj}$ such that $\sum_{j=1}^{\infty} |\nu(A_{nj})| > |\nu|(A_n) - \frac{\epsilon}{2^n}$. Hence,

$$\sum_{n=1}^{\infty}\sum_{j=1}^{\infty} |\nu(A_{nj})| > \sum_{n=1}^{\infty} |\nu|(A_n) - \epsilon.$$

Note that $\{A_{nj}\}_{n\geq 1, j\geq 1}$ is a countable family of disjoint sets in $\mathcal{F}$ such that $A = \cup_{n\geq 1} A_n = \cup_{n\geq 1} \cup_{j\geq 1} A_{nj}$. It follows from the definition of $|\nu|$ that

$$|\nu|(A) \geq \sum_{n=1}^{\infty}\sum_{j=1}^{\infty} |\nu(A_{nj})| > \sum_{n=1}^{\infty} |\nu|(A_n) - \epsilon.$$

Since this is true for for all $\epsilon > 0$, it follows that

$$|\nu|(A) \geq \sum_{n=1}^{\infty} |\nu|(A_n). \tag{2.5}$$

To get the opposite inequality, let $\{B_j\}_{j\geq 1}$ be a countable family of disjoint sets in $\mathcal{F}$ such that $\cup_{j\geq 1} B_j = A = \cup_{n\geq 1} A_n$. Since $B_j = B_j \cap A = \cup_{n\geq 1}(B_j \cap A_n)$ and ν satisfies (2.2),

$$\nu(B_j) = \sum_{n=1}^{\infty} \nu(B_j \cap A_n) \quad \text{for all} \quad j \geq 1.$$

Thus

$$\begin{aligned}\sum_{j=1}^{\infty} |\nu(B_j)| &\leq \sum_{j=1}^{\infty}\sum_{n=1}^{\infty} |\nu(B_j \cap A_n)| \\ &= \sum_{n=1}^{\infty}\sum_{j=1}^{\infty} |\nu(B_j \cap A_n)|. \end{aligned} \tag{2.6}$$

Note that for each A_n, $\{B_j \cap A_n\}_{j\geq 1}$ is a countable family of disjoint sets in $\mathcal{F}$ such that $A_n = \cup_{j\geq 1}(B_j \cap A_n)$. Hence from (2.4), it follows that $|\nu|(A_n) \geq \sum_{j=1}^{\infty} |\nu(B_j \cap A_n)|$ and hence, $\sum_{n=1}^{\infty} |\nu|(A_n) \geq \sum_{n=1}^{\infty}\sum_{j=1}^{\infty} |\nu(B_j \cap A_n)|$. From (2.6), it follows that $\sum_{n=1}^{\infty} |\nu|(A_n) \geq \sum_{j=1}^{\infty} |\nu(B_j)|$. This being true for every such family $\{B_j\}_{j\geq 1}$, it follows from (2.4) that

$$|\nu|(A) \leq \sum_{i=1}^{\infty} |\nu|(A_i) \tag{2.7}$$

and with (2.5), this completes the proof. □

Definition 4.2.2: The measure $|\nu|$ defined by (2.4) is called the *total variation measure* of the signed measure ν.

Next, define the set functions

$$\nu^+ \equiv \frac{|\nu| + \nu}{2}, \quad \nu^- \equiv \frac{|\nu| - \nu}{2}. \tag{2.8}$$

It can be verified that ν^+ and ν^- are both finite measures on $(\Omega, \mathcal{F})$.

Definition 4.2.3: The measures ν^+ and ν^- are called the *positive* and *negative variation measures* of the signed measure ν, respectively.

It follows from (2.8) that

$$\nu = \nu^+ - \nu^-. \tag{2.9}$$

Thus every finite signed measure ν on $(\Omega, \mathcal{F})$ is the difference of two finite measures, as claimed earlier.

Note that both ν^+ and ν^- are dominated by $|\nu|$ and all three measures are finite. By the Radon-Nikodym theorem (Theorem 4.1.1), there exist functions h_1 and h_2 in $L^1(\Omega, \mathcal{F}, |\nu|)$ such that

$$\frac{d\nu^+}{d|\nu|} = h_1 \quad \text{and} \quad \frac{d\nu^-}{d|\nu|} = h_2. \tag{2.10}$$

This and (2.9) imply that for any A in $\mathcal{F}$,

$$\nu(A) = \int_A h_1 d|\nu| - \int_A h_2 d|\nu| = \int_A h d|\nu|, \tag{2.11}$$

where $h = h_1 - h_2$. Thus every finite signed measure ν on $(\Omega, \mathcal{F})$ can be expressed as

$$\nu(A) = \int_A f d\mu, \ A \in \mathcal{F} \tag{2.12}$$

for some finite measure μ on $(\Omega, \mathcal{F})$ and some $f \in L^1(\Omega, \mathcal{F}, \mu)$.

Conversely, it is easy to verify that a set function ν defined by (2.12) for some finite measure μ on $(\Omega, \mathcal{F})$ and some $f \in L^1(\Omega, \mathcal{F}, \mu)$ is a finite signed measure (cf. Problem 4.6). This leads to the following:

Theorem 4.2.2:

(i) A set function ν on a measurable space $(\Omega, \mathcal{F})$ is a finite signed measure iff there exist two finite measures μ_1 and μ_2 on $(\Omega, \mathcal{F})$ such that $\nu = \mu_1 - \mu_2$.

(ii) *A set function ν on a measurable space $(\Omega, \mathcal{F})$ is a finite signed measure iff there exist a finite measure μ on $(\Omega, \mathcal{F})$ and an $f \in L^1(\Omega, \mathcal{F}, \mu)$ such that for all A in $\mathcal{F}$,*

$$\nu(A) = \int_A f \, d\mu.$$

Definition 4.2.4: Let ν be a finite signed measure on a measurable space on $(\Omega, \mathcal{F})$. A set $A \in \mathcal{F}$ is called a *positive set* for ν if for any $B \subset A, B \in \mathcal{F}$, $\nu(B) \geq 0$. A set $A \in \mathcal{F}$ is called a *negative set* for ν if for any $B \subset A$ with $B \in \mathcal{F}$, $\nu(B) \leq 0$.

Let h be as in (2.11). Let

$$\Omega^+ = \{\omega : h(\omega) \geq 0\} \quad \text{and} \quad \Omega^- = \{\omega : h(\omega) < 0\}. \qquad (2.13)$$

From (2.11), it follows that for all A in $\mathcal{F}$, $\nu(A \cap \Omega^+) \geq 0$ and $\nu(A \cap \Omega^-) \leq 0$. Thus Ω^+ is a positive set and Ω^- is a negative set for ν. Furthermore, $\Omega^+ \cup \Omega^- = \Omega$ and $\Omega^+ \cap \Omega^- = \emptyset$. Summarizing this discussion, one gets the following theorem.

Theorem 4.2.3: (*Hahn decomposition theorem*). *Let ν be a finite signed measure on a measurable space $(\Omega, \mathcal{F})$. Then there exist a positive set Ω^+ and a negative set Ω^- for ν such that $\Omega = \Omega^+ \cup \Omega^-$ and $\Omega^+ \cap \Omega^- = \emptyset$.*

Let Ω^+ and Ω^- be as in (2.13). It can be verified (Problem 4.8) that for any $B \in \mathcal{F}$, if $B \subset \Omega^+$, then $\nu(B) = |\nu|(B)$. By (2.11), this implies that for all A in $\mathcal{F}$,

$$\int_{A \cap \Omega^+} h d|\nu| = |\nu|(A \cap \Omega^+).$$

It follows that $h = 1$ a.e. ($|\nu|$) on Ω^+. Similarly, $h = -1$ a.e. ($|\nu|$) on Ω^-. Thus, the measures ν^+ and ν^-, defined in (2.8), reduce to

$$\begin{aligned} \nu^+(A) &= \int_A \frac{(1+h)}{2} d|\nu| \\ &= \int_{A \cap \Omega^+} \frac{(1+h)}{2} d|\nu| + \int_{A \cap \Omega^-} \frac{(1+h)}{d} |\nu| \\ &= |\nu|(A \cap \Omega^+), \end{aligned}$$

and similarly

$$\nu^-(A) = |\nu|(A \cap \Omega^-).$$

Note that ν^+ and ν^- are both finite measures that are mutually singular. This particular decomposition of ν as

$$\nu = \nu^+ - \nu^-$$

is known as the *Jordan decomposition* of ν. It will now be shown that this decomposition is minimal and that it is unique in the class of signed measures with mutually singular components. Suppose there exist two finite measures μ_1 and μ_2 on $(\Omega, \mathcal{F})$ such that $\nu = \mu_1 - \mu_2$. For any $A \in \mathcal{F}$, $\nu^+(A) = \nu(A \cap \Omega^+) = \mu_1(A \cap \Omega^+) - \mu_2(A \cap \Omega^+) \leq \mu_1(A \cap \Omega^+) \leq \mu_1(A)$ and $\nu^-(A) = -\nu(A \cap \Omega^-) = \mu_2(A \cap \Omega^-) - \mu_1(A \cap \Omega^-) \leq \mu_2(A \cap \Omega^-) \leq \mu_2(A)$. Thus, $\nu^+ \leq \mu_1$ and $\nu^- \leq \mu_2$. Clearly, since both μ_1 and ν^+ are finite measures on $(\Omega, \mathcal{F})$, $\mu_1 - \nu^+$ is also a finite measure. Similarly, $\mu_2 - \nu^-$ is also a finite measure. Also, since $\mu_1 - \mu_2 = \nu = \nu^+ - \nu^-$, it follows that $\mu_1 - \nu^+ = \mu_2 - \nu^- = \lambda$, say. Thus, for any decomposition of ν as $\mu_1 - \mu_2$ with μ_1, μ_2 finite measures, it holds that $\mu_1 = \nu^+ + \lambda$ and $\mu_2 = \nu^- + \lambda$, where λ is a measure on $(\Omega, \mathcal{F})$. Thus, $\nu = \nu^+ - \nu^-$ is a *minimal* decomposition in the sense that in this case $\lambda = 0$. Now suppose μ_1 and μ_2 are mutually singular, i.e., there exist $\Omega_1, \Omega_2 \in \mathcal{F}$ such that $\Omega_1 \cap \Omega_2 = \emptyset$, $\Omega_1 \cup \Omega_2 = \Omega$, and $\mu_1(\Omega_2) = 0 = \mu_2(\Omega_1)$. Since $\mu_1 \geq \lambda$ and $\mu_2 \geq \lambda$, it follows that $\lambda(\Omega_2) = 0 = \lambda(\Omega_1)$. Thus $\lambda = 0$ and $\mu_1 = \nu^+$ and $\mu_2 = \nu^-$.

Summarizing the above discussion yields:

Theorem 4.2.4: *Let ν be a finite signed measure on a measurable space $(\Omega, \mathcal{F})$ and let μ_1 and μ_2 be two finite measures on $(\Omega, \mathcal{F})$ such that $\nu = \mu_1 - \mu_2$. Then there exists a finite measure λ such that $\mu_1 = \nu^+ + \lambda$ and $\mu_2 = \nu^- + \lambda$ with $\lambda = 0$ iff μ_1 and μ_2 are mutually singular.*

Let

$$\mathbb{S} \equiv \{\nu : \nu \quad \text{is a finite signed measure on} \quad (\Omega, \mathcal{F})\}.$$

Also, for any $\alpha \in \mathbb{R}$, let $\alpha^+ = \max(\alpha, 0)$ and $\alpha^- = \max(-\alpha, 0)$. Note that for ν_1, ν_2 in $\mathbb{S}$ and $\alpha_1, \alpha_2 \in \mathbb{R}$,

$$\begin{aligned}
\alpha_1\nu_1 + \alpha_2\nu_2 &= (\alpha_1^+ - \alpha_1^-)(\nu_1^+ - \nu_1^-) + (\alpha_2^+ - \alpha_2^-)(\nu_2^+ - \nu_2^-) \\
&= (\alpha_1^+\nu_1^+ + \alpha_1^-\nu_1^- + \alpha_2^+\nu_2^+ + \alpha_2^-\nu_2^-) \\
&\quad - (\alpha_1^+\nu_1^- + \alpha_1^-\nu_1^+ + \alpha_2^+\nu_2^- + \alpha_2^-\nu_2^+) \\
&= \lambda_1 - \lambda_2, \quad \text{say},
\end{aligned}$$

where λ_1 and λ_2 are both finite measures. It now follows from Theorem 4.2.2 that $\alpha_1\nu_1 + \alpha_2\nu_2 \in \mathbb{S}$. Thus, $\mathbb{S}$ is a vector space over $\mathbb{R}$.

Now it will be shown that $\|\nu\| \equiv |\nu|(\Omega)$ is a norm on $\mathbb{S}$ and that $(\mathbb{S}, \|\cdot\|)$ is a Banach space.

Definition 4.2.5: For a finite signed measure ν on a measurable space $(\Omega, \mathcal{F})$, the *total variation norm* ν is defined by $\|\nu\| \equiv |\nu|(\Omega)$.

Proposition 4.2.5: *Let $\mathbb{S} \equiv \{\nu : \nu$ a finite signed measure on $(\Omega, \mathcal{F})\}$. Then, $\|\nu\| \equiv |\nu|(\Omega)$, $\nu \in \mathbb{S}$ is a norm on $\mathbb{S}$.*

Proof: Let $\nu_1, \nu_2 \in \mathbb{S}$, $\alpha_1, \alpha_2 \in \mathbb{R}$ and $\lambda = \alpha_1\nu_1 + \alpha_2\nu_2$. For any $A \in \mathcal{F}$ and any $\{A_i\}_{i\geq 1} \subset \mathcal{F}$ with $A = \cup_{i\geq 1} A_i$,

$$\begin{aligned} |\lambda(A_i)| &\leq |\alpha_1||\nu_1(A_i)| + |\alpha_2||\nu_2(A_i)| \quad \text{for all} \quad i \geq 1 \\ \Rightarrow \sum_{i\geq 1} |\lambda(A_i)| &\leq |\alpha_1| \sum_{i\geq 1} |\nu_1(A_i)| + |\alpha_2| \sum_{i\geq 1} |\nu_2(A_i)| \\ &\leq |\alpha_1||\nu_1|(A) + |\alpha_2||\nu_2|(A). \end{aligned}$$

Taking supremum over all $\{A_i\}_{i\geq 1}$ yields,

$$\begin{aligned} & |\lambda|(A) \leq |\alpha_1||\nu_1|(A) + |\alpha_2||\nu_2|(\cdot) \\ \text{i.e.,} \quad & |\lambda|(\cdot) \leq |\alpha_1| \, |\nu_1|(\cdot) + |\alpha_2||\nu_2|(\cdot), \\ \Rightarrow \quad & \|\lambda\| \equiv |\lambda|(\Omega) \leq |\alpha_1| \, |\nu_1|(\Omega) + |\alpha_2| \, |\nu_2|(\Omega) \\ & \qquad = |\alpha_1|\|\nu_1\| + |\alpha_2|\|\nu_2\|. \end{aligned}$$

Taking $\alpha_1 = \alpha_2 = 1$ yields

$$\|\nu_1 + \nu_2\| \leq \|\nu_1\| + \|\nu_2\|,$$

i.e., the triangle inequality holds.

Next taking $\alpha_2 = 0$ yields $\|\alpha_1\nu_1\| \leq |\alpha_1|\|\nu_1\|$. To get the opposite inequality, note that for $\alpha_1 \neq 0$, $\nu_1 = \frac{1}{\alpha_1}\alpha_1\nu_1$ and so $\|\nu_1\| \leq |\frac{1}{\alpha_1}|\|\alpha_1\nu_1\|$. Hence, $|\alpha_1| \, \|\nu_1\| \leq \|\alpha_1\nu_1\|$. Thus, for any $\alpha_1 \neq 0$, $\|\alpha_1\nu\| = |\alpha_1|\|\nu\|$. Finally, $\|\nu\| = 0 \Rightarrow |\nu|(\Omega) = 0 \Rightarrow |\nu|(A) = 0 \quad \text{for all} \quad A \in \mathcal{F} \Rightarrow \nu(A) = 0 \quad \text{for all} \quad A \in \mathcal{F}$, i.e., ν is the zero measure. Thus $\|\cdot\|$ is a norm on $\mathbb{S}$. □

Proposition 4.2.6: *$(\mathbb{S}, \|\cdot\|)$ is complete.*

Proof: Let $\{\nu_n\}_{n\geq 1}$ be a Cauchy sequence in $(\mathbb{S}, \|\cdot\|)$. Note that for each $A \in \mathcal{F}$, $|\nu_n(A) - \nu_m(A)| \leq |\nu_n - \nu_m|(A) \leq \|\nu_n - \nu_m\|$. Hence, for each $A \in \mathcal{F}$, $\{\nu_n(A)\}_{n\geq 1}$ is a Cauchy sequence in $\mathbb{R}$ and hence

$$\nu(A) \equiv \lim_{n\to\infty} \nu_n(A) \quad \text{exists.}$$

It will be shown that $\nu(\cdot)$ is a finite signed measure and $\|\nu_n - \nu\| \to 0$ as $n \to \infty$. Let $\{A_i\}_{i\geq 1} \subset \mathcal{F}$, $A_i \cap A_j = \emptyset$ for $i \neq j$, and $A = \cup_{i\geq 1} A_i$. Let $x_n \equiv \{\nu_n(A_i)\}_{i\geq 1}$, $n \geq 1$, and let $x_0 \equiv \{\nu(A_i)\}_{i\geq 1}$. Note that each $x_n \in \ell_1$ where $\ell_1 \equiv \{x : x = \{x_i\}_{i\geq 1} \in \mathbb{R}, \ \sum_{i\geq 1} |x_i| < \infty\}$. For $x \in \ell_1$, let $\|x\|_1 = \sum_{i=1}^{\infty} |x_i|$. Then $\|x_n - x_m\| = \sum_{i\geq 1} |\nu_n(A_i) - \nu_m(A_i)| \leq |\nu_n - \nu_m|(A) \leq |\nu_n - \nu_m|(\Omega) \to 0$ as $n, m \to \infty$. But ℓ_1 is complete under $\|\cdot\|_1$. So there exists $x^* \in \ell_1$ such that $\|x_n - x^*\|_1 \to 0$. Since $x_{ni} \equiv \nu_n(A_i) \to \nu(A_i)$ for all $i \geq 1$, it follows that $x_i^* = \nu(A_i)$ for all $i \geq 1$ and that $\sum_{i=1}^{\infty} |\nu(A_i)| < \infty$. Also, for all $n \geq 1$, $\nu_n(A) = \sum_{i=1}^{\infty} \nu_n(A_i)$. Since $\sum_{i\geq 1} |\nu_n(A_i) - \nu(A_i)| \to 0$, $\nu_n(A) \equiv \sum_{i\geq 1} \nu_n(A_i) \to \sum_{i\geq 1} \nu(A_i)$ as $n \to \infty$. But $\nu_n(A) \to \nu(A)$. Thus,

$\nu(A) = \sum_{i=1}^{\infty} \nu(A_i)$. Also for any countable partition $\{A_i\}_{i \geq 1} \subset \mathcal{F}$ of Ω,

$$\sum_{i=1}^{\infty} |\nu(A_i)| = \lim_{n \to \infty} \sum_{i=1}^{\infty} |\nu_n(A_i)| \leq \lim_{n \to \infty} \|\nu_n\| < \infty.$$

Thus $|\nu|(\Omega) < \infty$ and hence, $\nu \in \mathbb{S}$. Finally,

$$\|\nu_n - \nu\| = \sup\left\{ \sum_{i=1}^{\infty} |\nu_n(A_i) - \nu(A_i)| : \{A_i\}_{i \geq 1} \subset \mathcal{F} \right.$$
$$\left. \text{is a disjoint partition of } \Omega \right\}.$$

But for every countable partition $\{A_i\}_{i \geq 1} \subset \mathcal{F}$,

$$\sum_{i=1}^{\infty} |\nu_n(A_i) - \nu(A_i)| = \lim_{m \to \infty} \sum_{i=1}^{\infty} |\nu_n(A_i) - \nu_m(A_i)| \leq \lim_{m \to \infty} \|\nu_n - \nu_m\|.$$

Thus, $\|\nu_n - \nu\| \leq \lim_{m \to \infty} \|\nu_n - \nu_m\|$ and hence, $\lim_{n \to \infty} \|\nu_n - \nu\| \leq \lim_{n \to \infty} \lim_{m \to \infty} \|\nu_n - \nu_m\| = 0$. Hence, $\nu_n \to \nu$ in $\mathbb{S}$. $\square$

Definition 4.2.6: (*Integration w.r.t. signed measures*). Let μ be a finite signed measure on a measurable space $(\Omega, \mathcal{F})$ and $|\mu|$ be its total variation measure as in Definition 4.2.2. Then, for any $f \in L^1(\Omega, \mathcal{F}, |\mu|)$, $\int f d\mu$ is defined as

$$\int f d\mu = \int f d\mu^+ - \int f d\mu^-,$$

where μ^+ and μ^- are the positive and negative variations of μ as defined in (2.8).

Proposition 4.2.7: *Let μ be a signed measure on a measurable space $(\Omega, \mathcal{F}, P)$. Let $\mu = \lambda_1 - \lambda_2$ where λ_1 and λ_2 are finite measures. Let $f \in L^1(\Omega, \mathcal{F}, \lambda_1 + \lambda_2)$. Then $f \in L^1(\Omega, \mathcal{F}, |\mu|)$ and*

$$\int f d\mu = \int f d\lambda_1 - \int f d\lambda_2. \tag{2.14}$$

Proof: Left as an exercise (Problem 4.13).

4.3 Functions of bounded variation

From the construction of the Lebesgue-Stieltjes measures on $(\mathbb{R}, \mathcal{B}(\mathbb{R}))$ discussed in Chapter 1, it is seen that to every nondecreasing right continuous function $F : \mathbb{R} \to \mathbb{R}$, there is a (Radon) measure μ_F on $(\mathbb{R}, \mathcal{B}(\mathbb{R}))$ such that

$\mu_F((a,b]) = F(b) - F(a)$ for all $a < b$ and conversely. If μ_1 and μ_2 are two Radon measures and $\mu = \mu_1 - \mu_2$, let

$$
\begin{aligned}
G(x) &\equiv \begin{cases} \mu((0,x]) & \text{for} \quad x > 0, \\ -\mu((x,0]) & \text{for} \quad x < 0, \\ 0 & \text{for} \quad x = 0. \end{cases} \\
&= \begin{cases} F_1(x) - F_2(x) - (F_1(0) - F_2(0)) & \text{for} \quad x > 0, \\ (F_1(0) - F_2(0)) - (F_1(x) - F_2(x)) & \text{for} \quad x < 0, \\ 0 & \text{for} \quad x = 0. \end{cases}
\end{aligned}
$$

Thus to every finite signed measure μ on $(\mathbb{R}, \mathcal{B}(\mathbb{R}))$, there corresponds a function $G(\cdot)$ that is the difference of two right continuous nondecreasing and bounded functions. The converse is also easy to establish. A characterization of such a function $G(\cdot)$ without any reference to measures is given below.

Definition 4.3.1: Let $f : [a,b] \to \mathbb{R}$, where $-\infty < a < b < \infty$. Then for any partition $Q = \{a = x_0 < x_1 < x_2 < \ldots < x_n = b\}$, $n \in \mathbb{N}$, the *positive, negative and total variations of f with respect to Q* are respectively defined as

$$
\begin{aligned}
P(f,Q) &\equiv \sum_{i=1}^{n} (f(x_i) - f(x_{i-1}))^+ \\
N(f,Q) &\equiv \sum_{i=1}^{n} (f(x_i) - f(x_{i-1}))^- \\
T(f,Q) &\equiv \sum_{i=1}^{n} |f(x_i) - f(x_{i-1})|.
\end{aligned}
$$

It is easy to verify that (i) if f is nondecreasing, then

$$P(f,Q) = T(f,Q) = f(b) - f(a) \quad \text{and} \quad N(f,Q) = 0$$

and that (ii) for any f,

$$P(f,Q) + N(f,Q) = T(f,Q).$$

Definition 4.3.2: Let $f = [a,b] \to \mathbb{R}$, where $-\infty < a < b < \infty$. The *positive, negative and total variations of f over $[a,b]$* are respectively defined as

$$
\begin{aligned}
P(f,[a,b]) &\equiv \sup_Q P(f,Q) \\
N(f,[a,b]) &\equiv \sup_Q N(f,Q) \\
T(f,[a,b]) &\equiv \sup_Q T(f,Q),
\end{aligned}
$$

where the supremum in each case is taken over all finite partitions Q of $[a, b]$.

Definition 4.3.3: Let $f : [a, b] \to \mathbb{R}$, where $-\infty < a < b < \infty$. Then, f is said to be of *bounded variation on* $[a, b]$ if $T(f, [a, b]) < \infty$. The set of all such functions is denoted by $BV[a, b]$.

As remarked earlier, if f is nondecreasing, then $T(f, Q) = f(b) - f(a)$ for each Q and hence $T(f, [a, b]) = f(b) - f(a)$. It follows that if $f = f_1 - f_2$, where both f_1 and f_2 are nondecreasing, then $f \in BV[a, b]$. A natural question is whether the converse is true. The answer is yes, as shown by the following result.

Theorem 4.3.1: *Let $f \in BV[a, b]$. Let $f_1(x) \equiv P(f, [a, x])$ and $f_2(x) \equiv N(f, [a, x])$. Then f_1 and f_2 are nondecreasing in $[a, b]$ and for all $a \leq x \leq b$,*

$$f(x) = f_1(x) - f_2(x)$$

Proof: That f_1 and f_2 are nondecreasing follows from the definition. It is enough to verify that if $f \in BV[a, b]$, then

$$f(b) - f(a) = P(f, [a, b]) - N(f, [a, b]),$$

as this can be applied to $[a, x]$ for $a \leq x < b$. For each finite partition Q of $[a, b]$,

$$\begin{aligned} f(b) - f(a) &= \sum_{i=1}^{n} (f(x_i) - f(x_{i-1})) \\ &= P(f, Q) - N(f, Q). \end{aligned}$$

Thus $P(f, Q) = f(b) - f(a) + N(f, Q)$. By taking supremum over all finite partitions Q, it follows that

$$P(f, [a, b]) = f(b) - f(a) + N(f, [a, b]).$$

If $f \in BV[a, b]$, this yields $f(b) - f(a) = P(f, [a, b]) - N(f, [a, b])$. □

Remark 4.3.1: Since $T(f, Q) = P(f, Q) + N(f, Q) = 2P(f, Q) - (f(b) - f(a))$, it follows that if $f \in BV[a, b]$, then

$$\begin{aligned} T(f, [a, b]) &= 2P(f, [a, b]) - (f(b) - f(a)) \\ &= P(f, [a, b]) + N(f, [a, b]). \end{aligned}$$

Corollary 4.3.2: *A function $f \in BV[a, b]$ iff there exists a finite signed measure μ on $(\mathbb{R}, \mathcal{B}(\mathbb{R}))$ such that $f(x) = \mu([a, x])$, $a \leq x \leq b$.*

The proof of this corollary is left as an exercise.

Remark 4.3.2: Some observations on functions of bounded variations are listed below.

(a) Let $f = I_{\mathbb{Q}}$ where $\mathbb{Q}$ is the set of rationals. Then for any $[a, b], a < b, P(f, [a, b]) = N(f, [a, b]) = \infty$ and so $f \notin BV[a, b]$. This holds for $f = I_D$ for any set D such that both D and D^c are dense in $\mathbb{R}$.

(b) Let f be *Lipschitz* on $[a, b]$. That is, $|f(x) - f(y)| \leq K|x - y|$ for all x, y in $[a, b]$ where $K \in (0, \infty)$ is a constant. Then, $f \in BV[a, b]$.

(c) Let f be differentiable in (a, b) and continuous on $[a, b]$ and $f'(\cdot)$ be bounded in (a, b). Then by the mean value theorem, f is Lipschitz and hence, f is in $BV[a, b]$.

(d) Let $f(x) = x^2 \sin \frac{1}{x}$, $0 < x \leq 1$, and let $f(0) = 0$. Then f is continuous on $[0, 1]$, differentiable on $(0, 1)$ with f' bounded on $(0, 1)$, and hence $f \in BV[0, 1]$.

(e) Let $g(x) = x^2 \sin \frac{1}{x^2}$, $0 < x \leq 1$, $g(0) = 0$. Then g is continuous on $[0, 1]$, differentiable on $(0, 1)$ but g' is not bounded on $(0, 1)$. This by itself does not imply that $g \notin BV[0, 1]$, since being Lipschitz is only a sufficient condition. But it turns out that $g \notin BV[0, 1]$. To see this, let $x_n = \sqrt{\frac{1}{(2n+1)\frac{\pi}{2}}}$, $n = 0, 1, 2 \ldots$. Then $\sum\limits_{i=1}^{n} |g(x_i) - g(x_{i-1})| \geq \sum\limits_{i=1}^{n} \frac{1}{(2i+1)\frac{\pi}{2}}$ and hence $T(g, [0, 1]) = \infty$.

(f) It is known (see Royden (1988), Chapter 4) that if $f : [a, b] \to \mathbb{R}$ is nondecreasing, then f is differentiable a.e. (m) on (a, b) and $\int_{[a,b]} f' dm \leq f(b) - f(c)$, where (m) denotes the Lebesgue measure. This implies that if $f \in BV[a, b]$, then f is differentiable a.e. (m) on (a, b) and so, $\int_{[a,b]} |f'| dm \leq T(f, [a, b])$.

4.4 Absolutely continuous functions on $\mathbb{R}$

Definition 4.4.1: A function $F : \mathbb{R} \to \mathbb{R}$ is *absolutely continuous* (a.c.) if for all $\epsilon > 0$, there exists $\delta > 0$ such that if $I_j = [a_j, b_j]$, $j = 1, 2, \ldots, k$ $(k \in \mathbb{N})$ are disjoint and $\sum_{j=1}^{k} (b_j - a_j) < \delta$, then $\sum_{j=1}^{k} |F(b_j) - F(a_j)| < \epsilon$.

By the mean value theorem, it follows that if F is differentiable and $F'(\cdot)$ is bounded, then F is a.c. Also note that F is a.c. implies F is uniformly continuous.

Definition 4.4.2: A function $F : [a, b] \to \mathbb{R}$ is *absolutely continuous* if the function $\tilde{F}$, defined by

$$\tilde{F}(x) = \begin{cases} F(x) & \text{if } a \leq x \leq b, \\ F(a) & \text{if } x < a, \\ F(b) & \text{if } x > b, \end{cases}$$

is absolutely continuous.

Thus $F(x) = x$ is a.c. on $\mathbb{R}$. Any polynomial is a.c. on any bounded interval but not necessarily on all of $\mathbb{R}$. For example, $F(x) = x^2$ is a.c. on any bounded interval but not a.c. on $\mathbb{R}$, since it is not uniformly continuous on $\mathbb{R}$.

The main result of this section is the following result due to H. Lebesgue, known as the *fundamental theorem of Lebesgue integral calculus.*

Theorem 4.4.1: *A function $F : [a, b] \to \mathbb{R}$ is absolutely continuous iff there is a function $f : [a, b] \to \mathbb{R}$ such that f is Lebesgue measurable and integrable w.r.t. m and such that*

$$F(x) = F(a) + \int_{[a,x]} f \, dm, \quad \textit{for all} \quad a \leq x \leq b \tag{4.1}$$

where m is the Lebesgue measure.

Proof: First consider the "if part." Suppose that (4.1) holds. Since $\int_{[a,b]} |f| dm < \infty$, for any $\epsilon > 0$, there exists a $\delta > 0$ such that (cf. Proposition 2.5.8).

$$m(A) < \delta \Rightarrow \int_A |f| dm < \epsilon. \tag{4.2}$$

Thus, if $I_j = (a_j, b_j), \subset [a, b]$, $j = 1, 2, \ldots, k$ are such that $\sum_{j=1}^k (b_j - a_j) < \delta$, then

$$\sum_{j=1}^k |F(b_j) - F(a_j)| \leq \int_{\bigcup_{j=1}^k I_j} |f| dm < \epsilon,$$

since $m(\cup_{j=1}^k I_j) \leq \sum_{j=1}^k (b_j - a_j) < \delta$ and (4.2) holds. Thus, F is a.c.

Next consider the "only if part." It is not difficult to verify (Problem 4.18) that F a.c. implies F is of bounded variation on any finite interval $[a, b]$ and both the positive and the negative variations of F on $[a, b]$ are a.c. as well. Hence, it suffices to establish (4.1) assuming that F is a.c. and nondecreasing. Let μ_F be the Lebesgue Stieltjes measure generated by $\tilde{F}$ as in Definition 4.4.2. It will now be shown that μ_F is absolutely continuous w.r.t. the Lebesgue measure. Fix $\epsilon > 0$. Let $\delta > 0$ be chosen so that

$$(a_j, b_j) \subset [a, b], j = 1, 2, \ldots, k, \sum_{j=1}^k (b_j - a_j) < \delta \Rightarrow \sum_{j=1}^k |F(b_j) - F(a_j)| < \epsilon.$$

Let $A \in \mathcal{M}_m$, $A \subset (a,b)$, and $m(A) = 0$. Then, there exist a countable collection of disjoint open intervals $\{I_j = (a_j, b_j) : I_j \subset [a,b]\}_{j\geq 1}$ such that

$$A \subset \bigcup_{j\geq 1} I_j \quad \text{and} \quad \sum_{j\geq 1}(b_j - a_j) < \delta.$$

Thus

$$\begin{aligned} \mu_F\Big(A \cap \bigcup_{j=1}^{k} I_j\Big) &\leq \mu_F\Big(\bigcup_{i=1}^{k} I_j\Big) \\ &\leq \sum_{j=1}^{k} \mu_F(I_j) = \sum_{j=1}^{k}(F(b_j) - F(a_j)) < \epsilon \end{aligned}$$

for all $k \in \mathbb{N}$.

Since $A \subset \cup_{j\geq 1} I_j$, by the m.c.f.b. property of $\mu_F(\cdot)$, $\mu_F(A) = \lim_{k\to\infty} \mu_F(A \cap \cup_{j=1}^{k} I_j) \leq \epsilon$. This being true for any $\epsilon > 0$, it follows that $\mu_F(A) = 0$. Since F is continuous, $\mu_F(\{a,b\}) = 0$ and hence $\mu_F\big((a,b)^c\big) = 0$. Thus, $\mu_F \ll m$, i.e., μ_F is dominated by m. Now, by the Radon-Nikodym theorem (cf. Theorem 4.1.1 (ii)), there exists a nonnegative measurable function f such that $A \in \mathcal{M}_m$ implies that $\mu_F(A) = \int_{A\cap[a,b]} f\,dm$ and, in particular, for $a \leq x \leq b$,

$$\mu_F([a,x]) = F(x) - F(a) = \int_{[a,x]} f\,dm,$$

i.e., (4.1) holds. □

The representation (4.1) of an absolutely continuous F can be strengthened as follows:

Theorem 4.4.2: *Let $F : \mathbb{R} \to \mathbb{R}$ satisfy (4.1). Then F is differentiable a.e. (m) and $F'(\cdot) = f(\cdot)$ a.e. (m).*

For a proof of this result, see Royden (1988), Chapter 4.

The relation between the notion of absolute continuity of a distribution function $F : \mathbb{R} \to \mathbb{R}$ and that of the associated Lebesgue-Stieltjes measure μ_F w.r.t. Lebesgue measure m will be discussed now.

Let $F : \mathbb{R} \to \mathbb{R}$ be a distribution function, i.e., F is nondecreasing and right continuous. Let μ_F be the associated Lebesgue-Stieltjes measure such that $\mu_F((a,b]) = F(b) - F(a)$ for all $-\infty < a < b < \infty$. Recall that F is said to be *absolutely continuous on an interval* $[a,b]$ if for each $\epsilon > 0$, there exists a $\delta > 0$ such that for any finite collection of intervals $I_j = (a_j, b_j), j = 1, 2, \ldots, n$, contained in $[a,b]$,

$$\sum_{j=1}^{n}(b_j - a_j) < \delta \quad \Rightarrow \quad \sum_{j=1}^{n}(F(b_j) - F(a_j)) < \epsilon.$$

Recall also that μ_F is *absolutely continuous w.r.t. the Lebesgue measure* m if for $A \in \mathcal{B}(\mathbb{R})$, $m(A) = 0 \Rightarrow \mu_F(A) = 0$. A natural question is that if F is absolutely continuous on every interval $[a, b] \subset \mathbb{R}$, is μ_F absolutely continuous w.r.t. m and conversely? The answer is yes.

Theorem 4.4.3: *Let $F : \mathbb{R} \to \mathbb{R}$ be a nondecreasing function and let μ_F be the associated Lebesgue-Stieltjes measure. Then F is absolutely continuous on $[a, b]$ for all $-\infty < a < b < \infty$ iff $\mu_F \ll m$ where m is the Lebesgue measure on $(\mathbb{R}, \mathcal{B}(\mathbb{R}))$.*

Proof: Suppose that $\mu_F \ll m$. Then by Theorem 4.1.1, there exists a nonnegative measurable function h such that

$$\mu_F(A) = \int_A h dm \quad \text{for all} \quad A \quad \text{in} \quad \mathcal{B}(\mathbb{R}).$$

Hence, for any $a < b$ in $\mathbb{R}$ and $a < x < b$,

$$F(x) - F(a) \equiv \mu_F((a, x]) = \int_{(a,x]} h dm.$$

This implies the absolute continuity of F on $[a, b]$ as shown in Theorem 4.4.1.

Conversely, if F is absolutely continuous on $[a, b]$ for all $-\infty < a < b < \infty$, then as shown in the proof of the "only if" part of Theorem 4.4.1, for all $-\infty < a < b < \infty$, then $\mu_F(A \cap [a, b]) = 0$ if $m(A \cap [a, b]) = 0$. Thus, if $m(A) = 0$, then for all $-\infty < a < b < \infty$, $m(A \cap [a, b]) = 0$ and hence $\mu_F(A \cap [a, b]) = 0$ and hence $\mu_F(A) = 0$, i.e., $\mu_F \ll m$. □

Recall that a measure μ on $(\mathbb{R}^k, \mathcal{B}(\mathbb{R}^k))$ is a *Radon measure* if $\mu(A) < \infty$ for every bounded Borel set A. In the following, let $m(\cdot)$ denote the Lebesgue measure on $\mathbb{R}^k$.

Definition 4.4.3: A Radon measure μ on $(\mathbb{R}^k, \mathcal{B}(\mathbb{R}^k))$ is *differentiable* at $x \in \mathbb{R}^k$ with *derivative* $(D\mu)(x)$ if for any $\epsilon > 0$, there is a $\delta > 0$ such that

$$\left| \frac{\mu(A)}{m(A)} - (D\mu)(x) \right| < \epsilon$$

for every open ball A such that $x \in A$ and diam. $(A) \equiv \sup\{\|x - y\| : x, y \in A\}$, the diameter of A, is less than δ.

Theorem 4.4.4: *Let μ be a Radon measure on $(\mathbb{R}^k, \mathcal{B}(\mathbb{R}^k))$. Then*

(i) μ is differentiable a.e. (m), $D\mu(\cdot)$ is Lebesgue measurable, and ≥ 0 a.e. (m) and for all bounded Borel sets $A \in \mathcal{B}(\mathbb{R}^k)$,

$$\int_A D\mu(\cdot) dm \leq \mu(A).$$

(ii) *Let* $\mu_a(A) \equiv \int_A D\mu(\cdot)dm$, $A \in \mathcal{B}(\mathbb{R}^k)$. *Let* $\mu_s(\cdot)$ *be the unique measure on* $\mathcal{B}(\mathbb{R}^k)$ *such that for all bounded Borel sets* A

$$\mu_s(A) = \mu(A) - \mu_a(A).$$

Then

$$\mu_s \perp m \quad \text{and} \quad D\mu_s(\cdot) = 0 \quad \text{a.e. } (m).$$

For a proof, see Rudin (1987).

Remark 4.4.1: By the uniqueness of the Lebesgue decomposition, it follows that a Radon measure μ on $\mathcal{B}(\mathbb{R}^k)$ is $\perp m$ iff $D\mu(\cdot) = 0$ a.e. (m) and is $\ll m$ iff $\mu(A) = \int_A D\mu(\cdot)dm$ for all $A \in \mathcal{B}(\mathbb{R}^k)$.

Let $f : \mathbb{R}^k \to \mathbb{R}_+$ be integrable w.r.t. m on bounded sets. Let $\mu(A) \equiv \int_A f dm$ for $A \in \mathcal{B}(\mathbb{R}^k)$. Then $\mu(\cdot)$ is a Radon measure and that is $\ll m$ and hence by Theorem 4.4.4

$$D\mu(x) = f(x) \quad \text{for almost all} \quad x(m).$$

That is, for almost all $x(m)$, for each $\epsilon > 0$, there is a $\delta > 0$ such that

$$\left|\frac{1}{m(A)} \int_A f dm - f(x)\right| < \epsilon$$

for all open balls A such that $x \in A$ and diam. $(A) < \delta$.

It turns out that a stronger result holds.

Theorem 4.4.5: *For almost all* $x(m)$, *for each* $\epsilon > 0$, *there is a* $\delta > 0$ *such that*

$$\frac{1}{m(A)} \int_A |f - f(x)| dm < \epsilon$$

for all open balls A *such that* $x \in A$ *and diam.* $(A) < \delta$ *(see Problems 4.23, 4.24).*

Theorem 4.4.6: (*Change of variables formula in* $\mathbb{R}^k$, $k > 1$). *Let* V *be an open set in* $\mathbb{R}^k$. *Let* $T \equiv (T_1, T_2, \ldots, T_k)$ *be a map from* $\mathbb{R}^k \to \mathbb{R}^k$ *such that for each* i, $T_i : \mathbb{R}^k \to \mathbb{R}$ *and* $\frac{\partial T_i(\cdot)}{\partial x_j}$ *exists on* V *for all* $1 \le i, j \le k$. *Suppose that the Jacobian* $J_T(\cdot) \equiv det\big(\big(\frac{\partial T_i(\cdot)}{\partial x_j}\big)\big)$ *is continuous and positive on* V. *Suppose further that* $T(V)$ *is a bounded open set* W *in* $\mathbb{R}^k$ *and that* T *is* $(1-1)$ *and* T^{-1} *is continuous. Then*

(i) *For all Borel set* $E \subset V$, $T(E)$ *is a Borel set* $\subset W$.

(ii) $\nu(\cdot) \equiv m(T(\cdot))$ *is a measure on* $\mathcal{B}(W)$ *and* $\nu \ll m$ *with*

$$\frac{d\nu(\cdot)}{dm} = J_T(\cdot).$$

(iii) For any $h \in L^1(W, m)$

$$\int_W h dm = \int_V h(T(\cdot)) J_T(\cdot) dm.$$

(iv) $\lambda(\cdot) \equiv mT^{-1}(\cdot)$ *is a measure on* $\mathcal{B}(W)$ *and* $\lambda \ll m$ *with*

$$\frac{d\lambda}{dm} = |J(T^{-1}(\cdot))|^{-1}.$$

(v) For any $\mu \ll m$ *on* $\mathcal{B}(V)$ *the measure* $\psi(\cdot) \equiv \mu T^{-1}(\cdot)$ *is dominated by* m *with*

$$\frac{d\psi}{dm}(\cdot) = \left(\frac{d\mu}{dm}\right)(T^{-1}(\cdot))\Big(J_T(T^{-1}(\cdot))\Big)^{-1} \quad \text{on} \quad W.$$

For a proof see Rudin (1987), Chapter 7.

4.5 Singular distributions

4.5.1 Decomposition of a cdf

Recall that a *cumulative distribution function* (cdf) on $\mathbb{R}$ is a function $F : \mathbb{R} \to [0, 1]$ such that it is nondecreasing, right continuous, $F(-\infty) = 0$, $F(\infty) = 1$. In Section 2.2, it was shown that any cdf F on $\mathbb{R}$ can be written as

$$F = \alpha F_d + (1 - \alpha) F_c, \tag{5.1}$$

where F_d and F_c are discrete and continuous cdfs respectively. In this section, the cdf F_c will be further decomposed into a singular continuous and absolutely continuous cdfs.

Definition 4.5.1: A cdf F is *singular* if $F' \equiv 0$ almost everywhere w.r.t. the Lebesgue measure on $\mathbb{R}$.

Example 4.5.1: The cdfs of Binomial, Poisson or any integer valued random variables are singular.

It is known (cf. Royden (1988), Chapter 5) that a monotone function $F : \mathbb{R} \to \mathbb{R}$ is differentiable almost everywhere w.r.t. the Lebesgue measure and its derivative F' satisfies

$$\int_a^b F'(x)dx \leq F(b) - F(a), \tag{5.2}$$

for any $-\infty < a < b < \infty$.

For $x \in \mathbb{R}$, let $\tilde{F}_{ac}(x) \equiv \int_{-\infty}^{x} F_c'(t)dt$ and $\tilde{F}_{sc}(x) \equiv F_c(x) - \tilde{F}_{ac}(x)$. If $\tilde{\beta} \equiv \int_{-\infty}^{\infty} F_c'(t)dt = \tilde{F}_{ac}(\infty) = 0$, then $F_c'(t) = 0$ a.e. and so F_c is singular continuous. If $\tilde{\beta} = 1$, then $F_c = \tilde{F}_{ac}$ and so, F_c is absolutely continuous. If $0 < \alpha < 1$ and $0 < \tilde{\beta} < 1$, then F can be written as

$$F = \alpha F_d + \beta F_{ac} + \gamma F_{sc}, \tag{5.3}$$

where $\beta = (1-\alpha)\tilde{\beta}$, $\gamma = (1-\alpha)(1-\tilde{\beta})$, $F_{ac} = \tilde{\beta}^{-1}\tilde{F}_{ac}$, $F_{sc} = (1-\tilde{\beta})^{-1}\tilde{F}_{sc}$, and F_d is as in (5.1). Note that F_d, F_{ac}, F_{sc} are all cdfs and α, β, γ are nonnegative numbers adding up to 1. Summarizing the above discussions, one has the following:

Proposition 4.5.1: *Given any cdf F, there exist nonnegative constants α, β, γ and cdfs F_d, F_{ac}, F_{sc} satisfying (a) $\alpha + \beta + \gamma = 1$, and (b) F_d is discrete, F_{ac} is absolutely continuous, F_{sc} is singular continuous, such that the decomposition (5.3) holds.*

It can be shown that the constants α, β, and γ are uniquely determined, and that when $0 < \alpha < 1$, the decomposition (5.1) is unique, and that when $0 < \alpha, \beta, \gamma < 1$, the decomposition (5.3) is unique. The decomposition (5.3) also has a probabilistic interpretation. Any random variable X can be realized as a randomized choice over three random variables X_d, X_{ac}, and X_{sc} having cdfs F_d, F_{ac}, and F_{sc}, respectively, and with corresponding randomization probabilities α, β, and γ. For more details see Problem 6.15 in Chapter 6.

4.5.2 Cantor ternary set

Recall the construction of the Cantor set from Section 1.3.

Let $I_0 = [0, 1]$ denote the unit interval. If one deletes the open middle third of I_0, then one gets two disjoint closed intervals $I_{11} = \left[0, \frac{1}{3}\right]$ and $I_{12} = \left[\frac{2}{3}, 1\right]$. Proceeding similarly with the closed intervals I_{11} and I_{12}, one gets four disjoint intervals $I_{21} = \left[0, \frac{1}{9}\right]$, $I_{22} = \left[\frac{2}{9}, \frac{1}{3}\right]$, $I_{23} = \left[\frac{2}{3}, \frac{7}{9}\right]$, $I_{24} = \left[\frac{8}{9}, 1\right]$, and so on. Thus, at each step, deleting the open middle third of the closed intervals constructed in the previous step, one is left with 2^n disjoint closed intervals each of length 3^{-n} after n steps. Let $C_n = \cup_{j=1}^{2^n} I_{nj}$ and $C = \cap_{n=1}^{\infty} C_n$. By construction $C_{n+1} \subset C_n$ for each n and C_n's are closed sets. With $m(\cdot)$ denoting Lebesgue measure, one has $m(C_0) = 1$ and $m(C_n) = 2^n 3^{-n} = \left(\frac{2}{3}\right)^n$.

Definition 4.5.2: The set $C \equiv \cap_{n=1}^{\infty} C_n$ is called the *Cantor ternary* set or simply the *Cantor set.*

Since $m(C_0) = 1$, by m.c.f.a. $m(C) = \lim_{n\to\infty} m(C_n) = \lim_{n\to\infty} \left(\frac{2}{3}\right)^n = 0$. Thus, the Cantor set C has zero Lebesgue measure. Next, let $U_1 = U_{11} = \left(\frac{1}{3}, \frac{2}{3}\right)$ be the deleted interval at the first stage, $U_2 = U_{21} \cup U_{22} =$

$\left(\frac{1}{9}, \frac{2}{9}\right) \cup \left(\frac{7}{9}, \frac{8}{9}\right)$ be the union of the deleted intervals at the second stage and similarly, $U_n = \cup_{j=1}^{2^{n-1}} U_{nj}$ at stage n. Thus $C^c = U = \cup_{n=1}^{\infty}\left(\cup_{j=1}^{2^{n-1}} U_{nj}\right)$ is open and $m(C^c) = 1$. Since $C \cup C^c = [0,1]$ and C^c is open it follows that C is nonempty. In fact, C is uncountably infinite as will be shown now. To do this, one needs the concept of *p-nary expansion of numbers* in [0,1]. Fix a positive integer $p > 1$. For each x in [0,1), let $a_1(x) = \lfloor px \rfloor$ where $\lfloor t \rfloor = n$ if $n \leq t < n+1$. Thus $a_1(x) \leq px < a_1(x) + 1$ and $a_1(x) \in \{0, 1, \ldots, p-1\}$, i.e., $\frac{a_1(x)}{p} \leq x < \frac{a_1(x)}{p} + \frac{1}{p}$. Thus, if $\frac{k}{p} \leq x < \frac{k+1}{p}$ for some $k = 0, 1, 2, \ldots, p-1$, then $a_1(x) = k$. Next, let $x_1 \equiv x - \frac{a_1(x)}{p}$ and $a_2(x) = \lfloor p^2 x_1 \rfloor$. Then, $x_1 \in \left[0, \frac{1}{p}\right)$ and

$$\frac{a_2(x)}{p^2} \leq x_1 = x - \frac{a_1(x)}{p} < \frac{a_2(x)}{p^2} + \frac{1}{p^2}$$

and $a_2(x) \in \{0, 1, 2, p-1\}$. Next, let $0 \leq x_2 \equiv x - \frac{a_1(x)}{p} - \frac{a_2(x)}{p^2} < \frac{1}{p^2}$ and $a_3(x) = \lfloor p^3 x_2 \rfloor$ and so on. After k such iterations one gets

$$0 \leq x - \sum_{i=1}^{k} \frac{a_i(x)}{p^i} < \frac{1}{p^k}$$

where $a_i(x) \in \{0, 1, 2, \ldots, p-1\}$ for all i. Since $\frac{1}{p^k} \to 0$ as $k \to \infty$, one gets the p-nary expansion of x in [0,1) as

$$x = \sum_{i=1}^{\infty} \frac{a_i(x)}{p^i}, \quad a_i(x) \in \{0, 1, 2, \ldots, p-1\}. \tag{5.4}$$

Notice that if $x = \frac{k}{p}$ for some $k \in \{0, 1, 2, \ldots, p-1\}$, then in the above expansion $a_1(x) = k$ and $a_i(x) = 0$ for $i \geq 2$, and the expansion *terminates*. But since $\sum_{i=1}^{\infty} \frac{(p-1)}{p^i} = 1$, one may also write $x = \frac{k}{p} = \frac{k-1}{p} + \sum_{i=1}^{\infty} \frac{(p-1)}{p^i}$, this being an expansion which is *nonterminating* and *recurring*. It can be shown that for all x in [0,1) of the form $x = \frac{\ell}{p^m}$ for some positive integers ℓ and m, there are exactly two expansions such that one terminates and the other is nonterminating and recurring. For all other x in [0,1), the p-nary expansion is nonterminating and nonrecurring and is unique.

The *decimal expansion* corresponds to the case $p = 10$, the *binary expansion* to the case $p = 2$ and the *ternary expansion* to the case $p = 3$. Here the convention of choosing only a nonterminating expansion for each x in [0,1) is used. Thus, for example, for $p = 3$, $\frac{1}{3}$ will be replaced by $\sum_{i=2}^{\infty} \frac{2}{3^i}$ so that $a_1\left(\frac{1}{3}\right) = 0$, $a_i\left(\frac{1}{3}\right) = 2$ for $i \geq 2$. Similarly, for $p = 3$, $x = \frac{7}{9} = \frac{2}{3} + \frac{1}{9}$ will be replaced by $\frac{2}{3} + \frac{0}{3^2} + \sum_{i=3}^{\infty} \frac{2}{3^i}$ so that $a_1\left(\frac{7}{9}\right) = 2$, $a_2\left(\frac{7}{9}\right) = 0$, $a_i\left(\frac{7}{9}\right) = 2$ for $i \geq 3$. By taking $p = 2$, i.e. the binary expansion, it is seen that every x in [0,1) can be uniquely represented as $x = \sum_{i=1}^{\infty} \frac{\delta_i(x)}{2^i}$ where $\delta_i(x) \in \{0, 1\}$ for all i. Thus, the interval [0,1) is in one-to-one correspondence with the set of all sequences of 0's and 1's.

It is not difficult to prove the following (Problem 4.31).

Theorem 4.5.2: *A number x belongs to the Cantor set C iff in (5.4), for all $i \geq 1$, $a_i(x)$ is either 0 or 2.*

Corollary 4.5.3: *The Cantor set C is in one-to-one correspondence with the set of all sequences of 0's and 1's and hence is in one-to-one correspondence with the unit interval [0,1].*

Remark 4.5.1: Thus the Cantor ternary set C is a closed subset of [0,1], its Lebesgue measure $m(C) = 0$, and its cardinality is the same as that of [0,1]. Further, it is *nowhere dense*, i.e., its complement U is dense in the sense that for every open interval $(a, b) \subset [0, 1]$, $U \cap (a, b)$ is nonempty. It is also possible to get a Cantor like set C_α with (Lebesgue) measure α, $0 < \alpha < 1$, by following the above iterative procedure of deleting at each stage intervals of length that is a fraction $\frac{(1-\alpha)}{3}$ of the full interval (Problem 4.32).

4.5.3 Cantor ternary function

The *Cantor ternary function* $F : [0, 1] \to [0, 1]$ is defined as follows: For $n \geq 1$, let $\{U_{nj} : j = 1, \ldots, 2^{n-1}\}$ denote the set of "deleted" intervals at step n in the definition of the Cantor set C. Define F on $C^c = U$ by

$$\begin{aligned} F(x) &= \frac{1}{2} \quad \text{on} \quad U_{11} = \left(\frac{1}{3}, \frac{2}{3}\right) \quad \text{and} \\ &= \frac{1}{4} \quad \text{on} \quad U_{21} = \left(\frac{1}{9}, \frac{2}{9}\right) \\ &= \frac{3}{4} \quad \text{on} \quad U_{22} = \left(\frac{7}{9}, \frac{8}{9}\right) \end{aligned}$$

and so on. It can be checked that F is uniformly continuous on U and has a continuous extension to $I_0 = [0, 1]$. The extension of the function F (also denoted by F) maps [0,1] onto [0,1] and is continuous and nondecreasing. Further, on U, it is differentiable with derivative $F' \equiv 0$. Since $m(U^c) = 0$, F is a singular cdf (cf. see Definition 4.5.1). It can be shown that if $\sum_{i=1}^{\infty} \frac{a_i(x)}{3^i}$ is the ternary expansion of $x \in (0, 1)$, then

$$F(x) = \sum_{i=1}^{N(x)-1} \frac{a_i(x)}{2^{i+1}} + \frac{1}{2^{N(x)}} \qquad (5.5)$$

Where $N(x) = \inf\{i : i \geq 1, a_i(x) = 1\}$. For example, $x = \frac{1}{3} = \sum_{i=2}^{\infty} \frac{2}{3^i} \Rightarrow N(x) = \infty$, $a_1(x) = 0$, $a_i(x) = 2$ for $i \geq 2 \Rightarrow F(x) = \sum_{i=2}^{\infty} \frac{1}{2^i} = \frac{1}{2}$ while $x = \frac{4}{9} = \frac{1}{3} + \frac{0}{3^2} + \sum_{i=3}^{\infty} \frac{2}{3^i} \Rightarrow N(x) = 1 \Rightarrow F(x) = \frac{1}{2}$.

It can be shown that if $\{\delta_n\}_{n\geq 1}$ is a sequence of independent $\{0, 1\}$ valued random variables with $P(\delta_1 = 0) = \frac{1}{2} = P(\delta_1 = 1)$, then the above F is the cdf of the random variable $X = \sum_{i=1}^{\infty} \frac{2\delta_i}{3^i}$ which lies in the Cantor set w.p. 1.

4.6 Problems

4.1 Show that in the proof of Theorem 4.1.1, $\mu(A_3) = 0$ where $A_3 = \{g \notin [0,1]\}$ and g satisfies (1.9).

(**Hint:** Apply (1.9) separately to $A_{31} = \{g > 1\}$ and $A_{32} = \{g < 0\}$.)

4.2 Verify that μ_{1s}, defined in (1.14) and μ_2 are singular.

(**Hint:** For each n, by Case 1, there exists a g_n in $L^2(D_n, \mathcal{F}_n, \mu^{(n)})$ where $\mu^{(n)} = \mu_1^{(n)} + \mu_2^{(n)}$ and $\mathcal{F}_n \equiv \{A \cap D_n : A \in \mathcal{F}\}$, such that $0 \leq \int_A g_n d\mu^{(n)} \leq \mu^{(n)}(A)$ for all A in $\mathcal{F}_n$. Let $A_{1n} = \{w : w \in D_n, 0 \leq g_n(w) < 1\}$, $A_{2n} = \{w : w \in D_n, g_n(w) = 1\}$, and $A_2 = \cup_{n \geq 1} A_{2n}$. Show that $\mu_2(A_2) = 0$ and $\mu_{1s}(A) = \sum_{n=1}^{\infty} \mu_{1n}(A \cap A_{2n})$ and hence $\mu_{1s}(A_2^c) = 0$.)

4.3 Let ν be the counting measure on $[0,1]$ and $\int_{[0,1]} h d\nu < \infty$ for some nonnegative function h. Show that $B = \{x : h(x) > 0\}$ is countable.

(**Hint:** Let $B_n = \{x : h(x) > \frac{1}{n}\}$. Show that B_n is a finite set for each $n \in \mathbb{N}$.)

4.4 Prove Proposition 4.1.2.

4.5 Find the Lebesgue decomposition of μ w.r.t. ν and the Radon-Nikodym derivative $\frac{d\mu_a}{d\nu}$ in the following cases where μ_a is the absolutely continuous component of μ w.r.t. ν.

(a) $\mu = N(0,1)$, $\nu =$ Exponential(1)

(b) $\mu =$ Exponential(1), $\nu = N(0,1)$

(c) $\mu = \mu_1 + \mu_2$, where $\mu_1 = N(0,1)$, $\mu_2 =$ Poisson(1) and $\nu =$ Cauchy(0, 1).

(d) $\mu = \mu_1 + \mu_2$, $\nu =$ Geometric(p), $0 < p < 1$, where $\mu_1 = N(0,1)$ and $\mu_2 =$ Poisson(1).

(e) $\mu = \mu_1 + \mu_2$, $\nu = \nu_1 + \nu_2$ where $\mu_1 = N(0,1)$, $\mu_2 =$ Poisson(1), $\nu_1 =$ Cauchy(0, 1) and $\nu_2 =$ Geometric(p), $0 < p < 1$.

(f) $\mu =$ Binomial (10, 1/2), $\nu =$ Poisson (1).

The measures referred to above are defined in Tables 4.6.1 and 4.6.2, given at the end of this section.

4.6 Let $(\Omega, \mathcal{F}, \mu)$ be a measure space and $f \in L_1(\Omega, \mathcal{F}, \mu)$. Let

$$\nu_f(A) \equiv \int_A f d\mu \quad \text{for all} \quad A \in \mathcal{F}.$$

(a) Show that ν_f is a finite signed measure.

(b) Show that $\|\nu\| = \int_\Omega |f| d\mu$ and for $A \in \mathcal{F}$, $\nu_f^+(A) = \int_A f^+ d\mu$, $\nu_f^-(A) = \int_A f d\mu$, and $|\nu_f|(A) = \int_A |f|(d\mu)$.

4.7 (a) Let μ_1 and μ_2 be two finite measures such that both are dominated by a σ-finite measure ν. Show that the total variation measure of the signed measure $\mu \equiv \mu_1 - \mu_2$ is given by $|\mu|(A) = \int_A |h_1 - h_2| d\nu$ where for $i = 1, 2$, $h_i = \frac{d\mu_i}{d\nu}$.

(b) Conclude that if μ_1 and μ_2 are two measures on a countable set $\Omega \equiv \{\omega_i\}_{i \geq 1}$ with $\mathcal{F} \equiv \mathcal{P}(\Omega)$, then $|\mu|(A) = \sum_{i \in A} |\mu_1(\omega_i) - \mu_2(\omega_i)|$.

(c) Show that if μ_n is the Binomial (n, p_n) measure and μ is the Poisson (λ) measure, $0 < \lambda < \infty$, then as $n \to \infty$, $|\mu_n - \mu|(\cdot) \to 0$ uniformly on $\mathcal{P}(\mathbb{Z}_+)$ iff $np_n \to \lambda$.

(**Hint:** Show that for each $i \in \mathbb{Z}_+ \equiv \{0, 1, 2, \ldots\}$, $\mu_n(\{i\}) \to \mu(\{i\})$ and use Scheffe's theorem.)

4.8 Let ν be a finite signed measure on a measurable space $(\Omega, \mathcal{F})$ and let $|\nu|$ be the total variation measure corresponding to ν. Show that for any $B \in \Omega^+$, $B \subset \mathcal{F}$,

$$|\nu|(B) = \nu(B),$$

where Ω^+ is as defined in (2.13).

(**Hint:** For any set $A \subset \Omega^+$,

$$\nu(A) = \int_A h d|\nu| = \int_{A \cap \Omega +} h d|\nu| \geq 0.)$$

4.9 Show that the Banach space $\mathbb{S}$ of finite signed measures on $(\mathbb{N}, \mathcal{P}(\mathbb{N}))$ is isomorphic to ℓ_1, the Banach space of absolutely convergent sequences $\{x_n\}_{n \geq 1}$ in $\mathbb{R}$.

4.10 Let μ_1 and μ_2 be two probability measures on $(\Omega, \mathcal{F})$.

(a) Show that

$$\|\mu_1 - \mu_2\| = 2 \sup\{|\mu_1(A) - \mu_2(A)| : A \in \mathcal{F}\}.$$

(**Hint:** For any $A \in \mathcal{F}$, $\{A, A^c\}$ is a partition of Ω and so $\|\mu_1 - \mu_2\| \geq |\mu_1(A) - \mu_2(A)| + |\mu_1(A^c) - \mu_2(A^c)| = 2|\mu_1(A) - \mu_2(A)|$, since μ_1 and μ_2 are probability measures. For the opposite inequality, use the Hahn decomposition of Ω w.r.t. $\mu_1 - \mu_2$ and the fact $\|\mu_1 - \mu_2\| = |(\mu_1 - \mu_2)(\Omega^+)| + |(\mu_1 - \mu_2)(\Omega^-)|$.)

(b) Show that $\|\mu_1 - \mu_2\|$ is also equal to

$$\sup\left\{\left|\int f d\mu_1 - \int f d\mu_2\right| : f \in B(\Omega, \mathbb{R})\right\}$$

where $B(\Omega, \mathbb{R})$ is the collection of all $\mathcal{F}$-measurable functions from Ω to $\mathbb{R}$ such that $\sup\{|f(\omega)| : \omega \in \Omega\} \leq 1$.

4.11 Let $(\Omega, \mathcal{F})$ be a measurable space.

(a) Let $\{\mu_n\}_{n\geq 1}$ be a sequence of finite measures on $(\Omega, \mathcal{F})$. Show that there exists a probability measure λ such that $\mu_n \ll \lambda$.

(**Hint:** Consider $\lambda(\cdot) = \sum_{n=1}^{\infty} \frac{1}{2^n} \frac{\mu_n(\cdot)}{\mu_n(\Omega)} \cdot$)

(b) Extend (a) to the case where $\{\mu_n\}_{n\geq 1}$ are σ-finite.

(c) Conclude that for any sequence $\{\nu_n\}_{n\geq 1}$ of finite signed measures on $(\Omega, \mathcal{F})$, there exists a probability measure λ such that $|\nu_n| \ll \lambda$ for all $n \geq 1$.

4.12 Let $\{\mu_n\}_{n\geq 1}$ be a sequence of finite measures on a measurable space $(\Omega, \mathcal{F})$. Show that there exists a finite measure μ on $(\Omega, \mathcal{F})$ such that $\|\mu_n - \mu\| \to 0$ iff there is a finite measure λ dominating μ and μ_n, $n \geq 1$ such that the Radon-Nikodym derivatives $f_n \equiv \frac{d\mu_n}{d\lambda} \to f \equiv \frac{d\mu}{d\lambda}$ in measure on $(\Omega, \mathcal{F}, \lambda)$ and $\mu_n(\Omega) \to \mu(\Omega)$.

4.13 (a) Let μ_1 and μ_2 be two finite measures on $(\Omega, \mathcal{F})$. Let $\mu_1 = \mu_{1a} + \mu_{1s}$ be the Lebesgue-Radon-Nikodym decomposition of μ_1 w.r.t. μ_2 as in Theorem 4.1.1. Show that if $\mu = \mu_1 - \mu_2$, then for all $A \in \mathcal{F}$,

$$|\mu|(A) = \int_A |h - 1| d\mu_2 + \mu_{1s}(A) \quad \text{where} \quad h = \frac{d\mu_{1a}}{d\mu_2}$$

is the Radon-Nikodym derivative of μ_{1a} w.r.t. μ_2. Conclude that if $\mu_1 \perp \mu_2$, then $|\mu|(\cdot) = \mu_1(\cdot) + \mu_2(\cdot)$ and if $\mu_1 \ll \mu_2$, then $|\mu|(A) = \int_A \left|\frac{d\mu_1}{d\mu_2} - 1\right| d\mu_2$.

(b) Compute $|\mu|(\cdot)$, $\|\mu\|$ if $\mu = \mu_1 - \mu_2$ for the following cases

(i) $\mu_1 \sim N(0, 1)$, $\mu_2 \sim N(1, 1)$

(ii) $\mu_1 \sim$ Cauchy (0,1), $\mu_2 \sim N(0, 1)$

(iii) $\mu_1 \sim N(0, 1)$, $\mu_2 \sim$ Poisson (λ).

(c) Establish Proposition 4.2.7.

4.14 Give another proof of the completeness of $(\mathbb{S}, \|\cdot\|))$ by verifying the following steps.

(a) For any sequence $\{\nu_n\}_{n\geq 1}$ in $\mathbb{S}$, there is a finite measure λ and $\{f_n\}_{n\geq 1} \subset L^1(\Omega, \mathcal{F}, \lambda)$ such that

$$\nu_n(A) = \int_A f_n d\lambda \quad \text{for all} \quad A \in \mathcal{F}, \quad \text{for all} \quad n \geq 1.$$

(b) $\{\nu_n\}_{n\geq 1}$ Cauchy in $\mathbb{S}$ is the same as $\{f_n\}_{n\geq 1}$ Cauchy in $L^1(\Omega, \mathcal{F}, \lambda)$ and hence, the completeness of $(\mathbb{S}, \|\cdot\|))$ follows from the completeness of $L^1(\Omega, \mathcal{F}, \lambda)$.

4.15 Let $f, g \in BV[a, b]$.

(a) Show that $P(f + g; [a, b]) \leq P(f; [a, b]) + P(g; [a, b])$ and that the same is true for $N(\cdot;\cdot)$ and $T(\cdot;\cdot)$.

(b) Show that for any $c \in \mathbb{R}$,

$$P(cf; [a, b]) = |c| P(f; [a, b])$$

and do the same for $N(\cdot;\cdot)$ and $T(\cdot;\cdot)$.

(c) For any $a < c < b$, $P(f; [a, b]) = P(f; [a, c]) + P(f; [c, b])$.

4.16 Let $\{f_n\}_{n\geq 1} \subset BV[a, b]$ and let $\lim_n f_n(x) = f(x)$ for all x in $[a, b]$. Show that $P(f; [a, b]) \leq \lim_{n\to\infty} P(f_n; [a, b])$ and do the same for $N(\cdot;\cdot)$ and $T(\cdot;\cdot)$.

4.17 Let $f \in BV[a, b]$. Show that f is continuous except on an at most countable set.

4.18 Let $F : [a, b] \to \mathbb{R}$ be a.c. Show that it is of bounded variation.

(**Hint:** By the definition of a.c., for $\epsilon = 1$, there is a $\delta_1 > 0$ such that $\sum_{j=1}^k |a_j - b_j| < \delta_1 \Rightarrow \sum_{j=1}^k |F(a_j) - F(b_j)| < 1$. Let M be an integer $> \frac{b-a}{\delta} + 1$. Show that $T(F, [a, b]) \leq M$.)

4.19 Let F be an absolutely continuous nondecreasing function on $\mathbb{R}$. Let μ_F be the Legesgue-Stieltjes measure corresponding to F. Show that for any $h \in L^1(\mathbb{R}, \mathcal{M}_{\mu_F}, \mu_F)$,

$$\int_{\mathbb{R}} h d\mu_F = \int h f dm$$

where f is a nonnegative measurable function such that $F(b) - F(a) = \int_{[a,b]} f dm$ for any $a < b$.

4.20 Let $F : [a,b] \to \mathbb{R}$ be absolutely continuous with $F'(\cdot) > 0$ a.e. on $[a,b]$, where $-\infty < a < b < \infty$. Let $F(a) = c$ and $F(b) = d$. Let $m(\cdot)$ denote the Lebesgue measure on $\mathbb{R}$. Show the following:

(a) (*Change of variables formula*). For any $g : [c,d] \to \mathbb{R}$ and Lebesgue measurable and integrable w.r.t. m

$$\int_{[c,d]} g dm = \int_{[a,b]} g(F)F' dm.$$

(b) For any Borel set $E \subset [a,b]$, $F(E)$ is also a Borel set.

(c) $\nu(\cdot) \equiv m\big(F(\cdot)\big)$ is a measure on $\mathcal{B}([a,b])$ and $\nu \ll m$ with

$$\frac{d\nu}{dm}(\cdot) = F'(\cdot).$$

(d) $\lambda(\cdot) \equiv mF^{-1}(\cdot)$ is a measure on $\mathcal{B}([c,d])$ and $\lambda \ll m$ with

$$\frac{d\lambda}{dm}(\cdot) = \Big(F'\big(F^{-1}(\cdot)\big)\Big)^{-1}.$$

(e) For any measure $\mu \ll m$ on $\mathcal{B}([a,b])$ the measure $\psi(\cdot) \equiv \mu F^{-1}(\cdot)$ is dominated by m with

$$\frac{d\psi}{dm} = \frac{d\mu}{dm}\big(F^{-1}(\cdot)\big)\Big(F'\big(F^{-1}(\cdot)\big)\Big)^{-1}.$$

(f) Establish (a) assuming that g and F' are both continuous noting that both integrals reduce to Riemann integrals.

(**Hint:**

(i) Verify (a) for $g = I_{[a,b]}$, $c < \alpha < \beta < d$ and approximate by step functions.

(ii) Show that F is $(1-1)$ and $F^{-1}(\cdot)$ is continuous and hence Borel measurable.

(iii) Show that $\nu(\cdot) = \mu_F$, the Lebesgue-Stieltjes measure corresponding to F.

(iv) Use the fact that for any $c \le \alpha \le \beta \le d$,

$$\begin{aligned}
\psi([\alpha,\beta]) &= \mu([\gamma,\delta]), \quad \text{where} \quad \gamma = F^{-1}(\alpha),\, \delta = F^{-1}(\beta),\\
&= \int_{[\gamma,\delta]} g dm, \quad \text{where} \quad g = \frac{d\mu}{dm}\\
&= \int_{[\gamma,\delta]} \frac{g\Big(F^{-1}\big(F(\cdot)\big)\Big)}{F'\Big(F^{-1}\big(F(\cdot)\big)\Big)} F'(\cdot) dm\\
&= \int_{[\alpha,\beta]} \frac{g\big(F^{-1}(\cdot)\big)}{F'\big(F^{-1}(\cdot)\big)} dm \quad \text{by (a).}
\end{aligned}$$

)

4.21 Let $F : \mathbb{R} \to \mathbb{R}$ be absolutely continuous on every finite interval.

(a) Show that the f in (4.1) can be chosen independently of the interval $[a, b]$.

(b) Further, if f is integrable over $\mathbb{R}$, then $\lim_{x\to-\infty} F(x) \equiv F(-\infty)$ and $\lim_{x\to\infty} F(x) \equiv F(\infty)$ exist and $F(x) = F(-\infty) + \int_{(-\infty,x)} f d\mu_L$ for all x in $\mathbb{R}$.

(c) Give an example where $F : \mathbb{R} \to \mathbb{R}$ is a.c., but f is not integrable over $\mathbb{R}$.

4.22 Let $F : \mathbb{R} \to \mathbb{R}$ be absolutely continuous on bounded intervals. Let $\{I_j = 1 \le j \le k \le \infty\}$ be a collection of disjoint intervals such that $\cup_{j=1}^k I_j \equiv \mathbb{R}$ and on each I_j, $F'(\cdot) > 0$ a.e. or $F'(\cdot) < 0$ a.e. w.r.t. m.

(a) Show that for any $h \in L^1(\mathbb{R}, m)$,

$$\int_{\mathbb{R}} h dm = \int_{\mathbb{R}} h(F(\cdot))|F'(\cdot)| dm.$$

(b) Show that if μ is a measure on $(\mathbb{R}, \mathcal{B}(\mathbb{R}))$ dominated by m then the measure $\mu F^{-1}(\cdot)$ is also dominated by m and

$$\frac{d\mu F^{-1}}{dm}(y) = \sum_{x_j \in D(y)} \frac{f(x_j)}{|F'(x_j)|}$$

where $f(\cdot) = \frac{d\mu}{dm}$ and $D(y) = \{x_j : x_j \in I_j, F(x_j) = y\}$.

(c) Let μ be the $N(0, 1)$ measure on $(\mathbb{R}, \mathcal{B}(\mathbb{R}))$, i.e.,

$$\frac{d\mu}{dm}(x) = \frac{1}{\sqrt{2n}} e^{-\frac{x^2}{2}}, \quad -\infty < x < \infty.$$

Let $F(x) = x^2$. Find $\frac{d\mu F^{-1}}{dm}$.

4.23 Let $f : \mathbb{R} \to \mathbb{R}$ be integrable w.r.t. m on bounded intervals. Show that for almost all x_0 in $\mathbb{R}$ (w.r.t. m),

$$\lim_{\substack{a \uparrow x_0 \\ b \downarrow x_0}} \frac{1}{(b-a)} \int_a^b |f(x) - f(x_0)| dx = 0.$$

(**Hint:** For each rational r, by Theorem 4.4.4

$$\lim_{\substack{a \uparrow x_0 \\ b \downarrow x_0}} \frac{1}{(b-a)} \int_a^b |f(x) - r| dx = |f(x_0) - r|.$$

a.e. (m). Let A_r denote the set of x_0 for which this fails to hold. Let $A = \cup_{r\in\mathbb{Q}} A_r$. Then $m(A) = 0$. For any $x_0 \notin A$ and any $\epsilon > 0$, choose a rational r such that $|f(x_0) - r| < \epsilon$ and now show that

$$\lim_{\substack{a\uparrow x_0 \\ b\downarrow x_0}} \frac{1}{(b-a)} \int_a^b |f(x) - f(x_0)|dx < \epsilon.\ )$$

4.24 Use the hint to the above problem to establish Theorem 4.4.5.

4.25 Let $(\Omega, \mathcal{B})$ be a measurable space. Let $\{\mu_n\}_{n\geq 1}$ and μ be σ-finite measures on $(\Omega, \mathcal{B})$. Let for each $n \geq 1$, $\mu_n = \mu_{na} + \mu_{ns}$ be the Lebesgue decomposition of μ_n w.r.t. μ with $\mu_{na} \ll \mu$ and $\mu_{ns} \perp \mu$. Let $\lambda = \sum_{n\geq 1} \mu_n$, $\lambda_a = \sum_{n\geq 1} \mu_{na}$, $\lambda_s = \sum_{n\geq 1} \mu_{ns}$. Show that $\lambda_a \ll \mu$ and $\lambda_s \perp \mu$ and that $\lambda = \lambda_a + \lambda_s$ is the Lebesgue decomposition of λ w.r.t. μ.

4.26 Let $\{\mu_n\}_{n\geq 1}$ be Radon measures on $(\mathbb{R}^k, \mathcal{B}(\mathbb{R}^k))$ and m be the Lebesgue measure on $\mathbb{R}^k$. Show that if $\lambda = \sum_{n=1}^{\infty} \mu_n$ is also a Radon measure, then

$$D\lambda = \sum_{n=1}^{\infty} D\mu_n \quad \text{a.e.} \quad (m).$$

(**Hint:** Use Theorem 4.4.4 and the uniqueness of the Lebesgue decomposition.)

4.27 Let F_n, $n \geq 1$ be a sequence of nondecreasing functions from $\mathbb{R} \to \mathbb{R}$. Let $F(x) = \sum_{n\geq 1} F_n(x) < \infty$ for all $x \in \mathbb{R}$. Show that $F(\cdot)$ is nondecreasing and

$$F'(\cdot) = \sum_{n\geq 1} F_n'(\cdot) \quad \text{a.e.} \quad (m).$$

4.28 Let E be a Lebesgue measurable set in $\mathbb{R}$. The *metric density* of E at x is defined as

$$D_E(x) \equiv \lim_{\delta\downarrow 0} \frac{m\big(E \cap (x-\delta, x+\delta)\big)}{2\delta}$$

if it exists. Show that $D_E(\cdot) = I_E(\cdot)$ a.e. m.

(**Hint:** Consider the measure $\lambda_E(\cdot) \equiv m(E \cap \cdot)$ on the Lebesgue σ-algebra. Show that $\lambda_E \ll m$ and find $\lambda_E'(\cdot)$ (cf. Definition 4.4.2).)

4.29 Let $F, G : [a, b] \to \mathbb{R}$ be both absolutely continuous. Show that $H = FG$ is also absolutely continuous on $[a, b]$ and that

$$\int_{[a,b]} FdG + \int_{[a,b]} GdF = F(b)G(b) - F(a)G(a).$$

4.30 Let $(\Omega, \mathcal{F}, \mu)$ be a finite measure space. Fix $1 \le p < \infty$. Let $T : L^p(\mu) \to \mathbb{R}$ be a bounded linear functional as defined in (3.2.10) (cf. Section 3.2). Complete the following outline of a proof of Theorem 3.2.3 (Riesz representation theorem).

(a) Let $\nu(A) \equiv T(I_A)$, $A \in \mathcal{F}$. Verify that $\nu(\cdot)$ is a signed measure on $(\Omega, \mathcal{F})$.

(b) Verify that $|\nu| \ll \mu$.

(c) Let $g \equiv \frac{d\nu}{d\mu}$. Show that $g \in L^q(\mu)$ where $q = \frac{p}{p-1}$, $1 < p < \infty$ and $q = \infty$ if $p = 1$.

(d) Show that $T = T_g$.

4.31 Prove Theorem 4.5.2 and Corollary 4.5.3.

4.32 For $0 < \alpha < 1$, construct a Cantor like set C_α as described in Remark 4.5.1.

4.33 Show that the Cantor ternary function can be expressed as in (5.5).

4.34 Let $(\Omega, \mathcal{F}, \mu)$ be a σ-finite measure space.

(a) Let $\mathcal{G}$ be a σ-algebra $\subset \mathcal{F}$. Let $f : \Omega \to \mathbb{R}_+$ be $\langle \mathcal{F}, \mathcal{B}(\mathbb{R})\rangle$-measurable. Show that there exists a $g : \Omega \to \mathbb{R}_+$ that is $\langle \mathcal{G}, \mathcal{B}(\mathbb{R})\rangle$-measurable and $\nu(A) \equiv \int_A f d\mu = \int_A g d\mu$ for all A in $\mathcal{G}$.

(**Hint:** Apply Theorem 4.1.1 (b) to the measures ν and μ restricted to $\mathcal{G}$. When μ is a probability measure, g is called the conditional expectation of f given $\mathcal{G}$ (cf. Chapter 12).)

(b) Now suppose $\mathcal{G} = \sigma\langle \{A_i\}_{i \ge 1}\rangle$ where $\{A_i\}_{i \ge 1}$ is a partition of $\Omega \subset \mathcal{F}$. Determine $g(\cdot)$ explicitly on each A_i such that $0 < \mu(A_i) < \infty$.

TABLE 4.6.1. Some discrete univariate distributions.

Mean μ	$\mu(A),\ A \in \mathcal{B}(\mathbb{R})$
Bernoulli (p), $0 < p < 1$	$\sum_{i=0}^{1} p^i (1-p)^{1-i} I_A(i)$
Binomial (n, p), $0 < p < 1$, $n \in \mathbb{N}$	$\sum_{i=0}^{n} \binom{n}{i} p^i (1-p)^{n-i} I_A(i)$
Geometric (p), $0 < p < 1$	$\sum_{i=0}^{\infty} p(1-p)^{i-1} I_A(i)$
Poisson (λ), $0 < \lambda < \infty$	$\sum_{i=0}^{\infty} e^{-\lambda} \frac{\lambda^i}{i!} \cdot I_A(i)$

TABLE 4.6.2. Some standard absolutely continuous distributions. Here m denotes the Lebesgue measure on $(\mathbb{R}, \mathcal{B}(\mathbb{R}))$.

Measure μ	$\mu(A)$, $A \in \mathcal{B}(\mathbb{R})$
Uniform (a, b), $-\infty < a < b < \infty$	$m(A \cap [a, b])/(b - a)$
Exponential (β), $\beta \in (0, \infty)$	$\int_A \frac{1}{\beta} \exp(-x/\beta) I_{(0,\infty)}(x) m(dx)$
Gamma (α, β), $\alpha, \beta \in (0, \infty)$	$\int_A \frac{1}{\Gamma(\alpha)\beta^\alpha} x^{\alpha-1} \exp(-x/\beta) I_{(0,\infty)}(x) m(dx)$ where $\Gamma(a) = \int_0^\infty x^{a-1} e^{-x} dx$, $a \in (0, \infty)$
Beta (α, β), $\alpha, \beta \in (0, \infty)$	$\frac{1}{B(\alpha,\beta)} \int_A x^{\alpha-1} (1 - x)^{\beta-1} I_{(0,1)}(x) m(dx)$, where $B(a, b) = \Gamma(a)\Gamma(b)/\Gamma(a + b)$, $a, b \in (0, \infty)$
Cauchy (γ, σ) $\gamma \in \mathbb{R}$, $\sigma \in (0, \infty)$	$\int_A \frac{1}{\pi\sigma} \frac{\sigma^2}{\sigma^2 + (x - \gamma)^2} m(dx)$
Normal (γ, σ^2), $\gamma \in \mathbb{R}$, $\sigma \in (0, \infty)$	$\int_A \frac{1}{\sqrt{2\pi}\sigma} \exp(-(x - \gamma)^2/\sigma^2) m(dx)$
Lognormal (γ, σ^2), $\gamma \in \mathbb{R}$, $\sigma \in (0, \infty)$	$\int_A \frac{1}{\sqrt{2\pi}\sigma} \frac{e^{-(\log x - \gamma)^2/2\sigma^2}}{x} I_{(0,\infty)}(x) m(dx)$

5

Product Measures, Convolutions, and Transforms

5.1 Product spaces and product measures

Given two measure spaces $(\Omega_i, \mathcal{F}_i, \mu_i)$, $i = 1, 2$, is it possible to construct a measure $\mu \equiv \mu_1 \times \mu_2$ on a σ-algebra on the product space $\Omega_1 \times \Omega_2$ such that $\mu(A \times B) = \mu_1(A)\mu_2(B)$ for $A \in \mathcal{F}_1$ and $B \in \mathcal{F}_2$? This section is devoted to studying this question.

Definition 5.1.1: Let $(\Omega_i, \mathcal{F}_i)$, $i = 1, 2$ be measurable spaces.

(a) $\Omega_1 \times \Omega_2 \equiv \{(\omega_1, \omega_2) : \omega_1 \in \Omega_1, \omega_2 \in \Omega_2\}$, the set of all ordered pairs, is called the *(Cartesian) product of* Ω_1 *and* Ω_2.

(b) The set $A_1 \times A_2$ with $A_1 \in \mathcal{F}_1, A_2 \in \mathcal{F}_2$ is called a *measurable rectangle*. The collection of measurable rectangles will be denoted by $\mathcal{C}$.

(c) The *product σ-algebra of* $\mathcal{F}_1$ *and* $\mathcal{F}_2$ on $\Omega_1 \times \Omega_2$, denoted by $\mathcal{F}_1, \times \mathcal{F}_2$, is the smallest σ-algebra generated by $\mathcal{C}$, i.e.,

$$\mathcal{F}_1 \times \mathcal{F}_2 \equiv \sigma\langle\{A_1 \times A_2 : A_1 \in \mathcal{F}_1, A_2 \in \mathcal{F}_2\}\rangle.$$

(d) $(\Omega_1 \times \Omega_2, \mathcal{F}_1 \times \mathcal{F}_2)$ is called the *product measurable space.*

Starting with the definition of μ on the class $\mathcal{C}$ of measurable rectangles by

$$\mu(A_1 \times A_2) = \mu_1(A_1)\mu_2(A_2) \tag{1.1}$$

for all $A_1 \in \mathcal{F}_1, A_2 \in \mathcal{F}_2$, one can extend it to a measure on the algebra $\mathcal{A}$ of all finite unions of disjoint measurable rectangles in a natural way. Indeed, the extension to $\mathcal{A}$ is obtained simply by assigning the μ-measure of a finite union of disjoint measurable rectangles as the sum of the μ-measures of the corresponding individual measurable rectangles. Then, by the extension theorem, (cf. Theorem 1.3.3), it can be further extended to a complete measure μ on a σ-algebra containing $\mathcal{F}_1 \times \mathcal{F}_2$, defined in (c) above. However, to evaluate $\mu(A)$ for an arbitrary A in $\mathcal{F}_1 \times \mathcal{F}_2$ that is not a measurable rectangle and to evaluate $\int h d\mu$ for arbitrary measurable functions h, some further work is needed. For the case of the product of two measure spaces, here an alternate approach that yields a direct way of computing these quantities is presented. The extension theorem approach will be used for the more general case of products of finitely many measure spaces in Section 5.3. The case of products of infinitely many probability spaces will be discussed in Chapter 6.

Definition 5.1.2:

(a) Let $A \in \mathcal{F}_1 \times \mathcal{F}_2$. Then, for any $\omega_1 \in \Omega_1$, the set

$$A_{1\omega_1} \equiv \{\omega_2 \in \Omega_2 : (\omega_1, \omega_2) \in A\} \tag{1.2}$$

is called the ω_1*-section of* A and for any $\omega_2 \in \Omega_2$, the set $A_{2\omega_2} \equiv \{\omega_1 \in \Omega_1 : (\omega_1, \omega_2) \in A\}$ is called the ω_2*-section* of A.

(b) If $f : (\Omega_1 \times \Omega_2) \to \Omega_3$ is a $\langle \mathcal{F}_1 \times \mathcal{F}_2, \mathcal{F}_3 \rangle$ measurable mapping from $\Omega_1 \times \Omega_2$ into some measurable space $(\Omega_3, \mathcal{F}_3)$, then the ω_1*-section of* f is the function $f_{1\omega_1} : \Omega_2 \to \Omega_3$, given by

$$f_{1\omega_1}(\omega_2) = f(\omega_1, \omega_2), \ \omega_2 \in \Omega_2. \tag{1.3}$$

Similarly, one may define the ω_2*-sections* of f by $f_{2\omega_2}(\omega_1) = f(\omega_1, \omega_2)$, $\omega_1 \in \Omega_1$. The following result shows that the ω_1-sections of a $\mathcal{F}_1 \times \mathcal{F}_2$-measurable function is $\mathcal{F}_2$-measurable when considered as a function of $\omega_2 \in \Omega_2$.

Proposition 5.1.1: *Let* $(\Omega_1 \times \Omega_2, \mathcal{F}_1 \times \mathcal{F}_2)$ *be a product space,* $A \in \mathcal{F}_1 \times \mathcal{F}_2$ *and let* $f : \Omega_1 \times \Omega_2 \to \Omega_3$ *be a* $\langle \mathcal{F}_1 \times \mathcal{F}_2, \mathcal{F}_3 \rangle$*-measurable function.*

(i) *For every* $\omega_1 \in \Omega_1$, $A_{1\omega_1} \in \mathcal{F}_2$ *and for every* $\omega_2 \in \Omega_2$, $A_{2\omega_2} \in \mathcal{F}_1$.

(ii) *For every* $\omega_1 \in \Omega_1$, $f_{1\omega_1}$ *is* $\langle \mathcal{F}_2, \mathcal{F}_3 \rangle$*-measurable and for every* $\omega_2 \in \Omega_2$, $f_{2\omega_2}$ *is* $\langle \mathcal{F}_1, \mathcal{F}_3 \rangle$*-measurable.*

Proof: Fix $\omega_1 \in \Omega_1$. Define the function $g : \Omega_2 \to \Omega_1 \times \Omega_2$ by

$$g(\omega_2) = (\omega_1, \omega_2), \ \omega_2 \in \Omega_2.$$

Note that for any measurable rectangle $A \equiv A_1 \times A_2 \in \mathcal{F}_1 \times \mathcal{F}_2$,

$$A_{1\omega_1} = \begin{cases} A_2 & \text{if} \quad \omega_1 \in A_1 \\ \emptyset & \text{if} \quad \omega_1 \notin A_1 \end{cases}$$

and hence $g^{-1}(A_1 \times A_2) \in \mathcal{F}_2$. Since the class of all measurable rectangles generates $\mathcal{F}_1 \times \mathcal{F}_2$, for fixed ω_1 in Ω_1, g is $\langle \mathcal{F}_2, \mathcal{F}_1 \times \mathcal{F}_2 \rangle$-measurable. Therefore, for A in $\mathcal{F}_1 \times \mathcal{F}_2$, $A_{1\omega_1} = g^{-1}(A) \in \mathcal{F}_2$ and for f as given, $f_{1\omega_1} = f \circ g$ is $\langle \mathcal{F}_2, \mathcal{F}_3 \rangle$-measurable. This proves (i) and (ii) for the ω_1-sections. The proof for the ω_2-sections are similar. □

Now suppose μ_1 and μ_2 are measures on $(\Omega_1, \mathcal{F}_1)$ and $(\Omega_2, \mathcal{F}_2)$, respectively. Then, for any set $A \in \mathcal{F}_1 \times \mathcal{F}_2$, for all $\omega_1 \in \Omega_1$, $A_{1\omega_1} \in \mathcal{F}_2$ and hence, $\mu_2(A_{1\omega_1})$ is well defined. If this were an $\mathcal{F}_1$-measurable function, then one might define a set function on $\mathcal{F}_1 \times \mathcal{F}_2$ by

$$\mu_{12}(A) = \int_{\Omega_1} \mu_2(A_{1\omega_1}) \mu_1(d\omega_1). \tag{1.4}$$

And similarly, reversing the order of μ_1 and μ_2, one might define a second set function

$$\mu_{21}(A) = \int_{\Omega_2} \mu_1(A_{2\omega_2}) \mu_2(d\omega_2), \tag{1.5}$$

provided that $\mu_1(A_{2\omega_2})$ is $\mathcal{F}_2$-measurable. Note that for the measurable rectangles $A = A_1 \times A_2$, $\mu_{12}(A) = \mu_1(A_1)\mu_2(A_2) = \mu_{21}(A)$ and thus, both μ_{12} and μ_{21} coincide (with the *product* measure μ) on the class $\mathcal{C}$ of all measurable rectangles. This implies that if the product measure μ is *unique* on $\mathcal{F}_1 \times \mathcal{F}_2$, and μ_{12} and μ_{21} are measures on $\mathcal{F}_1 \times \mathcal{F}_2$, then μ_{12}, μ_{21} and μ must coincide on the σ-algebra $\mathcal{F}_1 \times \mathcal{F}_2$. Then, one can evaluate $\mu(A)$ for any set $A \in \mathcal{F}_1 \times \mathcal{F}_2$ using either of the relations (1.4) or (1.5). The following result makes this heuristic discussion rigorous.

Theorem 5.1.2: *Let $(\Omega_i, \mathcal{F}_i, \mu_i)$, $i = 1, 2$ be σ-finite measure spaces. Then,*

(i) for all $A \in \mathcal{F}_1 \times \mathcal{F}_2$, the functions $\mu_2(A_{1\omega_1})$ and $\mu_1(A_{2\omega_2})$ are $\mathcal{F}_1$- and $\mathcal{F}_2$-measurable, respectively.

(ii) The set functions μ_{12} and μ_{21}, given by (1.4) and (1.5) respectively, are measures on $\mathcal{F}_1 \times \mathcal{F}_2$, satisfying $\mu_{12}(A) = \mu_{21}(A)$ for all $A \in \mathcal{F}_1 \times \mathcal{F}_2$.

(iii) Further, $\mu_{12} = \mu_{21} \equiv \mu$ is σ-finite and it is the only measure satisfying

$$\mu(A_1 \times A_2) = \mu_1(A_1)\mu_2(A_2) \quad \textit{for all} \quad A_1 \times A_2 \in \mathcal{C}.$$

Proof: First suppose that μ_1 and μ_2 are finite measures. Define

$$\mathcal{L} = \big\{ A \in \mathcal{F}_1 \times \mathcal{F}_2 : \mu_2(A_{1\omega_1}) \quad \text{is a} \quad \langle \mathcal{F}_1, \mathcal{B}(\mathbb{R}) \rangle\text{-measurable function} \big\}.$$

For $A = \Omega_1 \times \Omega_2$, $\mu_2(A_{1\omega_1}) \equiv \mu_2(\Omega_2)$ for all $\omega_1 \in \Omega_1$ and hence, $\Omega_1 \times \Omega_2 \in \mathcal{L}$. Next, let $A, B \in \mathcal{L}$ with $A \subset B$. Then, it is easy to check that $(A \setminus B)_{1\omega_1} = A_{1\omega_1} \setminus B_{1\omega_1}$. Since μ_2 is a finite measure and $A, B \in \mathcal{L}$, it follows that $\mu_2((A \setminus B)_{1\omega_1}) = \mu_2(A_{1\omega_1} \setminus B_{1\omega_1}) = \mu_2(A_{1\omega_1}) - \mu_2(B_{1\omega_1})$ is $\langle \mathcal{F}_1, \mathcal{B}(\mathbb{R})\rangle$-measurable. Thus, $A \setminus B \in \mathcal{L}$. Finally, let $\{B_n\}_{n\geq 1} \subset \mathcal{L}$ be such that $B_n \subset B_{n+1}$ for all $n \geq 1$. Then for any $\omega_1 \in \Omega_1$, $(B_n)_{1\omega_1} \subset (B_{n+1})_{1\omega_1}$ for all $n \geq 1$. Hence by the MCT,

$$\infty > \mu_2\Big(\Big(\bigcup_{n\geq 1} B_n\Big)_{1\omega_1}\Big) = \mu_2\Big(\bigcup_{n\geq 1} (B_n)_{1\omega_1}\Big) = \lim_{n\to\infty} \mu_2((B_n)_{1\omega_1})$$

for all $\omega_1 \in \Omega_1$. By Proposition 2.1.5, this implies that $\mu_2((\cup_{n\geq 1} B_n)_{1\omega_1})$ is $\langle \mathcal{F}_1, \mathcal{B}(\mathbb{R})\rangle$-measurable, and hence, $\cup_{n\geq 1} B_n \in \mathcal{L}$. Thus, $\mathcal{L}$ is a λ-system. For $A = A_1 \times A_2 \in \mathcal{C}$, $\mu_2(A_{1\omega_1}) = \mu_2(A_2) I_{A_1}(\omega_1)$ and hence $\mathcal{C} \subset \mathcal{L}$. Since $\mathcal{C}$ is a π-system, by Corollary 1.1.3, it follows that $\mathcal{L} = \mathcal{F}_1 \times \mathcal{F}_2$. Thus, $\mu_2(A_{1\omega_1})$, considered as a function of ω_1, is $\langle \mathcal{F}_1, \mathcal{B}(\mathbb{R})\rangle$-measurable for all $A \in \mathcal{F}_1 \times \mathcal{F}_2$, proving (i).

Next, consider part (ii). By part (i), μ_{12} is a well-defined set function on $\mathcal{F}_1 \times \mathcal{F}_2$. It is easy to check that μ_{12} is a measure on $\mathcal{F}_1 \times \mathcal{F}_2$ (Problem 5.1). Similarly, μ_{21} is a well-defined measure on $\mathcal{F}_1 \times \mathcal{F}_2$. Since $\mu_{12}(A) = \mu_{21}(A) = \mu_1(A_1)\mu_2(A_2)$ for all $A = A_1 \times A_2 \in \mathcal{C}$ and $\mathcal{C}$ is a π-system generating $\mathcal{F}_1 \times \mathcal{F}_2$, it follows from Theorem 1.2.4 that $\mu_{12}(A) = \mu_{21}(A)$ for all $A \in \mathcal{F}_1 \times \mathcal{F}_2$. This proves (ii) for the case where $\mu_i(\Omega_i) < \infty$, $i = 1, 2$.

Next, suppose that μ_i's are σ-finite. Then, there exist disjoint sets $\{B_{in}\}_{n\geq 1} \subset \mathcal{F}_i$, such that $\cup_{n\geq 1} B_{in} = \Omega_i$ and $\mu_i(B_{in}) < \infty$ for all $n \geq 1$, $i = 1, 2$. Define the finite measures

$$\mu_{in}(D) = \mu_i(D \cap B_{in}), \quad D \in \mathcal{F}_i,$$

for $n \geq 1$, $i = 1, 2$. The arguments above with μ_i replaced by μ_{in} imply that for any $A \in \mathcal{F}_1 \times \mathcal{F}_2$, $\mu_{2n}(A_{1\omega_1})$ is $\langle \mathcal{F}_1, \mathcal{B}(\mathbb{R})\rangle$-measurable for all $n \geq 1$. Since μ_2 is a measure on $\mathcal{F}_2$, $\mu_2(A_{1\omega_1}) = \sum_{n=1}^{\infty} \mu_{2n}(A_{\omega_1})$ and hence, considered as a function of ω_1, it is $\langle \mathcal{F}_1, \mathcal{B}(\mathbb{R})\rangle$-measurable for all $A \in \mathcal{F}_1 \times \mathcal{F}_2$. Thus, the set function μ_{12} of (1.4) is well defined in the σ-finite case as well. Similarly, μ_{21} of (1.5) is a well-defined set function on $\mathcal{F}_1 \times \mathcal{F}_2$. Let $\mu_{12}^{(m,n)}$ and $\mu_{21}^{(m,n)}$, respectively, denote the set functions defined by (1.4) and (1.5) with μ_1 replaced by μ_{1m} and μ_2 replaced by μ_{2n}, $m \geq 1$, $n \geq 1$. Then, by repeated use of the MCT,

$$\begin{aligned} \mu_{12}(A) &= \int_{\Omega_1} \mu_2(A_{1\omega_1})\mu_1(d\omega_1) \\ &= \sum_{m=1}^{\infty} \Big(\int_{B_{1m}} \sum_{n=1}^{\infty} \mu_2(A_{1\omega_1} \cap B_{2n})\Big)\mu_1(d\omega_1) \\ &= \sum_{m=1}^{\infty}\sum_{n=1}^{\infty} \int_{B_{1m}} \mu_2(A_{1\omega_1} \cap B_{2n})\mu_1(d\omega_1) \end{aligned}$$

$$= \sum_{m=1}^{\infty} \sum_{n=1}^{\infty} \mu_{12}^{(m,n)}(A), \quad A \in \mathcal{F}_1 \times \mathcal{F}_2 \tag{1.6}$$

and similarly,

$$\mu_{21}(A) = \sum_{n=1}^{\infty} \sum_{m=1}^{\infty} \mu_{21}^{(m,n)}(A), \quad A \in \mathcal{F}_1 \times \mathcal{F}_2. \tag{1.7}$$

Since $\mu_{12}^{(m,n)}$ and $\mu_{21}^{(m,n)}$ are (finite) measures, it is easy to check that μ_{12} and μ_{21} are measures on $\mathcal{F}_1 \times \mathcal{F}_2$. Also, by the finite case, $\mu_{12}^{(m,n)}(A_1 \times A_2) = \mu_{21}^{(m,n)}(A_1 \times A_2)$ for all $n \geq 1$, $m \geq 1$ and hence

$$\mu_{12}(A_1 \times A_2) = \mu_{21}(A_1 \times A_2) \quad \text{for all} \quad A_1 \times A_2 \in \mathcal{C}.$$

Next note that $\{B_{1m} \times B_{2n} : m \geq 1, n \geq 1\}$ is a partition of $\Omega_1 \times \Omega_2$ by $\mathcal{F}_1 \times \mathcal{F}_2$ sets and by (1.6) and (1.7), for all $m \geq 1$, $n \geq 1$,

$$\mu_{12}(B_{1m} \times B_{2n}) = \mu_1(B_{1m})\mu_2(B_{2n}) = \mu_{21}(B_{1m} \times B_{2n}) < \infty.$$

Hence, μ_{12} and μ_{21} are σ-finite on $\mathcal{F}_1 \times \mathcal{F}_2$. Since μ_{12} and μ_{21} agree on $\mathcal{C}$ and $\mathcal{C}$ is a π-system generating the product σ-algebra, it follows that $\mu_{12}(A) = \mu_{21}(A)$ for all $A \in \mathcal{F}_1 \times \mathcal{F}_2$ and it is the unique measure satisfying $\mu(A_1 \times A_2) = \mu_1(A_1)\mu_2(A_2)$ for all $A_1 \times A_2 \in \mathcal{C}$. This completes the proof of the theorem. □

Remark 5.1.1: In the above theorem, the σ-finiteness condition on the measures μ_1 and μ_2 cannot be dropped. For example, let $\Omega_1 = \Omega_2 = [0, 1]$, $\mathcal{F}_1 = \mathcal{F}_2 = \mathcal{B}([0, 1])$, $\mu_1 =$ the Lebesgue measure and $\mu_2 =$ the counting measure. Clearly μ_2 is not σ-finite since $[0, 1]$ is uncountable. Let A be the diagonal set in the product space $[0, 1] \times [0, 1]$, i.e., $A = \{(\omega_1, \omega_2) \in \Omega_1 \times \Omega_2 : \omega_1 = \omega_2\}$. Then $A \in \mathcal{F}_1 \times \mathcal{F}_2$, $A_{1\omega_1} = \{\omega_1\}$ and $A_{2\omega_2} = \{\omega_2\}$. Further, $\mu_2(A_{1\omega_1}) = 1$ for all ω_1 and $\mu_1(A_{2\omega_2}) = 0$ for all ω_2. Thus $\mu_{12}(A) = \int_{\Omega_1} \mu_2(A_{1\omega_1})\mu_1(d\omega_1) = 1$, but $\mu_{21}(A) = \int_{\Omega_2} \mu_1(A_{2\omega_2})\mu_2(d\omega_2) = 0$, and hence, $\mu_{12}(A) \neq \mu_{21}(A)$.

Remark 5.1.2: Although this approach allows one to compute the product measure $\mu_1 \times \mu_2$ of a set $A \in \mathcal{F}_1 \times \mathcal{F}_2$, the measure space $(\Omega_1 \times \Omega_2, \mathcal{F}_1 \times \mathcal{F}_2, \mu_1 \times \mu_2)$ may not be complete even if both $(\Omega_i, \mathcal{F}_i, \mu_i)$, $i = 1, 2$ are complete (Problem 5.2). However, the approach based on the extension theorem yields a product measure space that is complete. See Remark 5.2.2 for further discussion on the topic.

Definition 5.1.3:

(a) The unique measure μ on $\mathcal{F}_1 \times \mathcal{F}_2$ defined in Theorem 5.1.2 (iii) is called the *product measure* and is denoted by $\mu_1 \times \mu_2$.

(b) The measure space $(\Omega_1 \times \Omega_2, \mathcal{F}_1 \times \mathcal{F}_2, \mu_1 \times \mu_2)$ is called the *product measure space.*

Formulas (1.4) and (1.5) give two different ways of evaluating $\mu_1 \times \mu_2(A)$ for an $A \in \mathcal{F}_1 \times \mathcal{F}_2$. In the next section, integration of a measurable function $f : \Omega_1 \times \Omega_2 \to \mathbb{R}$ w.r.t. the product measure $\mu_1 \times \mu_2$ is considered and the above approach is extended to justify evaluation of the integral iteratively.

5.2 Fubini-Tonelli theorems

Let $f : \Omega_1 \times \Omega_2 \to \mathbb{R}$ be a $\langle \mathcal{F}_1 \times \mathcal{F}_2, \mathcal{B}(\mathbb{R})\rangle$-measurable function. Relations (1.4) and (1.5) suggest that the integral of f w.r.t. $\mu_1 \times \mu_2$ may be evaluated as iterated integrals, using the formulas

$$\int_{\Omega_1 \times \Omega_2} f(\omega_1, \omega_2) \mu_1 \times \mu_2(d(\omega_1, \omega_2)) = \int_{\Omega_2} \left[\int_{\Omega_1} f(\omega_1, \omega_2) \mu_1(d\omega_1) \right] \mu_2(d\omega_2) \tag{2.1}$$

and

$$\int_{\Omega_1 \times \Omega_2} f(\omega_1, \omega_2) \mu_1 \times \mu_2(d(\omega_1, \omega_2)) = \int_{\Omega_1} \left[\int_{\Omega_2} f(\omega_1, \omega_2) \mu_2(d\omega_2) \right] \mu_1(d\omega_1). \tag{2.2}$$

Here, the left sides of both (2.1) and (2.2) are simply the integral of f on the space $\Omega = \Omega_1 \times \Omega_2$ w.r.t. the measure $\mu = \mu_1 \times \mu_2$. The expressions on the right sides of (2.1) and (2.2) are, however, *iterated integrals*, where integrals of sections of f are evaluated first and then the resulting *sectional integrals* are integrated again to get the final expression. Conditions for the validity of (2.1) and (2.2) are provided by the Fubini-Tonelli theorems stated below.

Theorem 5.2.1: *(Tonelli's theorem). Let $(\Omega_i, \mathcal{F}_i, \mu_i)$, $i = 1, 2$ be σ-finite measure spaces and let $f : \Omega_1 \times \Omega_2 \to \mathbb{R}_+$ be a nonnegative $\mathcal{F}_1 \times \mathcal{F}_2$-measurable function. Then*

$$g_1(\omega_1) \equiv \int_{\Omega_2} f(\omega_1, \omega_2) \mu_2(d\omega_2) : \Omega_1 \to \bar{\mathbb{R}} \quad \textit{is} \quad \langle \mathcal{F}_1, \mathcal{B}(\bar{\mathbb{R}})\rangle\textit{-measurable} \tag{2.3}$$

and

$$g_2(\omega_2) \equiv \int_{\Omega_1} f(\omega_1, \omega_2) \mu_1(d\omega_1) : \Omega_2 \to \bar{\mathbb{R}} \quad \textit{is} \quad \langle \mathcal{F}_2, \mathcal{B}(\bar{\mathbb{R}})\rangle\textit{-measurable.} \tag{2.4}$$

Further

$$\int_{\Omega_1 \times \Omega_2} f d\mu = \int_{\Omega_1} g_1 d\mu_1 = \int_{\Omega_2} g_2 d\mu_2, \tag{2.5}$$

where $\mu = \mu_1 \times \mu_2$.

Proof: If $f = I_A$ for some A in $\mathcal{F}_1 \times \mathcal{F}_2$, the result follows from Theorem 5.1.2. By the linearity of integrals, the result now holds for all simple nonnegative functions f. For a general nonnegative function f, there exist a sequence $\{f_n\}_{n \geq 1}$ of nonnegative simple functions such that $f_n(\omega_1, \omega_2) \uparrow f(\omega_1, \omega_2)$ for all $(\omega_1, \omega_2) \in \Omega_1 \times \Omega_2$. Write $g_{1n}(\omega_1) = \int_{\Omega_1} f_n(\omega_1, \omega_2)\mu_2(d\omega_2)$. Then, g_{1n} is $\mathcal{F}_1$-measurable for all $n \geq 1$, g_{1n}'s are nondecreasing, and by the MCT,

$$\begin{aligned} g_1(\omega_1) &\equiv \int_{\Omega_1} f(\omega_1, \omega_2)\mu_2(d\omega_2) \\ &= \lim_{n\to\infty} \int f_n(\omega_1, \omega_2)\mu_2(d\omega_2) \\ &= \lim_{n\to\infty} g_{1n}(\omega_1) \end{aligned} \tag{2.6}$$

for all $\omega_1 \in \Omega_1$. Thus, by Proposition 2.1.5, g_1 is $\langle \mathcal{F}_1, \mathcal{B}(\bar{\mathbb{R}}) \rangle$-measurable. Since (2.5) holds for simple functions, $\int f_n d\mu = \int g_{1n} d\mu_1$ for all $n \geq 1$. Hence, by repeated applications of the MCT, it follows that

$$\begin{aligned} \int f d\mu &= \lim_{n\to\infty} \int f_n d\mu = \lim_{n\to\infty} \int g_{1n} d\mu_1 \\ &= \int (\lim_{n\to\infty} g_{1n}) d\mu_1 = \int g_1 d\mu_1. \end{aligned}$$

The proofs of (2.4) and the second equality in (2.5) are similar. □

Theorem 5.2.2: *(Fubini's theorem). Let* $(\Omega_i, \mathcal{F}_i, \mu_i)$, $i = 1, 2$ *be* σ*-finite measure spaces and let* $f \in L^1(\Omega_1 \times \Omega_2, \mathcal{F}_1 \times \mathcal{F}_2, \mu_1 \times \mu_2)$. *Then there exist sets* $B_i \in \mathcal{F}_i$, $i = 1, 2$ *such that*

(i) $\mu_i(\Omega_i \setminus B_i) = 0$ *for* $i = 1, 2$,

(ii) for $\omega_1 \in B_1$, $f(\omega_1, \cdot) \in L^1(\Omega_2, \mathcal{F}_2, \mu_2)$,

(iii) the function

$$g_1(\omega_1) \equiv \begin{cases} \int_{\Omega_2} f(\omega_1, \omega_2)\mu_2(d\omega_2) & \text{for } \omega_1 \text{ in } B_1 \\ 0 & \text{for } \omega_1 \text{ in } B_1^c \end{cases}$$

is $\mathcal{F}_1$*-measurable and*

$$\int_{\Omega_1} g_1 d\mu_1 = \int_{\Omega_1 \times \Omega_2} f d(\mu_1 \times \mu_2), \tag{2.7}$$

(iv) for $\omega_2 \in B_2$, $f(\cdot, \omega_2) \in L^1(\Omega_1, \mathcal{F}_1, \mu_1)$,

(v) the function

$$g_2(\omega_2) \equiv \begin{cases} \int_{\Omega_1} f(\omega_1, \omega_2)\mu_1(d\omega_1) & \text{for} \quad \omega_2 \text{ in } B_2 \\ 0 & \text{for} \quad \omega_2 \text{ in } B_2^c \end{cases}$$

is $\mathcal{F}_2$-measurable and

$$\int_{\Omega_2} g_2 d\mu_2 = \int_{\Omega_1 \times \Omega_2} f d(\mu_1 \times \mu_2). \tag{2.8}$$

Remark 5.2.1: An informal statement of the above theorem is as follows. If f is integrable on the product space then the sectional integrals $\int_{\Omega_2} f(\omega_1, \cdot) d\mu_2$ and $\int_{\Omega_1} f(\cdot, \omega_2) d\mu_1$ are well defined a.e., and their integrals w.r.t. μ_1 and μ_2, respectively, are equal to the integral of f w.r.t. the product measure $\mu_1 \times \mu_2$.

Proof: By Tonelli's theorem

$$\int_{\Omega_1 \times \Omega_2} |f| d(\mu_1 \times \mu_2) = \int_{\Omega_1} \left(\int_{\Omega_2} |f(\omega_1, \omega_2)| \mu_2(d\omega_2) \right) \mu_1(d\omega_1).$$

So $\int_{\Omega_1 \times \Omega_2} |f| d(\mu_1 \times \mu_2) < \infty$ implies that $\mu_1(B_1^c) = 0$ where $B_1 = \{\omega_1 : \int |f(\omega_1, \cdot)| d\mu_2 < \infty\}$. Also, by Tonelli's theorem

$$g_{11}(\omega_1) \equiv \int_{\Omega_2} f^+(\omega_1, \cdot) d\mu_2 \quad \text{and} \quad g_{12}(\omega_1) \equiv \int_{\Omega_2} f^-(\omega_1, \cdot) d\mu_2$$

are both $\mathcal{F}_1$-measurable and

$$\int_{\Omega_1} g_{11} d\mu_1 = \int_{\Omega_1 \times \Omega_2} f^+ d(\mu_1 \times \mu_2), \quad \int_{\Omega_1} g_{12} d\mu_1 = \int_{\Omega_1 \times \Omega_2} f^- d(\mu_1 \times \mu_2). \tag{2.9}$$

Since g_1 defined in (iii) can be written as $g_1 = (g_{11} - g_{12}) I_{B_1}$, g_1 is $\mathcal{F}_1$-measurable. Also,

$$\begin{aligned} \int_{\Omega_1} |g_1| d\mu_1 &\leq \int_{\Omega_1} g_{11} d\mu_1 + \int_{\Omega_1} g_{12} d\mu_1 \\ &= \int_{\Omega_1 \times \Omega_2} f^+ d(\mu_1 \times \mu_2) + \int_{\Omega_1 \times \Omega_2} f^- d(\mu_1 \times \mu_2) \\ &< \infty. \end{aligned}$$

Further, as $\int_{\Omega_1 \times \Omega_2} |f| d\mu_1 \times \mu_2 < \infty$, by (2.9), g_{11} and $g_{12} \in L^1(\Omega_1, \mathcal{F}_1, \mu_1)$. Noting that $\mu_1(B_1^c) = 0$, one gets

$$\begin{aligned} \int_{\Omega_1} g_1 d\mu_1 &= \int_{\Omega_1} (g_{11} - g_{12}) I_{B_1} d\mu_1 \\ &= \int_{\Omega_1} g_{11} I_{B_1} d\mu_1 - \int_{\Omega_1} g_{12} I_{B_1} d\mu_1 \\ &= \int_{\Omega_1} g_{11} d\mu_1 - \int_{\Omega_1} g_{12} d\mu_1 \end{aligned}$$

which, by (2.9), equals $\int_{\Omega_1\times\Omega_2} f^+ d(\mu_1 \times \mu_2) - \int_{\Omega_1\times\Omega_2} f^- d(\mu_1 \times \mu_2) = \int_{\Omega_1\times\Omega_2} f d(\mu_1 \times \mu_2)$. Thus, (ii) and (iii) of the theorem have been established as well as (i) for $i = 1$. The proofs of (iv) and (v) and that of (i) for $i = 2$ are similar. □

An application of the Fubini-Tonelli theorems gives an *integration by parts* formula. Let F_1 and F_2 be two nondecreasing right continuous functions on an interval $[a, b]$. Let μ_i be the Lebesgue-Stieltjes measure on $\mathcal{B}([a, b])$ corresponding to F_i, $i = 1, 2$. The 'integration by parts' formula allows one to write $\int_{(a,b]} F_1(x)dF_2(x) \equiv \int_{(a,b]} F_1 d\mu_2$ in terms of $\int_{(a,b]} F_2(x)dF_1(x) \equiv \int_{(a,b]} F_2 d\mu_1$.

Theorem 5.2.3: *Let F_1, F_2 be two nondecreasing right continuous functions on $[a, b]$ with no common points of discontinuity in $(a, b]$. Then*

$$\int_{(a,b]} F_1(x)dF_2(x) = F_1(b)F_2(b) - F_1(a)F_2(a) - \int_{(a,b]} F_2(x)dF_1(x). \quad (2.10)$$

Proof: Note that $\big((a, b], \mathcal{B}(a, b], \mu_i\big)$, $i = 1, 2$ are finite measure spaces. Consider the product space $\big((a, b] \times (a, b], \mathcal{B}((a, b] \times (a, b]), \mu_1 \times \mu_2\big)$. Define the sets

$$\begin{aligned} A &= \{(x, y) : a < x \leq y \leq b\} \\ B &= \{(x, y) : a < y \leq x \leq b\} \end{aligned}$$

and

$$C = \{(x, y) : a < x = y \leq b\}.$$

For notational simplicity, write A_x and A_y for the x-section and the y-section of A, respectively, and similarly, for the sets B and C. Since F_1 and F_2 have no common points of discontinuity, by Theorem 5.1.2

$$\begin{aligned} \mu_1 \times \mu_2(C) &= \int_{(a,b]} \mu_2(C_x)\mu_1(dx) \\ &= \int_{(a,b]} \mu_2(\{x\})\mu_1(dx) \\ &= 0. \end{aligned} \quad (2.11)$$

(see Problem 5.3). And by Theorem 5.1.2,

$$\begin{aligned} \mu_1 \times \mu_2(A) &= \int_{(a,b]} \mu_1(A_y)\mu_2(dy) \\ &= \int_{(a,b]} \mu_1((a, y])\mu_2(dy) \\ &= \int_{(a,b]} [F_1(y) - F_1(a)]dF_2(y) \end{aligned} \quad (2.12)$$

and similarly,

$$\begin{aligned}\mu_1 \times \mu_2(B) &= \int_{(a,b]} \mu_2(B_x)\mu_1(dx) \\ &= \int_{(a,b]} [F_2(x) - F_2(a)]dF_1(x). \qquad (2.13)\end{aligned}$$

Next note that $(\mu_1 \times \mu_2)((a,b] \times (a,b]) = \mu_1((a,b]) \cdot \mu_2((a,b]) = [F_1(b) - F_1(a)][F_2(b) - F_2(a)]$. Hence, by (2.11)–(2.13),

$$\begin{aligned}& [F_1(b) - F_1(a)][F_2(b) - F_2(a)] \\ = & (\mu_1 \times \mu_2)((a,b] \times (a,b]) \\ = & \mu_1 \times \mu_2(A) + \mu_1 \times \mu_2(B) - \mu_1 \times \mu_2(C) \\ = & \int_{(a,b]} [F_1(y) - F_1(a)]dF_2(y) + \int_{(a,b]} [F_2(x) - F_2(a)]dF_1(x),\end{aligned}$$

which yields (2.10), thereby completing the proof of Theorem 5.2.3. □

If F_1 and F_2 are absolutely continuous with nonnegative densities f_1 and f_2 w.r.t. the Lebesgue measure on $(a,b]$, then (2.10) yields

$$\int_a^b F_1(x)f_2(x)dx = F_1(b)F_2(b) - F_1(a)F_2(a) - \int_a^b F_2(x)f_1(x)dx. \quad (2.14)$$

If f_1 and f_2 are any two Lebesgue integrable function on $(a,b]$ that are not necessarily nonnegative, one can decompose f_i as $f_i = f_i^+ - f_i^-$ and apply (2.14) to f_i^+'s and f_i^-'s separately. Then, by linearity, it follows that the relation (2.14) also holds for the given f_1 and f_2. Thus, the standard 'integration by parts' formula is a special case of Theorem 5.2.3. Relations (2.10) and (2.14) can be extended to unbounded intervals under suitable conditions on F_1 and F_2 (Problem 5.5).

Remark 5.2.2: The measure space $(\Omega_1 \times \Omega_2, \mathcal{F}_1 \times \mathcal{F}_2, \mu_1 \times \mu_2)$ constructed using the integrals in (1.4) and (1.5) is not necessarily complete. As mentioned at the beginning of this section, the approach based on the extension theorem (applied to the set function defined in (1.1) on the algebra $\mathcal{A}$ of finite disjoint unions of measurable rectangles) does yield a complete measure space $(\Omega_1 \times \Omega_2, \mathcal{M}, \lambda)$ such that $\mathcal{F}_1 \times \mathcal{F}_2 \subset \mathcal{M}$ and $\lambda(A) = (\mu_1 \times \mu_2)(A)$ for all A in $\mathcal{F}_1 \times \mathcal{F}_2$. To compute the integral $\int f d\lambda$ of an $\mathcal{M}$-measurable function f w.r.t. λ, formula (2.5) in Tonelli's theorem and formulas (2.7) and (2.8) in Fubini's theorem continue to be valid with some modifications. In the following, a modified statement of Fubini's theorem is presented. Similar modification also holds for Tonelli's theorem (cf. Royden (1988), Chapter 12, Section 4).

Theorem 5.2.4: *Let $(\Omega_i, \mathcal{F}_i, \mu_i)$, $i = 1,2$ be σ-finite measure spaces and let $f \in L^1(\Omega_1 \times \Omega_2, \mathcal{M}, \mu_1 \times \mu_2)$. Let $(\Omega_i, \bar{\mathcal{F}}_i, \bar{\mu}_i)$ be the completion of $(\Omega_i, \mathcal{F}_i, \mu_i)$, $i = 1,2$. Then there exist sets B_i in $\bar{\mathcal{F}}_i$, $i = 1,2$ such that*

(i) $\bar{\mu}_i(B_i^c) = 0$, $i = 1, 2$,

(ii) for $\omega_1 \in B_1$, $f(\omega_1, \cdot)$ *is* $\bar{\mathcal{F}}_2$*-measurable,*

(iii) $\int_{\Omega_2} |f(\omega_1, \cdot)| d\bar{\mu}_2 < \infty$ *for all* $\omega_1 \in B_1$,

(iv) $g_1(\omega_1) \equiv \begin{cases} \int_{\Omega_2} f(\omega_1, \cdot) d\bar{\mu}_2 & \textit{for} \quad \omega_1 \textit{ in } B, \\ 0 & \textit{for} \quad \omega_1 \textit{ in } B_1^c \end{cases}$
is $\bar{\mathcal{F}}_1$*-measurable,*

(v) $\int_{\Omega_1} g_1 d\bar{\mu}_1 = \int_{\Omega_1 \times \Omega_2} f d\lambda$.

Further, a similar statement holds for $i = 2$.

5.3 Extensions to products of higher orders

Definition 5.3.1: Let $(\Omega_i, \mathcal{F}_i)$, $i = 1, \ldots, k$ be measurable spaces, $2 \leq k < \infty$. Then

(a) $\times_{i=1}^k \Omega_i = \Omega_1 \times \ldots \times \Omega_k = \{(\omega_1, \ldots, \omega_i) : \omega :\in \Omega_i \text{ for } i = 1, \ldots, k\}$ is called the *product set* of $\Omega_1, \ldots, \Omega_k$.

(b) A set of the form $A_1 \times \ldots \times A_k$ with $A_i \in \mathcal{F}_i, 1 \leq i \leq k$ is called a *measurable rectangle.* The *product* σ*-algebra* on $\times_{i=1}^k \Omega_i$, denoted by $\times_{i=1}^k \mathcal{F}_i$, is the σ-algebra generated by the collection of all measurable rectangles, i.e.,

$$\times_{i=1}^k \mathcal{F}_i = \sigma\langle\{A_1 \times \ldots \times A_k : A_i \in \mathcal{F}_i, 1 \leq i \leq k\}\rangle.$$

(c) $(\times_{i=1}^k \Omega_i, \times_{i=1}^k \mathcal{F}_i)$ is called the *product space* or the *product measurable space.* When $(\Omega_i, \mathcal{F}_i) = (\Omega, \mathcal{F})$ for all $1 \leq i \leq k$, the product space will be denoted by $(\Omega^k, \mathcal{F}^k)$.

To define the product measure, one starts with a natural set function on $\mathcal{C}$, the class of all measurable rectangles, extends it to the algebra $\mathcal{A}$ of finite disjoint unions of measurable rectangles by linearity, and then uses the extension theorem (Theorem 1.3.3) to obtain a measure on the product space. The details of this construction are now described below.

Define a set function μ on $\mathcal{C}$ by

$$\mu(A) = \prod_{i=1}^k \mu(A_i) \tag{3.1}$$

where $A = A_1 \times A_2 \times \quad \times A_k$ with $A_i \in \mathcal{F}_i$, $i = 1, 2, \ldots, k$. Next extend it to the algebra $\mathcal{A}$ by linearity. If $B \in \mathcal{A}$ is of the form $B = \cup_{j=1}^m B_j$ where

$\{B_j : 1, \ldots, m\} \subset \mathcal{C}$ are disjoint, then define $\mu(B)$ by

$$\mu(B) = \sum_{j=1}^{m} \mu(B_j). \tag{3.2}$$

If the set $B \in \mathcal{A}$ admits two different representations as finite unions of measurable rectangles then it is not difficult to verify that the value of $\mu(B)$ remains unchanged. Next it is shown that μ is a measure on $\mathcal{A}$.

Proposition 5.3.1:. *Let μ be as defined by (3.1) and (3.2) above on the algebra $\mathcal{A}$. Then, it is countably additive on $\mathcal{A}$.*

Proof: Let $\{B_n\}_{n=1}^{\infty} \subset \mathcal{A}$ be disjoint such that $B = \cup_{n\geq 1} B_n$ also belongs to $\mathcal{A}$. It is enough to show that

$$\mu(B) = \sum_{n=1}^{\infty} \mu(B_n). \tag{3.3}$$

Let B and $\{B_n\}_{n=1}^{\infty}$ admit the representations $B = \cup_{i=1}^{\ell} A_i$ and $B_j = \cup_{r=1}^{\ell_j} A_{jr}$ where ℓ and ℓ_j are (finite) integers, A_i, $A_{jr} \in \mathcal{C}$ and each of the collections $\{A_i\}_{i=1}^{\ell}$ and $\{A_{jr}\}_{r=1}^{\ell_j}$, $j \geq 1$ is disjoint. Then each A_i can be written as

$$A_i = A_i \cap \Big[\bigcup_{n\geq 1} B_n \Big] = \bigcup_{n\geq 1} \bigcup_{r=1}^{\ell_n} (A_i \cap A_{nr}).$$

Suppose it is shown that for each $i \geq 1$,

$$\mu(A_i) = \sum_{j=1}^{\infty} \sum_{r=1}^{\ell_j} \mu(A_i \cap A_{jr}). \tag{3.4}$$

Then, by (3.2),

$$\begin{aligned}
\mu(B) &= \sum_{i=1}^{\ell} \mu(A_i) \\
&= \sum_{i=1}^{\ell} \sum_{j=1}^{\infty} \sum_{r=1}^{\ell_j} \mu(A_i \cap A_{jr}) \\
&= \sum_{j=1}^{\infty} \sum_{r=1}^{\ell_j} \sum_{i=1}^{\ell} \mu(A_i \cap A_{jr}) \\
&= \sum_{j=1}^{\infty} \sum_{r=1}^{\ell_j} \mu(B \cap A_{jr}),
\end{aligned}$$

where the last equality follows from the representation of $B \cap A_{jr}$ as $\cup_{i=1}^{\ell}(A_i \cap A_{jr})$ and the fact that $A_i \cap A_{jr} \in \mathcal{C}$ for all i, j, r. Since $A_{jr} \subset B_j \subset B$, the above yields

$$\sum_{j=1}^{\infty}\sum_{r=1}^{\ell_j} \mu(A_{jr}) = \sum_{j=1}^{\infty} \mu(B_j) \qquad \text{(by (3.2))}$$

which establishes (3.3).

Thus, it remains to show (3.4). This is implied by the following: *Let $C = \cup_{n\geq 1} C_n$ where $\{C_n\}_{n=1}^{\infty}$ is a collection of disjoint measurable rectangles and C is also a measurable rectangle. Then*

$$\mu(C) = \sum_{i=1}^{\infty} \mu(C_i). \tag{3.5}$$

Let $C = A_1 \times A_2 \times \quad \times A_k$ and $C_i = A_{i1} \times A_{i2} \times \quad \times A_{ik}$, $i = 1, 2, \ldots$. Since $C = \cup_{n\geq 1} C_n$ and $\{C_n\}_{n=1}^{\infty}$ are disjoint, this implies $I_C(\omega_1, \ldots, \omega_k) = \sum_{i=1}^{\infty} I_{C_i}(\omega_1, \ldots, \omega_k)$, for all $(\omega_1, \ldots, \omega_k) \in \Omega_1 \times \ldots \times \Omega_k$. That is,

$$\prod_{j=1}^{k} I_{A_j}(\omega_j) = \sum_{i=1}^{\infty}\prod_{j=1}^{k} I_{A_{ij}}(\omega_j) \tag{3.6}$$

for all $(\omega_1, \ldots, \omega_k) \in \Omega_1 \times \ldots \times \Omega_k$. Integration of both sides of (3.6) w.r.t. μ_1 over Ω_1 yields

$$\mu_1(A_1)\Big(\prod_{j=2}^{k} I_{A_j}(\omega_j)\Big) = \sum_{i=1}^{\infty} \mu_1(A_{i1}) \prod_{j=2}^{k} I_{A_{ij}}(\omega_j),$$

and by iteration

$$\prod_{j=1}^{k} \mu_j(A_j) = \sum_{i=1}^{\infty}\prod_{j=1}^{k} \mu_j(A_{ij}).$$

Hence, (3.5) follows. □

By the extension theorem (Theorem 1.3.3), there exists a σ-algebra $\mathcal{M}$ and a measure $\bar{\mu}$ such that

(i) $(\times_{i=1}^{k}\Omega_i, \mathcal{M}, \bar{\mu})$ is complete,

(ii) $\times_{i=1}^{k}\mathcal{F}_i \subset \mathcal{M}$, and

(iii) $\bar{\mu}(A) = \mu(A)$ for all $A \in \mathcal{A}$.

Thus, the above procedure yields an extension $\bar{\mu}$ of μ on $\mathcal{A}$ and this extension is unique when all the μ_i's are σ-finite. Further, under the hypothesis that $\mu_1, \ldots, \mu_k$ are σ-finite, the analogs of formulas (1.4) and (1.5) for computing the product measure of a set via the iterated integrals are valid. More generally, the Tonelli-Fubini theorems extend to the k-fold product spaces in an obvious manner. For example, if $(\Omega_i, \mathcal{F}_i, \mu_i)$ are σ-finite measure spaces for $i = 1, \ldots, k$ and f is a nonnegative measurable function on $(\times_{i=1}^k \Omega_i, \times_{i=1}^k \mathcal{F}_i, \bar{\mu})$, then

$$\int f d\bar{\mu} = \int_{\Omega_{i_1}} \int_{\Omega_{i_2}} \cdots \int_{\Omega_{i_k}} f(\omega_1, \omega_2, \quad \omega_k)\mu_{i_1}(d\omega_{i_1}) \ldots \mu_{i_k}(d\omega_{i_k})$$

for any permutation $(i_1, i_2, \ldots, i_k)$ of $(1, 2, \ldots, k)$.

Definition 5.3.2: The measure space $(\times_{i=1}^k \Omega_i, \mathcal{M}, \bar{\mu})$ is called the *complete product measure space* and $\bar{\mu}$ the *complete product measure.*

Remark 5.3.1: If $(\Omega_i, \mathcal{F}_i, \mu_i) \equiv (\mathbb{R}, \mathcal{L}, m)$ where $\mathcal{L}$ is the Lebesgue σ-algebra and m is the Lebesgue measure, then the above extension coincides with the $(\mathbb{R}^k, \mathcal{L}^k, m^k)$, where $\mathcal{L}^k$ is the Lebesgue σ-algebra in $\mathbb{R}^k$ and m^k is the k dimensional Lebesgue measure.

5.4 Convolutions

5.4.1 Convolution of measures on $(\mathbb{R}, \mathcal{B}(\mathbb{R}))$

In this section convolution of measures on $\big(\mathbb{R}, \mathcal{B}(\mathbb{R})\big)$ is discussed. From this, one can easily obtain convolution of sequences, convolution of functions in $L^1(\mathbb{R})$ and convolution of functions with measures.

Proposition 5.4.1: *Let μ and λ be two σ-finite measures on $\big(\mathbb{R}, \mathcal{B}(\mathbb{R})\big)$. For any Borel set A in $\mathcal{B}(\mathbb{R})$, let*

$$(\mu * \lambda)(A) \equiv \int \int I_A(x + y)\mu(dx)\lambda(dy). \tag{4.1}$$

Then $(\mu * \lambda)(\cdot)$ is a measure on $\big(\mathbb{R}, \mathcal{B}(\mathbb{R})\big)$.

Proof: Let $h(x, y) \equiv x + y$ for $(x, y) \in \mathbb{R} \times \mathbb{R}$. Then $h : \mathbb{R} \times \mathbb{R} \to \mathbb{R}$ is continuous and hence is $\langle \mathcal{B}(\mathbb{R}) \times \mathcal{B}(\mathbb{R}), \mathcal{B}(\mathbb{R})\rangle$-measurable. Consider the measure $(\mu \times \lambda)h^{-1}(\cdot)$ on $\langle \mathbb{R}, \mathcal{B}(\mathbb{R})\rangle$ induced by the map h and the measure $\mu \times \lambda$ on $\langle \mathbb{R} \times \mathbb{R}, \mathcal{B}(\mathbb{R}) \times \mathcal{B}(\mathbb{R})\rangle$. Clearly

$$(\mu * \lambda)(\cdot) = (\mu \times \lambda)h^{-1}(\cdot)$$

and hence the proposition is proved. □

Definition 5.4.1: For any two σ-finite measures μ and λ on $\big(\mathbb{R}, \mathcal{B}(\mathbb{R})\big)$, the measure $(\mu * \lambda)(A)$ defined in (4.1) is called the *convolution of μ and λ.*

The following proposition is easy to verify.

Proposition 5.4.2: *Let μ_1, μ_2, μ_3 be σ-finite measures on $\big(\mathbb{R}, \mathcal{B}(\mathbb{R})\big)$. Then*

(i) (Commutativity) $\mu_1 * \mu_2 = \mu_2 * \mu_1$,

(ii) (Associativity) $\mu_1 * (\mu_2 * \mu_3) = (\mu_1 * \mu_2) * \mu_3$,

(iii) (Distributive) $\mu_1 * (\mu_2 + \mu_3) = \mu_1 * \mu_2 + \mu_1 * \mu_3$,

(iv) (Identity element) $\mu_1 * \delta_{\{0\}} = \mu_1$

where $\delta_{\{0\}}(\cdot)$ is the delta measure at 0, i.e.,

$$\delta_{\{0\}}(A) = \begin{cases} 1 & \text{if } \ 0 \in A, \\ 0 & \text{if } \ 0 \notin A \end{cases}$$

for all $A \in \mathcal{B}(\mathbb{R})$.

Remark 5.4.1: (*Extension to $\mathbb{R}^k$*). Definition 5.4.1 extends to measures on $\big(\mathbb{R}^k, \mathcal{B}(\mathbb{R}^k)\big)$ for any integer $k \geq 1$ as well as to any space that is a commutative group under addition such as the sequence space $\mathbb{R}^\infty$, the function spaces $\mathcal{C}[0,1]$ and $\mathcal{C}[0,\infty)$. These are relevant in the study of stochastic processes.

Remark 5.4.2: (*Sums of independent random variables*). It will be seen in Chapter 7 that if X and Y are two independent random variables on a probability space $(\Omega, \mathcal{F}, P)$, then $P_X * P_Y = P_{X+Y}$ where for any random variable Z on $(\Omega, \mathcal{F}, P)$, P_Z is the distribution of Z, i.e., the probability measure on $\big(\mathbb{R}, \mathcal{B}(\mathbb{R})\big)$ induced by P and Z, i.e., $P_Z(\cdot) = PZ^{-1}(\cdot)$ on $\mathcal{B}(\mathbb{R})$.

Remark 5.4.3: (*Extension to signed measures*). Let μ and λ be two finite signed measures on $\big(\mathbb{R}, \mathcal{B}(\mathbb{R})\big)$ as defined in Section 4.2. Let $\mu = \mu^+ - \mu^-$, $\lambda = \lambda^+ - \lambda^-$ be the Jordan decomposition of μ and λ, respectively. Then the *product measure* $\mu \times \lambda$ can be defined as the signed measure

$$\begin{aligned} \mu \times \lambda &\equiv \big[(\mu^+ \times \lambda^+) + (\mu^- \times \lambda^-)\big] - \big[(\mu^+ \times \lambda^-) + (\mu^- + \lambda^+)\big] \\ &\equiv \gamma_1 - \gamma_2, \quad \text{say.} \end{aligned} \tag{4.2}$$

This is well defined since the measures $(\mu^+ \times \lambda^+) + (\mu^- \times \lambda^-)$ and $(\mu^+ \times \lambda^-) + (\mu^- \times \lambda^+)$ are both finite measures. Now the definition of convolution of measures in (4.1) carries over to signed measures using the definition of integration w.r.t. signed measures discussed in Section 4.2.

Definition 5.4.2: Let μ and ν be (finite) signed measures on a measurable space $\big(\mathbb{R}, \mathcal{B}(\mathbb{R})\big)$. The *convolution of μ and ν*, denoted by $\mu * \nu$ is the signed measure defined by

$$(\mu * \nu)(A) \quad = \quad \int\int I_A(x+y)\, d(\mu \times \nu)$$

$$= \int\int I_A(x+y)d\gamma_1 - \int\int I_A(x+y)d\gamma_2 \qquad (4.3)$$

where γ_1 and γ_2 are as in (4.2).

5.4.2 *Convolution of sequences*

Definition 5.4.3: Let $\tilde{a} \equiv \{a_n\}_{n\geq 0}$ and $\tilde{b} \equiv \{b_n\}_{n\geq 0}$ be two sequences of real numbers. Then the *convolution of* $\tilde{a}$ *and* $\tilde{b}$ denoted by $\tilde{a} * \tilde{b}$ is the sequence

$$(\tilde{a} * \tilde{b})(n) = \sum_{j=0}^{n} a_j b_{n-j}, \quad n \geq 0. \qquad (4.4)$$

If $\sum_{j=0}^{\infty} |a_j| < \infty$ and $\sum_{j=0}^{\infty} |b_j| < \infty$, then $\tilde{a} * \tilde{b}$ corresponds to the convolution of the signed measures $\tilde{a}$ and $\tilde{b}$ on $\mathbb{Z}_+$, defined by $\tilde{a}(i) = a_i$, $\tilde{b}(i) = b_i$, $i \in \mathbb{Z}_+$.

Example 5.4.1:

(a) Let $b_n \equiv 1$ for $n \geq 0$. Then for any $\{a_n\}_{n\geq 0}$, $(\tilde{a} * \tilde{b})(n) = \sum_{j=0}^{n} a_j$.

(b) Fix $0 < p < 1$, k_1, k_2 positive integers. For $n \in \mathbb{Z}_+$, let

$$\begin{aligned} a_n &\equiv \binom{k_1}{n} p^n (1-p)^{k_1-n} I_{[0,k_1]}(n) \\ b_n &\equiv \binom{k_2}{n} p^n (1-p)^{k_2-n} I_{[0,k_2]}(n). \end{aligned}$$

Then $(\tilde{a} * \tilde{b})(n) = \binom{k_1+k_2}{n} p^n (1-p)^{k_2-n} I_{[0,k_1+k_2]}(n)$.

(c) Fix $0 < \lambda_1, \lambda_2 < \infty$. Let

$$\begin{aligned} a_n &\equiv e^{-\lambda_1} \frac{\lambda_1^n}{n!}, \quad n = 0, 1, 2, \ldots \\ b_n &\equiv e^{-\lambda_2} \frac{\lambda_2^n}{n!}, \quad n = 0, 1, 2, \ldots . \end{aligned}$$

Then $(\tilde{a} * \tilde{b})(n) = e^{-(\lambda_1+\lambda_2)} \frac{(\lambda_1+\lambda_2)^n}{n!}$, $n = 0, 1, 2, \ldots$.

The verification of these claims is left as an exercise (Problem 5.20).

A useful technique to determine $\tilde{a} * \tilde{b}$ is the use of generating functions. This will be discussed in Section 5.5.

5.4.3 *Convolution of functions in* $L^1(\mathbb{R})$

Let $L^1(\mathbb{R}) \equiv L^1(\mathbb{R}, \mathcal{B}(\mathbb{R}), m)$ where $m(\cdot)$ is the Lebesgue measure. In the following, for $f \in L^1(\mathbb{R})$, $\int f dm$ will also be written as $\int f(x)dx$.

Proposition 5.4.3: *Let $f,\ g \in L^1(\mathbb{R})$. Then for almost all x (w.r.t. m),*

$$\int |f(x-u)|\,|g(u)|du < \infty. \tag{4.5}$$

Proof: Let $k(x,u) = f(x-u)g(u)$. Since $\pi(x,u) \equiv x-u$ is a continuous map from $\mathbb{R}\times\mathbb{R} \to \mathbb{R}$, it is $\langle \mathcal{B}(\mathbb{R}) \times \mathcal{B}(\mathbb{R}), \mathcal{B}(\mathbb{R})\rangle$-measurable. Also since f and g are Borel measurable it follows that $k: \mathbb{R}\times\mathbb{R} \to \mathbb{R}$ is $\langle \mathcal{B}(\mathbb{R}) \times \mathcal{B}(\mathbb{R}), \mathcal{B}(\mathbb{R})\rangle$-measurable. Also note that by the translation invariance of m, $\int |f(x-u)|dx = \int |f(x)|dx$ for any Borel measurable $f: \mathbb{R} \to \mathbb{R}$. Hence, by Tonelli's theorem,

$$\begin{aligned} \int\int |k(x,u)|dxdu &= \int\Big(\int |k(x,u)|dx\Big)du \\ &= \int |g(u)|\Big(\int |f(x)|dx\Big)du \\ &= \|g\|_1\,\|f\|_1 < \infty. \end{aligned} \tag{4.6}$$

Also by Tonelli's theorem

$$\int\int |k(x,u)|dxdu = \int\Big(\int |k(x,u)|du\Big)dx.$$

By (4.6), this yields

$$\int |k(x,u)|du < \infty \quad \text{a.e.} \quad (m)$$

which is the same as (4.5). □

Definition 5.4.4: Let $f,\ g \in L^1(\mathbb{R})$. Then the *convolution of f and g*, denoted by $f*g$ is the function defined a.e. (m) by

$$(f*g)(x) \equiv \int f(x-u)g(u)du. \tag{4.7}$$

Note that by (4.5) this is well defined.

Proposition 5.4.4: *Let $f,\ g \in L^1(\mathbb{R})$. Then*

(i) $f*g = g*f$

(ii) $f*g \in L^1(\mathbb{R})$ *and* $\|f*g\|_1 \le \|f\|_1\,\|g\|_1$

(iii) $\int f*g\,dm = \Big(\int f dm\Big)\Big(\int g dm\Big)$.

Proof: For (i), use the change of variable $u \to x-u$ and the translation invariance of m. For (ii) use (4.6). For (iii), use Fubini's theorem. □

It is easy to see that if μ and λ denote the signed measures with Radon-Nikodym derivatives f and g, respectively, w.r.t. m, then $\mu * \lambda$ is the signed measure with Radon-Nikodym derivative $f * g$ w.r.t. m.

Example 5.4.2: Here are two examples from probability theory.

(a) (*Convolutions of Uniform* $[0,1]$ *densities*). Let $f = g = I_{[0,1]}$. Then

$$(f * g)(x) = \begin{cases} x & 0 \le x \le 1 \\ 2 - x & 1 \le x \le 2. \end{cases}$$

(b) (*Convolutions of* $N(0,1)$ *densities*). Let $f(x) \equiv g(x) \equiv \frac{1}{\sqrt{2\pi}} e^{-\frac{x^2}{2}}$, $-\infty < x < \infty$. Then $(f * g)(x) = \frac{1}{\sqrt{2\pi}} \frac{1}{\sqrt{2}} e^{-\frac{x^2}{4}}$, $-\infty < x < \infty$.

5.4.4 Convolution of functions and measures

Let $f \in L^1(\mathbb{R})$ and λ be a signed measure. Let $\mu(A) \equiv \int_A f dm$, $A \in \mathcal{B}(\mathbb{R})$. Then it can be shown that $\mu * \lambda$ is dominated by m and its Radon-Nikodym derivative is given by

$$f * \lambda(x) \equiv \int f(x - u)\lambda(du), \quad x \in \mathbb{R}. \tag{4.8}$$

Note that $f * \lambda$ in (4.8) is well defined for any nonnegative Borel measurable f and any measure λ on $\big(\mathbb{R}, \mathcal{B}(\mathbb{R})\big)$.

5.5 Generating functions and Laplace transforms

Definition 5.5.1: Let $\{a_n\}_{n \ge 0}$ be a sequence in $\mathbb{R}$ and let $\rho = \big(\overline{\lim_{n\to\infty}} |a_n|^{1/n}\big)^{-1}$. The power series

$$A(s) \equiv \sum_{n=0}^{\infty} a_n s^n, \tag{5.1}$$

defined for all s in $(-\rho, \rho)$, is called the *generating function of* $\{a_n\}_{n \ge 0}$.

Note that the power series in (5.1) converges absolutely for $|s| < \rho$ and ρ is called the *radius of convergence* of $A(s)$ (cf. Appendix A.2).

The usefulness of generating functions is given by the following:

Proposition 5.5.1: *Let* $\{a_n\}_{n \ge 0}$ *and* $\{b_n\}_{n \ge 0}$ *be two sequences in* $\mathbb{R}$. *Let* $\{c_n\}_{n \ge 0}$ *be the convolution of* $\{a_n\}$ *and* $\{b_n\}$. *That is,*

$$c_n = \sum_{j=0}^{n} a_j b_{n-j}, \ n \ge 0.$$

Then

$$C(s) = A(s)B(s) \tag{5.2}$$

for all s in $(-\rho, \rho)$, where $A(\cdot)$, $B(\cdot)$, and $C(\cdot)$ are, respectively, the generating function of the sequences $\{a_n\}_{n\geq 0}$, $\{b_n\}_{n\geq 0}$ and $\{c_n\}_{n\geq 0}$ and $\rho = \min(\rho_a, \rho_b)$ with $\rho_a = \big(\overline{\lim_{n\to\infty}} |a_n|^{1/n}\big)^{-1}$ and $\rho_b = \big(\overline{\lim_{n\to\infty}} |b_n|^{1/n}\big)^{-1}$.

Proof: By Tonelli's theorem applied to the counting measure on the product space $(\mathbb{Z}_+ \times \mathbb{Z}_+)$,

$$\begin{aligned}
&\sum_{n=0}^{\infty} |s|^n \Big(\sum_{j=0}^{n} |a_j|\,|b_{n-j}|\Big) \\
&= \sum_{j=0}^{\infty} |a_j|\,|s|^j \Big(\sum_{n=j}^{\infty} |s|^{n-j} |b_{n-j}|\Big) \\
&= A(|s|)B(|s|) < \infty \quad \text{if} \quad |s| < \rho.
\end{aligned}$$

Now by Fubini's theorem, (5.2) follows. □

It is easy to verify the claims in Example 5.4.1 by using the above proposition and the uniqueness of the power series coefficients (Problem 5.20).

Proposition 5.5.2: (*Renewal sequences*). *Let $\{a_n\}_{n\geq 0}$, $\{b_n\}_{n\geq 0}$, $\{p_n\}_{n\geq 0}$ be sequences in $\mathbb{R}$ such that*

$$a_n = b_n + \sum_{j=0}^{n} a_{n-j} p_j, \; n \geq 0 \tag{5.3}$$

and $p_0 = 0$. Then

$$A(s) = \frac{B(s)}{1 - P(s)}$$

for all s such that $|s| < \rho$ and $P(s) \neq 1$, where $\rho = \min(\rho_b, \rho_p)$, $\rho_b = \big(\overline{\lim_{n\to\infty}} |b_n|^{1/n}\big)^{-1}$, $\rho_p = \big(\overline{\lim_{n\to\infty}} |p_n|^{1/n}\big)^{-1}$.

For applications of this to renewal theory, see Chapter 8.

Definition 5.5.2: (*Laplace transform*). Let $f : [0, \infty) \to \mathbb{R}$ be Borel measurable. The function

$$(Lf)(s) \equiv \int_{[0,\infty)} e^{sx} f(x)dx, \tag{5.4}$$

defined for all s in $\mathbb{R}$ such that

$$\int_{[0,\infty)} e^{sx} |f(x)|dx < \infty, \tag{5.5}$$

is called the *Laplace transform of f*.

It is easily seen that if (5.5) holds for some $s = s_0$ then it does for all $s < s_0$. The analog of Proposition 5.5.1 is the following.

Proposition 5.5.3: *Let $f,\, g \in L^1(\mathbb{R})$. Then*

$$L(f * g)(s) = Lf(s)Lg(s) \tag{5.6}$$

for all s such that (5.5) holds for both f and g.

Definition 5.5.3: (*Laplace-Stieltjes transform*). Let μ be a measure on $(\mathbb{R}, \mathcal{B}(\mathbb{R}))$ such that $\mu((-\infty, 0)) = 0$. The function

$$\mu^*(s) \equiv L\mu(s) \equiv \int_{[0,\infty)} e^{sx}\mu(dx)$$

is called the *Laplace-Stieltjes transform* of μ. Clearly, $\mu^*(s)$ is well defined for all s in $\mathbb{R}$. However, $\mu^*(s_0) = \infty$ implies $\mu^*(s) = \infty$ for $s \geq s_0$.

Proposition 5.5.4: *Let μ and λ be measures on $(\mathbb{R}, \mathcal{B}(\mathbb{R}))$ such that $\mu((-\infty, 0)) = 0 = \lambda((-\infty, 0))$. Then*

$$L(\mu * \lambda)(s) = L\mu(s)L\lambda(s) \quad \textit{for all} \quad s \text{ in } \mathbb{R}.$$

For an inversion formula to obtain a probability measure μ from $L\mu(\cdot)$, see Feller (1966), Section 13.4.

5.6 Fourier series

In this section, $L^p[0, 2\pi]$ stands for $L^p\big([0, 2\pi], \mathcal{B}([0, 2\pi]), m\big)$ where $m(\cdot)$ is Lebesgue measure and $0 < p < \infty$.

Definition 5.6.1: For $f \in L^1[0, 2\pi]$, $n \geq 0$, let

$$\begin{aligned} a_n &\equiv \frac{1}{2\pi}\int_0^{2\pi} f(x)\cos nx\, dx \\ b_n &\equiv \frac{1}{2\pi}\int_0^{2\pi} f(x)\sin nx\, dx, \end{aligned} \tag{6.1}$$

$$s_n(f, x) \equiv a_0 + \sum_{j=1}^{n}(a_j \cos jx + b_j \sin jx). \tag{6.2}$$

Then $\{a_n, b_n, n = 0, 1, 2, \ldots\}$ are called the *Fourier coefficients of f* and the sequence $\{s_n(f, x) : n = 0, 1, 2, \ldots\}$ is called the *partial sum sequence* of the *Fourier series expansion of f*.

Since $\mathcal{C}[0, 2\pi] \subset L^2[0, 2\pi] \subset L^1[0, 2\pi]$, the Fourier coefficients and the partial sum sequence of the Fourier series are well defined for f in $\mathcal{C}[0, 2\pi]$

and f in $L^2[0, 2\pi]$. J. Fourier introduced these series in early 19th Century to approximate certain functions exhibiting a periodic behavior that arose in the study of physics and mechanics and in particular in the theory of heat conduction (see Körner (1989) and Bhatia (2003)).

A natural question is: *In what sense does $s_n(f, x)$ approximate $f(x)$?* It turns out that one can prove the strongest results for f in $C[0, 2\pi]$, less stronger results for f in $L^2[0, 2\pi]$ and finally, for f in $L^1[0, 2\pi]$.

In the early 19th Century it was believed that for any f in $C[0, 2\pi]$, $s_n(f, x)$ converged to $f(x)$ for all x in $[0, 2\pi]$. But in 1876, D. Raymond constructed a continuous function f for which $\overline{\lim}_{n\to\infty}|s_n(f; x_0)| = \infty$ for some x_0. However, Fejer showed in 1903 that for f in $C[0, 2\pi]$, $s_n(f, \cdot)$ does converge to f uniformly in the *Cesaro sense.*

Theorem 5.6.1: (*Fejer*). *Let $f \in C[0, 2\pi]$ and $s_n(f, \cdot)$, $n \geq 0$ be as in Definition 3.4.1. Let*

$$\mathcal{D}_n(f, \cdot) = \frac{1}{n} \sum_{j=0}^{n-1} s_j(f, \cdot), \ n \geq 1. \tag{6.3}$$

Then

$$\mathcal{D}_n(f, \cdot) \to f(\cdot) \quad \textit{uniformly on} \quad [0, 2\pi] \quad \textit{as} \quad n \to \infty. \tag{6.4}$$

For the proof of this result the following result of some independent interest is needed.

Lemma 5.6.2: (*Fejer*). *Let for $m \geq 1$*

$$K_m(x) \equiv 1 + \frac{2}{m} \sum_{j=1}^{m-1} (m - j) \cos jx, \ x \in \mathbb{R}. \tag{6.5}$$

Then

(i)

$$K_m(x) = \begin{cases} \frac{1}{m}\left(\frac{\sin \frac{mx}{2}}{\sin \frac{x}{2}}\right)^2 & \text{if} \quad x \neq 0 \\ m & \text{if} \quad x = 0 \ . \end{cases} \tag{6.6}$$

(ii) For $\delta > 0$,

$$\sup\{K_m(x) : \delta \leq |x| \leq 2\pi - \delta\} \to 0 \quad \textit{as} \quad m \to \infty. \tag{6.7}$$

(iii)

$$\frac{1}{2\pi} \int_0^{2\pi} K_m(x) dx = 1. \tag{6.8}$$

Proof: Clearly, (iii) follows from (6.5) on integration. Also, (ii) follows from (i) since for $\delta \leq x \leq 2\pi - \delta$,

$$K_m(x) \leq \frac{1}{m} \frac{1}{(\sin \frac{\delta}{2})^2} \to 0 \quad \text{as} \quad m \to \infty.$$

To establish (i) note that using Euler's formula (cf. Section A.3, Appendix) $e^{\iota x} = \cos x + \iota \sin x$, $K_m(x)$ can be written as

$$\begin{aligned} K_m(x) &= \frac{1}{m} \sum_{j=-(m-1)}^{(m-1)} (m - |j|) e^{\iota j x} \\ &= \frac{1}{m} \sum_{j=0}^{2(m-1)} (m - |j - (m-1)|) e^{\iota (j-(m-1))x} \\ &= \frac{1}{m} \left(\sum_{k=0}^{m-1} e^{\iota (k - \frac{m-1}{2})x} \right)^2. \end{aligned}$$

For $x \neq 0$,

$$\begin{aligned} K_m(x) &= \frac{1}{m} \left(e^{-\frac{\iota(m-1)x}{2}} \frac{1 - e^{\iota m x}}{1 - e^{\iota x}} \right)^2, \\ &= \frac{1}{m} \left(\frac{e^{-\frac{\iota m x}{2}} - e^{\frac{\iota m x}{2}}}{e^{-\frac{\iota x}{2}} - e^{\frac{\iota x}{2}}} \right)^2, \\ &= \frac{1}{m} \left(\frac{\sin \frac{mx}{2}}{\sin \frac{x}{2}} \right)^2. \end{aligned}$$

For $x = 0$, (6.5) yields

$$\begin{aligned} K_m(0) &= 1 + \frac{2}{m} \sum_{j=1}^{m-1} (m - j) \\ &= 1 + \frac{2}{m} \frac{(m-1)m}{2} = m. \end{aligned}$$

□

Proof of Theorem 5.6.1: Let $f \in \mathcal{C}[0, 2\pi]$. Let $\{a_n, b_n\}_{n \geq 0}$, $s_n(f, \cdot)$, $\mathcal{D}_m(f, \cdot)$ be as in (6.1), (6.2), and (6.3). Then from the definition of $a_n, b_n, n \geq 0$, it follows that

$$\begin{aligned} \mathcal{D}_m(f, \cdot) &= \frac{1}{2\pi} \int_0^{2\pi} K_m(x - u) f(u) du \\ &= \frac{1}{2\pi} \int_0^{2\pi} f(x - u) K_m(u) du \qquad (6.9) \end{aligned}$$

where $K_m(\cdot)$ is defined in (6.5) and $f(\cdot)$ is extended to all of $\mathbb{R}$ periodically with period 2π. Since $f \in \mathcal{C}[0, 2\pi]$, given $\epsilon > 0$, there exists a $\delta > 0$ such that

$$x, y \in [0, 2\pi],\ |x - y| < \delta \Rightarrow |f(x) - f(y)| < \epsilon.$$

Also from (6.7), for this $\delta > 0$, there exist an m_0 such that $m \geq m_0 \Rightarrow \sup\{K_m(x) : \delta \leq x \leq 2\pi - \delta\} < \epsilon$. Now (6.9) yields, for $x \in [0, 2\pi]$, $m \geq m_0$

$$\begin{aligned} |\mathcal{D}_m(f, x) - f(x)| &\leq \frac{1}{2\pi} \int_0^{2\pi} |f(x - u) - f(x)| K_m(u) du \\ &= \frac{1}{2\pi} \int_{(\delta \leq u \leq 2\pi - \delta)^c} |f(x - u) - f(x)| K_m(u) du \\ &\quad + \frac{1}{2\pi} \int_{(\delta \leq u \leq 2\pi - \delta)} |f(x - u) - f(x)| K_m(u) du \\ &\leq \frac{1}{2\pi}(\epsilon + 2||f||\epsilon 2\pi) \end{aligned}$$

where $||f|| = \sup\{|f(x)| : x \in [0, 2\pi]\}$. Thus, $m \geq m_0 \Rightarrow \sup\{|\mathcal{D}_m(f, x) - f(x)| : x \in [0, 2\pi]\} \leq \epsilon \frac{(1+4\pi||f||)}{2\pi}$. Since $\epsilon > 0$ is arbitrary, the proof of Theorem 5.6.1 is complete. □

An immediate consequence of Theorem 5.6.1 is the completeness of the trigonometric functions.

Theorem 5.6.3: *The collection* $\mathcal{T}_0 \equiv \{\cos nx : n = 0, 1, 2, \ldots\} \cup \{\sin nx : n = 1, 2, \ldots\}$ *is a complete orthogonal system for* $L^2[0, 2\pi]$.

Proof: By Theorem 5.6.1 for each $f \in \mathcal{C}[0, 2\pi]$ and $\epsilon > 0$, there exists a finite linear combination $\mathcal{D}_m(f, \cdot)$ of $\{\cos nx : n = 0, 1, 2, \ldots\}$ and $\{\sin nx : n = 1, 2, \ldots\}$ such that

$$\begin{aligned} &\int_0^{2\pi} |f - \mathcal{D}_m(f, \cdot)|^2 dx \\ &\quad \leq 2\pi(\sup\{|f(x) - \mathcal{D}_m(f, x)| : x \in [0, 2\pi]\})^2 < \epsilon^2. \end{aligned}$$

Also, from Theorem 2.3.14 it is known that given any $g \in L^2[0, 2\pi]$, and any $\epsilon > 0$, there is a $f \in \mathcal{C}[0, 2\pi]$ such that $\int_0^{2\pi} |f - g|^2 dx < \epsilon^2$. Thus, for any $g \in L^2[0, 2\pi]$, $\epsilon > 0$, there is a $f \in \mathcal{C}[0, 2\pi]$ and an $m \geq 1$ such that

$$||g - \mathcal{D}_m(f, \cdot)||_2 < 2\epsilon.$$

That is, the set $\mathcal{T}$ of all finite linear combinations of the functions in the class $\mathcal{T}_0$ is dense in $L^2[0, 2\pi]$. Further, it is easy to verify that $h_1, h_2 \in \mathcal{T}_0$, $h_1 \neq h_2$ implies

$$\int_0^{2\pi} h_1(x) h_2(x) dx = 0,$$

i.e., $\mathcal{T}_0$ is an orthogonal family. Since $\mathcal{T}_0$ is orthogonal and $\mathcal{T}$ is dense in $L^2[0, 2\pi]$, $\mathcal{T}_0$ is complete. □

Definition 5.6.2: A function in $\mathcal{T}$ is called a *trigonometric polynomial.*

Thus, the above theorem says that trigonometric polynomials are dense in $L^2[0, 2\pi]$. Completeness of $\mathcal{T}$ in $L^2[0, 2\pi]$ and the results of Section 3.3 lead to

Theorem 5.6.4: *Let $f \in L^2[0, 2\pi]$. Let $\{(a_n, b_n), n = 0, 1, 2, \ldots\}$ and $\{s_n(f, \cdot)\}_{n \geq 0}$ be the associated Fourier coefficient sequences and partial sum sequence of the Fourier series for f as in Definition 5.6.1. Then*

(i) $s_n(f, \cdot) \to f$ in $L^2[0, 2\pi]$,

(ii) $\sum_{n=0}^{\infty} (\breve{a}_n^2 + \breve{b}_n^2) = \frac{1}{2\pi} \int_0^{2\pi} |f|^2 dx$ where for $n \geq 0$, $\breve{a}_n = a_n / c_n$, with $c_n^2 = \frac{1}{2\pi} \int_0^{2\pi} (\cos nx)^2 dx = \frac{1}{2}$, and for $n \geq 1$, $\breve{b}_n = \frac{b_n}{d_n}$ with $d_n^2 = \frac{1}{2\pi} \int_0^{2\pi} (\sin nx)^2 dx = \frac{1}{2}$.

(iii) Further, if f, $g \in L^2[0, 2\pi]$, then $\frac{1}{2\pi} \int_0^{2\pi} fg\, dx = a_0 \alpha_0 + \sum_{n=1}^{\infty} \left(\frac{a_n \alpha_n}{c_n^2} + \frac{b_n \beta_n}{d_n^2} \right)$, where $\{(a_n, b_n) : n = 0, 1, 2, \ldots\}$ and $\{(\alpha_n, \beta_n) : n = 0, 1, 2, \ldots\}$ are, respectively, the Fourier coefficients of f and g.

Clearly (ii) above is a restatement of Bessel's equality. Assertion (iii) is known as the *Parseval identity.* As for convergence pointwise or almost everywhere, A. N. Kolmogorov showed in 1926 (see Körner (1989)) that there exists an $f \in L^1[0, 2\pi]$ such that $\overline{\lim}_{n \to \infty} |s_n(f, x)| = \infty$ everywhere on $[0, 2\pi]$. This led to the belief that for $f \notin C[0, 2\pi]$, the mean square convergence of (i) in Theorem 5.6.4 cannot be improved upon. But L. Carleson showed in 1964 (see Körner (1989)) that for f in $L^2[0, 2\pi]$, $s_n(f, \cdot) \to f(\cdot)$ almost everywhere. Finally, turning to $L^1[0, 2\pi]$, one has the following:

Theorem 5.6.5: *Let $f \in L^1[0, 2\pi]$. Let $\{(a_n, b_n) : n \geq 0\}$ be as in (6.1) and satisfy $\sum_{n=0}^{\infty} (|a_n| + |b_n|) < \infty$. Let $s_n(f, \cdot)$ be as in (6.2). Then $s_n(f, \cdot)$ converges uniformly on $[0, 2\pi]$ and the limit coincides with f almost everywhere.*

Proof: Note that $\sum_{n=0}^{\infty} (|a_n| + |b_n|) < \infty$ implies that the sequence $\{s_n(f, \cdot)\}_{n \geq 0}$ is a Cauchy sequence in the Banach space $C[0, 2\pi]$ with the sup-norm. Thus, there exists a g in $C[0, 2\pi]$ such that $s_n(f, \cdot) \to g$ uniformly on $[0, 2\pi]$. It is easy to check that this implies that g and f have the same Fourier coefficients. Set $h = g - f$. Then $h \in L^1[0, 2\pi]$ and the Fourier coefficients of h are all zero. This implies that h is orthogonal to the members of the class $\mathcal{T}$ which in turn yields that h is orthogonal to all continuous functions in $C[0, 2\pi]$, i.e.,

$$\int_0^{2\pi} h(x)k(x)dx = 0$$

for all $k \in C[0, 2\pi]$. Since $h \in L^1[0, 2\pi]$ and for any interval $A \subset [0, 2\pi]$, there exists a sequence $\{k_n\}_{n \geq 1}$ of uniformly bounded continuous functions, such that $k_n \to I_A$ a.e. (m), by the DCT,

$$\int h(x)I_A(x)dx = \lim_{n\to\infty} \int h(x)k_n(x)dx = 0.$$

This in turn implies that $h = 0$ a.e., i.e., $g = f$ a.e. □

Remark 5.6.1: If $f \in L^2[0, 2\pi]$, the Fourier coefficients $\{a_n, b_n\}$ are square summable and hence go to zero as $n \to \infty$. What if $f \in L^1[0, 2\pi]$?

If $f \in L^1[0, 2\pi]$, one can assert the following:

Theorem 5.6.6: (*Riemann-Lebesgue lemma*). *Let* $f \in L^1[0, 2\pi]$. *Then*

$$\lim_{n\to\infty} \int_0^{2\pi} f(x)\cos nx\, dx = 0 = \lim_{n\to\infty} \int_0^{2\pi} f(x)\sin nx\, dx.$$

Proof: The lemma holds if $f = I_A$ for any interval $A \subset [0, 2\pi]$ and since step functions (i.e., linear combinations of indicator functions of intervals) is dense in $L^1[0, 2\pi]$, the lemma is proved. □

It can be shown that the mapping $f \to \{(a_n, b_n)\}_{n\geq 0}$ from $L^1[0, 2\pi]$ to bivariate sequences that go to zero as $n \to \infty$ is one-to-one but not onto (Rudin (1987), Chapter 5).

Remark 5.6.2: (*The complex case*). Let $T \equiv \{z : z = e^{\iota\theta}, 0 \leq \theta \leq 2\pi\}$ be the unit circle in the complex plane $\mathbb{C}$. Every function $g : T \to \mathbb{C}$ can be identified with a function f on $\mathbb{R}$ by $f(t) = g(e^{\iota t})$. Clearly, $f(\cdot)$ is periodic on $\mathbb{R}$ with period 2π. In the rest of this section, for $0 < p < \infty$, $L^p(T)$ will stand for the collection of all Borel measurable functions $f : [0, 2\pi]$ to $\mathbb{C}$ such that $\int_{[0,2\pi]} |f|^p dm < \infty$ where $m(\cdot)$ is the Lebesgue measure. A *trigonometric polynomial* is a function of form

$$f(\cdot) \equiv \sum_{n=-k}^{k} \alpha_n e^{\iota nx} \equiv a_0 + \sum_{n=1}^{k} (a_n \cos nx + b_n \sin nx),$$

$k < \infty$, where $\{\alpha_n\}$, $\{a_n\}_{n\geq 0}$, and $\{b_n\}_{n\geq 0}$ are sequences of complex numbers.

The completeness of the trigonometric polynomials proved in Theorem 5.6.3 implies that the family $\{e^{\iota nx} : n = 0, \pm 1, \pm 2, \ldots\}$ is a complete orthonormal basis for $L^2(T)$ which is a complex Hilbert space.

Thus Theorem 5.6.4 carries over to this case.

Theorem 5.6.7:

(i) *Let $f \in L^2(T)$. Then,*

$$\sum_{n=-k}^{k} \alpha_n e^{\iota n x} \to f \quad in \quad L^2(T)$$

where

$$\alpha_n \equiv \frac{1}{2\pi} \int_0^{2\pi} f(x) e^{-\iota n s} dx, \ n \in \mathbb{Z}.$$

Further,

$$\sum_{n=-\infty}^{\infty} |\alpha_n|^2 = \frac{1}{2\pi} \int_0^{2\pi} |f|^2 dm.$$

(ii) *For any sequence $\{\alpha_n\}_{n\in\mathbb{Z}}$ of complex numbers such that $\sum_{n=-\infty}^{\infty} |\alpha_n|^2 < \infty$, the sequence $\{f_k(x) \equiv \sum_{n=-k}^{k} \alpha_n e^{\iota n x}\}_{k \geq 1}$ converges in $L^2(T)$ to a unique f such that*

$$\alpha_n = \frac{1}{2\pi} \int_0^{2\pi} f(x) e^{-\iota n x} dx.$$

(iii) *For any $f, g \in L^2(T)$,*

$$\sum_{n=-\infty}^{\infty} \alpha_n \bar{\beta}_n = \frac{1}{2\pi} \int_0^{2\pi} f(x) \overline{g(x)} dx$$

where

$$\begin{aligned} \alpha_n &= \hat{f}(n) = \frac{1}{2\pi} \int_0^{2\pi} f(x) e^{-\iota n x} dx, \\ \beta_n &= \hat{g}(n) = \frac{1}{2\pi} \int_0^{2\pi} g(x) e^{-\iota n x} dx, \ n \in \mathbb{Z}. \end{aligned}$$

Further,

$$\sum_{n=-\infty}^{\infty} |\alpha_n \beta_n| < \infty.$$

(iv) *$L^2(T)$ is isomorphic to $\ell_2(\mathbb{Z})$, the Hilbert space of all square summable sequences of complex numbers on $\mathbb{Z}$.*

Similarly, Theorem 5.6.5 carries over to the complex case.

Theorem 5.6.8: *Let $f \in L^1(T)$. Suppose*

$$\sum_{n=-\infty}^{\infty} |\hat{f}(n)| < \infty$$

where

$$\hat{f}(n) = \frac{1}{2\pi}\int_0^{2\pi} f(x)e^{-\iota nx}dx, \ n \in \mathbb{Z}.$$

Then

$$s_n(f,x) \equiv \sum_{j=-n}^{n} \hat{f}(j)e^{-\iota jx}$$

converges uniformly on $[0, 2\pi]$ and the limit coincides with f a.e. and hence f is continuous a.e.

5.7 Fourier transforms on $\mathbb{R}$

In this section and in Section 5.8, let $L^p(\mathbb{R})$ stand for

$$L^p(\mathbb{R}) \equiv \{f : f : \mathbb{R} \to \mathbb{C}, \text{ Borel measurable, } \int_{\mathbb{R}} |f|^p dm < \infty\} \qquad (7.1)$$

where $m(\cdot)$ is Lebesgue measure. Also, for $f \in L^1(\mathbb{R})$, $\int_{\mathbb{R}} f dm$ will often be written as $\int f(x)dx$. Let

$$C_0 \equiv \{f : f : \mathbb{R} \to \mathbb{C}, \text{ continuous and } \lim_{|x|\to\infty} f(x) = 0\}. \qquad (7.2)$$

Definition 5.7.1: For $f \in L^1(\mathbb{R})$, $t \in \mathbb{R}$,

$$\hat{f}(t) \equiv \int f(x)e^{-\iota tx}dx \qquad (7.3)$$

is called the *Fourier transform of f*.

Proposition 5.7.1: *Let $f \in L^1(\mathbb{R})$ and $\hat{f}(\cdot)$ be as in (7.3). Then*

(i) $\hat{f}(\cdot) \in C_0$.

(ii) If $f_a(x) \equiv f(x-a)$, $a \in \mathbb{R}$, then $\hat{f}_a(t) = e^{\iota ta}\hat{f}(t)$, $t \in \mathbb{R}$.

Proof:

(i) For any $t \in \mathbb{R}$, $t_n \to t \Rightarrow e^{\iota t_n x}f(x) \to e^{\iota tx}f(x)$ for all $x \in \mathbb{R}$ and since $|e^{\iota t_n x}f(x)| \le |f(x)|$ for all n and x, by the DCT $\hat{f}(t_n) \to \hat{f}(t)$. To show that $\hat{f}(t) \to 0$ as $|t| \to \infty$, the same proof as that of Theorem 5.6.6 works. Thus, it holds if $f = I_{[a,b]}$, for $a, b \in \mathbb{R}$, $a < b$ and since the step functions are dense in $L^1(\mathbb{R})$, it holds for all $f \in L^1(\mathbb{R})$.

(ii) This is a consequence of the translation invariance of $m(\cdot)$, i.e., $m(A+a) = m(A)$ for all $A \in \mathcal{B}(\mathbb{R})$, $a \in \mathbb{R}$. □

The continuity of $\hat{f}(\cdot)$ can be strengthened to differentiability if, in addition to $f \in L^1(\mathbb{R})$, $xf(x) \in L^1(\mathbb{R})$. More generally, if $f \in L^1(\mathbb{R})$ and $x^k f(x) \in L^1(\mathbb{R})$ for some $k \geq 1$, then $\hat{f}(\cdot)$ is differentiable k-times with all derivatives $\hat{f}^{(r)}(t) \to 0$ as $|t| \to \infty$ for $r \leq k$ (Problem 5.22).

Proposition 5.7.2: *Let $f, g \in L^1(\mathbb{R})$ and $f * g$ be their convolution as defined in (4.7). Then*

$$\widehat{f * g} = \hat{f}\hat{g}. \tag{7.4}$$

Proof:

$$\begin{aligned} \widehat{f * g}(t) &= \int_{\mathbb{R}} e^{-\iota tx} \Big(\int_{\mathbb{R}} f(x-u) g(u) du \Big) dx \\ &= \int_{\mathbb{R}} \Big(\int_{\mathbb{R}} e^{-\iota t(x-u)} f(x-u) e^{-\iota tu} g(u) du \Big) dx \\ &= \int_{\mathbb{R}} (f_t * g_t)(x) dx \end{aligned} \tag{7.5}$$

where $f_t(x) = e^{-\iota tx} f(x)$, $g_t(x) = e^{-\iota tx} g(x)$. Thus, by Proposition 5.4.4

$$\begin{aligned} \widehat{f * g}(t) &= \Big(\int_{\mathbb{R}} f_t(x) dx \Big) \Big(\int_{\mathbb{R}} g_t(x) dx \Big) \\ &= \hat{f}(t) \hat{g}(t). \end{aligned}$$

□

The process of recovering f from $\hat{f}$, i.e., an inversion formula can be developed along the lines of Fejer's theorem, i.e., Theorem 5.6.1 of Section 5.6.

Theorem 5.7.3: *(Fejer's theorem). Let $f \in L^1(\mathbb{R})$, $\hat{f}(\cdot)$ be as in (7.3) and*

$$S_T(f, x) \equiv \frac{1}{2\pi} \int_{-T}^{T} \hat{f}(t) e^{\iota tx} dt, \ T \geq 0, \tag{7.6}$$

$$D_R(f, x) \equiv \frac{1}{R} \int_0^R S_T(f, x) dT, \ R \geq 0. \tag{7.7}$$

(i) If f is continuous at x_0 and f is bounded on $\mathbb{R}$, then

$$\lim_{R \to \infty} D_R(f, x_0) = f(x_0). \tag{7.8}$$

(ii) If f is uniformly continuous and bounded on $\mathbb{R}$, then

$$\lim_{R \to \infty} D_R(f, \cdot) = f(\cdot) \text{ uniformly on } \mathbb{R}. \tag{7.9}$$

(iii) As $R \to \infty$,

$$D_R(f, \cdot) \to f(\cdot) \quad in \quad L^1(\mathbb{R}). \tag{7.10}$$

(iv) If $f \in L^p(\mathbb{R})$, $1 \le p < \infty$, *then as* $R \to \infty$,

$$D_R(f, \cdot) \to f(\cdot) \quad in \quad L^p(\mathbb{R}). \tag{7.11}$$

Corollary 5.7.4: (*Uniqueness theorem*). *If* f *and* $g \in L^1(\mathbb{R})$ *and* $\hat{f}(\cdot) = \hat{g}(\cdot)$, *then* $f = g$ *a.e.* (m).

Proof: Let $h = f - g$. Then $h \in L^1(\mathbb{R})$ and $\hat{h}(\cdot) \equiv 0$. Thus, $S_T(h, \cdot) \equiv 0$ and $D_R(h, \cdot) \equiv 0$ where $S_T(h, \cdot)$ and $D_R(h, \cdot)$ are as in (7.6) and (7.7). Hence by Theorem 5.7.3 (iii), $h = 0$ a.e. (m), i.e., $f = g$ a.e. (m). □

Corollary 5.7.5: (*Inversion formula*). *Let* $f \in L^1(\mathbb{R})$ *and* $\hat{f} \in L^1(\mathbb{R})$. *Then*

$$f(x) = \frac{1}{2\pi} \int_{\mathbb{R}} \hat{f}(t) e^{\iota t x} dx \quad a.e. \quad (m). \tag{7.12}$$

Proof: Since $\hat{f} \in L^1(\mathbb{R})$, by the DCT,

$$S_T(f, x) \to \frac{1}{2\pi} \int_{\mathbb{R}} \hat{f}(t) e^{\iota t x} dt \quad \text{as} \quad T \to \infty$$

for all x in $\mathbb{R}$ and hence $D_R(f, x)$ has the same limit as $R \to \infty$. Now (7.12) follows from (7.10). □

The following results, i.e., Lemma 5.7.6 and Lemma 5.7.7, are needed for the proof of Theorem 5.7.3. The first one is an analog of Lemma 5.6.2.

Lemma 5.7.6: (*Fejer*). *For* $R > 0$, *let*

$$K_R(x) \equiv \frac{1}{2\pi} \frac{1}{R} \int_0^R \Big(\int_{-T}^{T} e^{\iota t x} dt \Big) dT. \tag{7.13}$$

Then

(i)

$$K_R(x) = \begin{cases} \frac{1}{\pi} \frac{(1 - \cos Rx)}{x^2}, & x \neq 0 \\ \frac{R}{2\pi} & x = 0, \end{cases} \tag{7.14}$$

and hence $K_R(\cdot) \ge 0$.

(ii) For $\delta > 0$,

$$\int_{|x| \ge \delta} K_R(x) dx \to 0 \quad as \quad R \to \infty. \tag{7.15}$$

(iii)

$$\int_{-\infty}^{\infty} K_R(x)dx = 1. \tag{7.16}$$

Proof:

(i) $K_R(0) = \frac{1}{2\pi R}\int_0^R (2T)dT = \frac{1}{2\pi}R$. For $x \neq 0$ and $R > 0$,

$$\begin{aligned} K_R(x) &= \frac{1}{2\pi R}\int_0^R \frac{2\sin Tx}{x} dT \\ &= \frac{1}{2\pi R}\frac{2(1-\cos Rx)}{x^2}. \end{aligned}$$

(ii) For $\delta > 0$,

$$\begin{aligned} 0 \le \int_{|x|\ge\delta} K_R(x)dx &\le \frac{2}{2\pi R}\int_{|x|\ge\delta}\frac{1}{x^2}\,dx \\ &= \frac{2}{\pi R}\frac{1}{\delta} \to 0 \quad \text{as} \quad R \to \infty. \end{aligned}$$

(iii)

$$\begin{aligned} \int_{-\infty}^{\infty} K_R(x)dx &= \frac{2}{\pi}\frac{1}{R}\int_0^\infty \Big(\frac{1-\cos Rx}{x^2}\Big)dx \\ &= \frac{2}{\pi}\int_0^\infty \Big(\frac{1-\cos u}{u^2}\Big)du. \end{aligned}$$

Now

$$\begin{aligned} &\int_0^\infty \frac{1-\cos u}{u^2}\,du \\ &= \lim_{L\to\infty}\int_0^L \Big(\frac{1-\cos u}{u^2}\Big)du \quad \text{(by the MCT)} \\ &= \lim_{L\to\infty}\int_0^L \Big(\int_0^u \sin x\,dx\Big)\frac{1}{u^2}\,du \\ &= \lim_{L\to\infty}\int_0^L \sin x\Big(\int_x^L \frac{1}{u^2}\,du\Big)dx \quad \text{(by Fubini's theorem)} \\ &= \lim_{L\to\infty}\Big(\int_0^L \frac{\sin x}{x}\,dx - \frac{1}{L}\int_0^L \sin x\,dx\Big) \\ &= \lim_{L\to\infty}\int_0^L \frac{\sin x}{x}\,dx \quad \text{since} \quad \Big|\int_0^L \sin x\,dx\Big| \le 1. \end{aligned}$$

Thus,

$$\int_0^\infty \frac{1-\cos u}{u^2}du = \int_0^\infty \frac{\sin x}{x}dx = \frac{\pi}{2} \tag{7.17}$$

(cf. Problem 5.9). Hence (iii) follows. □

Lemma 5.7.7: *Let $f \in L^p(\mathbb{R})$, $0 < p < \infty$. Then*

$$\int_{-\infty}^{\infty} |f(x-u) - f(x)|^p dx \to 0 \quad \text{as} \quad |u| \to 0. \tag{7.18}$$

Proof: The lemma holds if $f \in C_K$, i.e., if f is continuous on $\mathbb{R}$ (with values in $\mathbb{C}$) and vanishes outside a bounded interval. By Theorem 2.3.14, such functions are dense in $L^p(\mathbb{R})$. So given $f \in L^p(\mathbb{R})$, $0 < p < \infty$ and $\epsilon > 0$, let $g \in C_K$ be such that

$$\int |f - g|^p dm < \epsilon.$$

For any $0 < p < \infty$, there is a $0 < c_p < \infty$ such that for all $x, y, z \in (0, \infty)$, $|x + y + z|^p \leq c_p(|x|^p + |y|^p + |z|^p)$. Then,

$$\begin{aligned}
&\int |f(x-u) - f(x)|^p dx \\
&\leq c_p \int \big(|f(x-u) - g(x-u)|^p + |g(x-u) - g(x)|^p \\
&\quad + \int |g(x) - f(x)|^p\big) dx \\
&= c_p\Big(2\epsilon + \int |g(x-u) - g(x)|^p\Big) du.
\end{aligned}$$

So

$$\overline{\lim_{|u|\to 0}} \int |f(x-u) - f(x)|^p dx \leq c_p\, 2\epsilon.$$

Since $\epsilon > 0$ is arbitrary, the lemma is proved. □

Proof of Theorem 5.7.3: From (7.7)

$$\begin{aligned}
D_R(f, x) &\equiv \frac{1}{2\pi R} \int_0^R \Big(\int_{-T}^{T} e^{\iota t x} \Big(\int_{-\infty}^{\infty} e^{-\iota t y} f(y) dy\Big) dt\Big) dT \\
&= \frac{1}{2\pi} \frac{1}{R} \int_0^R \Big(\int_{-T}^{T} \Big(\int_{-\infty}^{\infty} e^{\iota t u} f(x-u) du\Big) dt\Big) dT.
\end{aligned}$$

Now Fubini's theorem yields

$$\begin{aligned}
D_R(f, x) &= \int_{-\infty}^{\infty} f(x-u) \Big(\frac{1}{2\pi} \frac{1}{R} \int_0^R \Big(\int_{-T}^{T} e^{\iota t u} dt\Big) dR\Big) du \\
&= \int_{-\infty}^{\infty} f(x-u) K_R(u) du \qquad (7.19)
\end{aligned}$$

where $K_R(\cdot)$ is as in (7.13).

Now let f be continuous at x_0 and bounded on $\mathbb{R}$ by M_f. Fix $\epsilon > 0$ and choose $\delta > 0$ such that $|x - x_0| < \delta \Rightarrow |f(x) - f(x_0)| < \epsilon$. From (7.16) and (7.19),

$$D_R(f, x_0) - f(x_0) = \int \big(f(x_0 - u) - f(x_0)\big) K_R(u) du$$

implying

$$\begin{aligned} |D_R(f, x_0) - f(x_0)| &\leq \int_{|u|<\delta} |f(x_0 - u) - f(x_0)| K_R(u) du \\ &\quad + \int_{|u|\geq\delta} |f(x_0 - u) - f(x_0)| K_R(u) du \\ &< \epsilon + 2M_f \int_{|u|\geq\delta} K_R(u) du. \end{aligned}$$

Now from (7.15), it follows that

$$\varlimsup_{R\to\infty} |D_R(f, x_0) - f(x_0)| \leq \epsilon,$$

proving (i).

The proof of (ii) is similar to this and is omitted.

Clearly (iv) implies (iii). To establish (iv), note that for $1 \leq p < \infty$, by Jensen's inequality (which applies since $K_R(u) \geq 0$ and $\int K_R(u) du = 1$), for x in $\mathbb{R}$,

$$|D_R(f, x) - f(x)|^p \leq \int \big|\big(f(x - u) - f(x)\big)\big|^p K_R(u) du$$

and hence

$$\int |D_R(f, x) - f(x)|^p dx \leq \int \Big(\int |f(x - u) - f(x)|^p dx\Big) K_R(u) du.$$

Now (7.18) and the arguments in the proof of (i) yield (iv). □

5.8 Plancherel transform

If $f \in L^2(\mathbb{R})$, it need not be in $L^1(\mathbb{R})$ and so the Fourier transform is not defined. However, it is possible to extend the definition using an approximation procedure due to Plancherel.

Proposition 5.8.1: *Let $f \in L^1(\mathbb{R}) \cap L^2(\mathbb{R})$ and let $\hat{f}(t) \equiv \int f(x) e^{-\iota tx} dx$, $t \in \mathbb{R}$. Then, $\hat{f} \in L^2(\mathbb{R})$ and*

$$\int |f|^2 dm = \frac{1}{2\pi} \int |\hat{f}|^2 dm. \tag{8.1}$$

Proof: Let $\tilde{f}(x) = \overline{f(-x)}$ and $g = f * \tilde{f}$. Since f and $\tilde{f}$ are in $L^1(\mathbb{R})$, g is well defined and is in $L^1(\mathbb{R})$. Further by Cauchy-Schwarz inequality,

$$\begin{aligned} |g(x_1) - g(x_2)| &\leq \int |f(x_1 - u) - f(x_2 - u)|\,|\tilde{f}(u)|du \\ &\leq \Big(\int |f(x_1 - u) - f(x_2 - u)|^2 du\Big)^{1/2}\Big(\int |\tilde{f}(u)|^2 du\Big). \end{aligned}$$

By Lemma 5.7.7, the right side goes to zero uniformly as $|x_1 \quad x_2| \to 0$. Thus $g(\cdot)$ is uniformly continuous. Also, by Cauchy-Schwarz inequality,

$$\begin{aligned} |g(x)| &= \Big|\int f(x-u)\tilde{f}(u)du\Big| \\ &\leq \Big(\int |f(u)|^2 du\Big)^{1/2}\Big(\int |\tilde{f}(u)|^2 du\Big)^{1/2} \\ &= \int |f(u)|^2 du, \end{aligned}$$

and hence $g(\cdot)$ is bounded. Thus, g is continuous, bounded and integrable on $\mathbb{R}$. By Theorem 5.7.3 (Fejer's theorem)

$$D_R(g, 0) \to g(0) \quad \text{as} \quad R \to \infty. \tag{8.2}$$

But

$$g(0) = \int f(u)\overline{f(-u)}du = \int |f|^2 dm. \tag{8.3}$$

By Proposition 5.7.2, $\hat{g}(t) = \hat{f}(t)\hat{\tilde{f}}(t)$. Also, note that

$$\begin{aligned} \hat{\tilde{f}}(t) &= \int \tilde{f}(x)e^{-\iota tx}dx \\ &= \int \overline{f(-x)}e^{-\iota tx}dx \\ &= \overline{\hat{f}(t)} \end{aligned}$$

and hence $\hat{g}(t) = |\hat{f}(t)|^2 \geq 0$. But

$$\begin{aligned} D_R(g, 0) &= \frac{1}{2\pi}\frac{1}{R}\int_0^R \Big(\int_{-T}^T \hat{g}(t)dt\Big)dT \\ &= \frac{1}{\pi}\int_0^R \hat{g}(t)\Big(1 - \frac{|t|}{R}\Big)dt. \end{aligned}$$

Since $\hat{g}(\cdot) \geq 0$, by the MCT,

$$\int_0^R \hat{g}(t)\Big(1 - \frac{|t|}{R}\Big)dt \uparrow \int_0^\infty \hat{g}(t)dt \quad \text{as} \quad R \uparrow \infty.$$

Thus, $\lim_{R\to\infty} D_R(g,0) = \frac{1}{\pi}\int_0^\infty \hat{g}(t)dt$. Since $\hat{g}(-t) = \hat{g}(t)$ for all t in $\mathbb{R}$,

$$\lim_{R\to\infty} D_R(g,0) = \frac{1}{2\pi}\int_{-\infty}^{\infty} \hat{g}(t)dt = \frac{1}{2\pi}\int |\hat{f}(t)|^2 dt. \tag{8.4}$$

Clearly, (8.2)–(8.4) imply (8.1). □

Proposition 5.8.2: *Let $f \in L^2(\mathbb{R})$ and $f_n(\cdot) \equiv fI_{[-n,n]}(\cdot)$, $n \geq 1$. Then, $\{f_n\}_{n\geq 1}$, $\{\hat{f}_n\}_{n\geq 1}$ are both Cauchy in $L^2(\mathbb{R})$ and hence, convergent in $L^2(\mathbb{R})$.*

Proof: Since for each $n \geq 1$, $f_n \in L^2(\mathbb{R}) \cap L^1(\mathbb{R})$, by Proposition 5.8.1,

$$\|f_{n_1} - f_{n_2}\|_2^2 = \frac{1}{2\pi}\int |\hat{f}_{n_1} - \hat{f}_{n_2}|^2 dm. \tag{8.5}$$

Since $f \in L^2(\mathbb{R})$, $f_n \to f$ in $L^2(\mathbb{R})$, $\{f_n\}_{n\geq 1}$ is Cauchy in $L^2(\mathbb{R})$. By (8.5) $\{\hat{f}_n\}_{n\geq 1}$ is also Cauchy in $L^2(\mathbb{R})$. □

Definition 5.8.1: Let $f \in L^2(\mathbb{R})$. The *Plancherel transform of f*, denoted by $\hat{f}$, is defined as $\lim_{n\to\infty} \hat{f}_n$, where

$$\hat{f}_n(t) = \int_{-n}^{n} e^{-\iota tx} f(x)dx \tag{8.6}$$

and the limit is taken in $L^2(\mathbb{R})$.

Theorem 5.8.3: *Let $f \in L^2(\mathbb{R})$ and $\hat{f}$ be its Plancherel transform. Then*

(i)

$$\int |f|^2 dm = \frac{1}{2\pi}\int |\hat{f}|^2 dm. \tag{8.7}$$

(ii) For $f \in L^1(\mathbb{R}) \cap L^2(\mathbb{R})$, the Plancherel transform coincides with the Fourier transform.

Proof: From Propositions 5.8.1 and 5.8.2 and the definition of $\hat{f}$,

$$\frac{1}{2\pi}\int |\hat{f}|^2 dm = \lim_{n\to\infty}\frac{1}{2\pi}\int |\hat{f}_n|^2 dm = \lim_{n\to\infty}\int |f_n|^2 dm = \int |f|^2 dm$$

proving (i). If $f \in L^1(\mathbb{R}) \cap L^2(\mathbb{R})$, $\hat{f}_n$ defined in (8.6) converges as $n \to \infty$ pointwise to the Fourier transform of f (by the DCT). But by Definition 5.7.1, $\hat{f}_n$ converges in $L^2(\mathbb{R})$ to the Plancherel transform of f. So (ii) follows. □

It can also be shown that for f, $\hat{f}$ as in the above theorem, the following *inversion formula* holds:

$$\frac{1}{2\pi}\int_{-n}^{n} \hat{f}(t)e^{\iota tx}dt \to f(x) \quad \text{in} \quad L^2(\mathbb{R}) \tag{8.8}$$

and that the map $f \to \hat{f}$ is a Hilbert space isomorphism of $L^2(\mathbb{R})$ onto $L^2(\mathbb{R})$. For a proof, see Rudin (1987), Chapter 9. This in turn implies that if $f, g \in L^2(\mathbb{R})$ with Plancherel transforms $\hat{f}$ and $\hat{g}$, respectively, then

$$\int f\,\bar{g}\,dm = \frac{1}{2\pi}\int \hat{f}\,\bar{\hat{g}}\,dm. \tag{8.9}$$

This is known as the *Parseval identity.*

5.9 Problems

5.1 Verify that if μ_1 and μ_2 are finite measures on $(\Omega_1, \mathcal{F}_1)$ and $(\Omega_2, \mathcal{F}_2)$ respectively, then $\mu_{12}(\cdot)$ and $\mu_{21}(\cdot)$, defined by (1.4) and (1.5) respectively, are measures on $(\Omega_1 \times \Omega_2, \mathcal{F}_1 \times \mathcal{F}_2)$.

(**Hint:** Use the MCT.)

5.2 Let $(\Omega, \mathcal{F}, \mu)$ be a complete measure space such that $\mathcal{F} \neq \mathcal{P}(\Omega)$ and for some $A_0 \in \mathcal{F}$ with $A_0 \neq \emptyset$, $\mu(A_0) = 0$. Let $B \in \mathcal{P}(\Omega) \setminus \mathcal{F}$. Then show that $(\mu \times \mu)(\Omega \times A_0) = 0$ but $B \times A_0 \notin \mathcal{F} \times \mathcal{F}$. Conclude that $(\Omega \times \Omega, \mathcal{F} \times \mathcal{F}, \mu \times \mu)$ is not complete. (An example of such a measure space $(\Omega, \mathcal{F}, \mu)$ is the space $([0,1], \mathcal{M}_L, m)$ where $\mathcal{M}_L$ is the Lebesgue σ-algebra on $[0,1]$ and m is the Lebesgue measure.)

5.3 Let μ_i, $i = 1, 2$ be two σ-finite measures on $(\mathbb{R}, \mathcal{B}(\mathbb{R}))$. Let $D_i = \{x : \mu_i(\{x\}) > 0\}, i = 1, 2$.

(a) Show that $D_1 \cup D_2$ is countable.

(b) Let $\phi_i(x) = \mu_i(\{x\})$ $x \in \mathbb{R}$, $i = 1, 2$. Show that ϕ_i is Borel measurable for $i = 1, 2$.

(c) Show that $\int \phi_1 d\mu_2 = \sum_{z \in D_1 \cap D_2} \phi_1(z)\phi_2(z)$.

(d) Deduce (2.11) from (c).

5.4 Extend Theorem 5.2.3 as follows. Let F_1, F_2 be two nondecreasing right continuous functions on $[a, b]$. Then

$$\int_{(a,b]} F_1 dF_2 + \int_{(a,b]} F_2 dF_1 = F_1(b)F_2(b) - F_1(a)F_2(a) + \int_{(a,b]} \phi_1 d\mu_2,$$

where ϕ_1 is as in Problem 5.3.

5.5 Let $F_i : \mathbb{R} \to \mathbb{R}$ be nondecreasing and right continuous, $i = 1, 2$. Show that if $\lim_{b \uparrow \infty} F_1(b)F_2(b) = \lambda_1$ and $\lim_{a \downarrow -\infty} F_1(a)F_2(a) = \lambda_2$ exist and are finite, then

$$\int_{\mathbb{R}} F_1 dF_2 + \int_{\mathbb{R}} F_2 dF_1 = \lambda_1 - \lambda_2 + \int_{\mathbb{R}} \phi_1 d\mu_2$$

where ϕ_1 is as in Problem 5.3.

5.6 Let $(\Omega, \mathcal{F}, \mu)$ be a σ-finite measure space and f be a nonnegative measurable function. Then, $\int_\Omega f d\mu = \int_{[0,\infty)} \mu(\{f \geq t\})dt$.

(**Hint:** Consider the product space of $(\Omega, \mathcal{F}, \mu)$ with $(\mathbb{R}_+, \mathcal{B}(\mathbb{R}_+), m)$ and apply Tonelli's theorem to the function $g(\omega, t) = I(f(\omega) \geq t)$, after showing that g is $\mathcal{F} \times \mathcal{B}(\mathbb{R}_+))$-measurable.)

5.7 Let $(\Omega, \mathcal{F}, P)$ be a probability space and $X : \Omega \to \mathbb{R}_+$ be a random variable.

(a) Show that for any $h : \mathbb{R}_+ \to \mathbb{R}_+$ that is absolutely continuous,

$$\begin{aligned} \int_\Omega h(X)dP &= h(0) + \int_{[0,\infty)} h'(t)P(X \geq t)dt \\ &= h(0) + \int_{(0,\infty)} h'(t)P(X > t)dt. \end{aligned}$$

(b) Show that for any $0 < p < \infty$,

$$\int_\Omega X^p dP = \int_{[0,\infty)} pt^{p-1}P(X \geq t)dt.$$

(c) Show that for any $0 < p < \infty$,

$$\int_\Omega X^{-p} dP = \frac{1}{\Gamma(p)} \int_{[0,\infty)} \psi_X(t)t^{p-1}dt,$$

where $\Gamma(p) = \int_{[0,\infty)} e^{-t}t^{p-1}dt, p > 0$, and $\psi_X(t) = \int_\Omega e^{-tX}dP$, $t \in \mathbb{R}_+$.

(**Hint:** (a) Apply Tonelli's theorem to the function $f(t, \omega) \equiv h'(t)I(X(w) \geq t)$ on the product measure space $([0,\infty) \times \Omega, \mathcal{B}([0,\infty)) \times \mathcal{F}, m \times P)$, where m is Lebesgue measure on $(\mathbb{R}_+, \mathcal{B}(\mathbb{R}_+))$.)

5.8 Let $g : \mathbb{R}_+ \to \mathbb{R}_+$ and $f : \mathbb{R}^2 \to \mathbb{R}_+$ be Borel measurable. Let $A = \{(x, y) : x \geq 0, 0 \leq y \leq g(x)\}$.

(a) Show that $A \in \mathcal{B}(\mathbb{R}^2)$.

(b) Show that

$$\int_{\mathbb{R}_+} \Big(\int_{[0,g(x)]} f(x,y)m(dy) \Big) m(dx) = \int f I_A dm^{(2)},$$

where $m^{(2)}$ is Lebesgue measure on $(\mathbb{R}^2, \mathcal{B}(\mathbb{R}^2))$ and $m(\cdot)$ is Lebesgue measure on $\mathbb{R}$.

(c) If g is continuous and strictly increasing show that the two integrals in (b) equal

$$\int_{\mathbb{R}_+}\Big(\int_{[g^{-1}(y),\infty)} f(x,y)m(dx)\Big)m(dy).$$

5.9 (a) For $1 < A < \infty$, let

$$h_A(t) = \int_0^A e^{-xt}\sin x\,dx,\ t \geq 0.$$

Use integration by parts to show that

$$|h_A(t)| \leq \frac{1}{1+t^2} + e^{-t}$$

and

$$h_A(t) \to \frac{1}{1+t^2} \quad \text{as} \quad A \to \infty.$$

(b) Show using Fubini's theorem that for $0 < A < \infty$,

$$\int_0^\infty h_A(t)dt = \int_0^A \frac{\sin x}{x}\,dx.$$

(c) Conclude using the DCT that

$$\lim_{A\to\infty}\int_0^A \frac{\sin x}{x}\,dx = \int_0^\infty \frac{1}{1+t^2}\,dt.$$

(d) Using Theorem 4.4.1 and the fact that $\phi(x) \equiv \tan x$ is a (1–1) strictly monotone map from $(0, \frac{\pi}{2})$ to $(0, \infty)$ having the inverse map $\psi(\cdot)$ with derivative $\psi'(t) \equiv \frac{1}{1+t^2}$, $0 < t < \infty$, conclude that

$$\int_0^\infty \frac{1}{1+t^2}dt = \frac{\pi}{2}\ .$$

5.10 Show that $I \equiv \int_0^\infty e^{-x^2/2}dx = \sqrt{\frac{\pi}{2}}$.

(**Hint:** By Tonelli's theorem, $I^2 = \int_0^\infty\int_0^\infty e^{-\frac{(x^2+y^2)}{2}}dxdy$. Now use the change of variables $x = r\cos\theta$, $y = r\sin\theta$, $0 < r < \infty$, $0 < \theta < \frac{\pi}{2}$.)

5.11 Let μ be a finite measure on $\big(\mathbb{R}, \mathcal{B}(\mathbb{R})\big)$. Let $f,\ g : \mathbb{R} \to \mathbb{R}_+$ be nondecreasing. Show that

$$\mu(\mathbb{R})\int fg\,d\mu \geq \Big(\int f d\mu\Big)\Big(\int g d\mu\Big).$$

(**Hint:** Consider $h(x_1, x_2) = \big(f(x_1) - f(x_2)\big)\big(g(x_1) - g(x_2)\big)$ on $\mathbb{R}^2$ and integrate w.r.t. $\mu \times \mu$.)

5.12 Let μ and λ be σ-finite measures on $\big(\mathbb{R}, \mathcal{B}(\mathbb{R})\big)$. Recall that

$$\nu(A) \equiv (\mu * \lambda)(A) \equiv \int\int I_A(x+y)d\mu d\lambda, \ A \in \mathcal{B}(\mathbb{R}).$$

(a) Show that for any Borel measurable $f : \mathbb{R} \to \mathbb{R}_+$, $f(x+y)$ is $\langle \mathcal{B}(\mathbb{R}) \times \mathcal{B}(\mathbb{R}), \mathcal{B}(\mathbb{R})\rangle$-measurable from $\mathbb{R} \times \mathbb{R} \to \mathbb{R}$ and

$$\int f d\nu = \int\int f(x+y)d\mu d\lambda.$$

(b) Show that

$$\nu(A) = \int_{\mathbb{R}} \mu(A-t)\lambda(dt), \ A \in \mathcal{B}(\mathbb{R}).$$

(c) Suppose there exist countable sets B_λ, B_μ such that $\mu(B_\mu^c) = 0 = \lambda(B_\lambda^c)$. Show that there exists a countable set B_ν such that $\nu(B_\nu^c) = 0$.

(d) Suppose that $\mu(\{x\}) = 0$ for all x in $\mathbb{R}$. Show that $\nu(\{x\}) = 0$ for all x in $\mathbb{R}$.

(e) Suppose that $\mu \ll m$ with $\frac{d\mu}{dm} = h$. Show that $\nu \ll m$ and find $\frac{d\nu}{dm}$ in terms of h, μ and λ.

(f) Suppose that $\mu \ll m$ and $\lambda \ll m$. Show that

$$\frac{d\nu}{dm} = \frac{d\mu}{dm} * \frac{d\lambda}{dm}.$$

5.13 (*Convolution of cdfs*). Let F_i, $i = 1, 2$ be cdfs on $\mathbb{R}$. Recall that a cdf F on $\mathbb{R}$ is a function from $\mathbb{R} \to \mathbb{R}_+$ such that it is nondecreasing, right continuous with $F(x) \to 0$ as $x \to -\infty$ and $F(x) \to 1$ as $x \to \infty$.

(a) Show that $(F_1 * F_2)(x) \equiv \int_{\mathbb{R}} F_1(x-u)dF_2(u)$ is well defined and is a cdf on $\mathbb{R}$.

(b) Show also that $(F_1 * F_2)(\cdot) = (F_2 * F_1)(\cdot)$.

(c) Suppose $t \in \mathbb{R}$ is such that $\int e^{tx} dF_i(x) < \infty$ for $i = 1, 2$. Show that $\int e^{tx} d(F_1 * F_2)(x) = \Big(\int e^{tx} dF_1(x)\Big)\Big(\int e^{tx} dF_2(x)\Big)$.

5.14 Let f, $g \in L^1(\mathbb{R}, \mathcal{B}(\mathbb{R}), m)$.

(a) Show that if f is continuous and bounded on $\mathbb{R}$, then so is $f * g$.

(b) Show that if f is differentiable with a bounded derivative on $\mathbb{R}$, then so is $f * g$.

(**Hint:** Use the DCT.)

5.15 Let $f \in L^1(\mathbb{R})$, $g \in L^p(\mathbb{R})$, $1 \le p \le \infty$.

(a) Show that if $1 \le p < \infty$, then for all x in $\mathbb{R}$

$$\Big|\int |f(u)g(x-u)|du\Big|^p \le \Big(\int |f|dm\Big)^{p-1}\Big(\int |g(x-u)|^p|f(u)|du\Big)$$

and hence that

$$(f * g)(x) \equiv \int f(u)g(x-u)du$$

is well defined a.e. (m) and

$$\|f * g\|_p \le \|f\|_1 \, \|g\|_p$$

with "=" holding iff either $f = 0$ a.e. or $g = 0$ a.e.

(b) Show that if $p = \infty$ then

$$\|f * g\|_\infty \le \|f\|_1 \, \|g\|_\infty$$

and "=" can hold for some nonzero f and g.

(**Hint for (a):** Use Jensen's inequality with probability measure $d\mu = \frac{|f|dm}{\|f\|_1}$ if $\|f\|_1 > 0$.)

5.16 Let $1 \le p \le \infty$ and $q = 1 - \frac{1}{p}$. Let $f \in L^p(\mathbb{R})$, $g \in L^q(\mathbb{R})$.

(a) Show that $f * g$ is well defined and uniformly continuous.

(b) Show that if $1 < p < \infty$,

$$\lim_{|x|\to\infty} (f * g)(x) = 0.$$

(**Hint:** For (a) use Hölder's inequality and Lemma 5.7.7. For (b) approximate g by simple functions.)

5.17 Let $g : \mathbb{R} \to \mathbb{R}$ be infinitely differentiable and be zero outside a bounded interval.

(a) Let $f : \mathbb{R} \to \mathbb{R}$ be Borel measurable and $\int_A |f|dm < \infty$ for all bounded intervals A in $\mathbb{R}$. Show that $f * g$ is well defined and infinitely differentiable.

(b) Show that for any $f \in L^1(\mathbb{R})$, there exist a sequence $\{g_n\}_{n\ge1}$ of such functions such that $f * g_n \to f$ in $L^1(\mathbb{R})$.

5.18 For $f \in L^1(\mathbb{R})$, let $f_\sigma = f * \phi_\sigma$ where $\phi_\sigma(x) = \frac{1}{\sqrt{2\pi}\sigma}e^{-\frac{x^2}{2\sigma^2}}$, $0 < \sigma < \infty$, $x \in \mathbb{R}$.

(a) Show that f_σ is infinitely differentiable.

(b) Show that if f is continuous and zero outside a bounded interval, then f_σ converges to g uniformly on $\mathbb{R}$ as $\sigma \to 0$.

(c) Show that if $f \in L^p(\mathbb{R})$, $1 \leq p < \infty$, then $f_\sigma \to f$ in $L^p(\mathbb{R})$ as $\sigma \to 0$.

5.19 Let $f \in L^p(\mathbb{R})$, $1 \leq p < \infty$ and $h(x) = \int_{A+x} f(u)du$ where A is a bounded Borel set and

$$A + x \equiv \{y : y = a + x, a \in A\}.$$

(a) Show that $h = f * g$ for some g bounded and with bounded support.

(b) Show that $h(\cdot)$ is continuous and that $\lim\limits_{|x|\to\infty} h(x) = 0$.

(**Hint:** For $1 < p < \infty$, use Hölder's inequality and show that $m\big((A+x_1) \triangle (A + x_2)\big) \to 0$ as $|x_1 - x_2| \to 0$.)

5.20 (a) Verify the claims in Examples 5.4.1 directly.

(b) Verify the same using generating functions.

5.21 Let f be a probability density on $\mathbb{R}$, i.e., f is nonnegative, Borel measurable and $\int f dm = 1$. Show that $|\hat{f}(t)| < 1$ for all $t \neq 0$.

(**Hint:** If $|\hat{f}(t_0)| = 1$ for some $t_0 \neq 0$, show that $\int \big(1 - \cos t_0(x - \theta)\big) f(x)dx = 0$ for some θ.)

5.22 (a) Let $f \in L^1(\mathbb{R})$ and $x^k f(x) \in L^1(\mathbb{R})$ for some $k \geq 1$. Show that $\hat{f}(\cdot)$ is k-times differentiable on $\mathbb{R}$ with all derivatives $\hat{f}^{(r)}(t) \to 0$ as $|t| \to \infty$ for $r \leq k$.

(b) Let $f \in L^1(\mathbb{R})$. Suppose $\int |t\hat{f}(t)|dt < \infty$. Show that there exists a function $g : \mathbb{R} \to \mathbb{R}$ such that it is differentiable, $\lim_{|x|\to\infty}(|g(x)| + |g'(x)|) = 0$ and $g = f$ a.e. (m). Extend this to the case where $\int |t^k \hat{f}(t)|dt < \infty$ for some $k > 1$.

(**Hint:** Consider $g(x) = \frac{1}{2\pi} \int e^{-\iota tx} \hat{f}(t)dt$.)

5.23 Let $f(x) = \frac{1}{\sqrt{2\pi}} e^{-\frac{x^2}{2}}$, $x \in \mathbb{R}$.

(a) Show that $\hat{f}(\cdot)$ is real valued, differentiable and satisfies the ordinary differential equation $\hat{f}'(t) + t\hat{f}(t) = 0$, $t \in \mathbb{R}$ and $\hat{f}(0) = 1$. Find $\hat{f}(t)$.

(b) For μ in $\mathbb{R}$, $\sigma > 0$, let

$$f_{\mu,\sigma}(x) = \frac{1}{\sqrt{2\pi}\sigma} e^{-\frac{1}{2}(\frac{x-u}{\sigma})^2}.$$

Find $\hat{f}_{\mu,\sigma}$ and verify that for any (μ_i, σ_i), $i = 1, 2$,

$$f_{\mu_1,\sigma_1} * f_{\mu_2,\sigma_2} = f_{\mu_1+\mu_2,\sigma_1^2+\sigma_2^2}.$$

(**Hint for (b):** Use Fourier transforms and uniqueness.)

5.24 (*Rate of convergence of Fourier series*). Consider the function $h(x) = \frac{\pi}{2} - |x|$ in $-\pi \le x \le \pi$.

(a) Find $\hat{h}(n) \equiv \frac{1}{2\pi}\int_{-\pi}^{\pi} e^{-\iota nx} h(x)dx$, $n = 0, \pm 1, \pm 2$.

(b) Show that $\sum_{n=-\infty}^{\infty} |\hat{h}(n)| < \infty$.

(c) Show that $S_n(h, x) \equiv \sum_{j=-n}^{+n} \hat{h}(j)e^{\iota jx}$ converges to $h(x)$ uniformly on $[-\pi, \pi]$.

(d) Verify that

$$\sup\{|S_n(h, x) - h(x)| : -\pi \le x \le \pi\} \le \frac{2}{\pi}\frac{1}{(n-1)}, \quad n \ge 2$$

and

$$|S_n(h, 0) - h(0)| \ge \frac{2}{\pi}\frac{1}{(n+2)}.$$

(**Remark:** This example shows that the Fourier series of a function can converge very slowly such as in this example where the rate of decay is $\frac{1}{n}$.)

5.25 Using Fejer's theorem (Theorem 5.6.1) prove Wierstrass' theorem on uniform approximation of a continuous function on a bounded closed interval by a polynomial.

(**Hint:** Show that on bounded intervals a trigonometric polynomial can be approximated uniformly by a polynomial using the power series representation of sine and cosine functions (see Section A.3).

5.26 Evaluate

$$\lim_{n\to\infty} \int_{-n}^{n} \frac{\sin \lambda x}{x} e^{\iota tx} dx, \quad 0 < \lambda < \infty, \ 0 < x < \infty.$$

(**Hint:** For $0 < \lambda < \infty$, $\frac{\sin \lambda x}{x} \in L^2(\mathbb{R})$ and it is the Fourier transform of $f(t) = I_{[-\lambda,\lambda]}(\cdot)$. Now apply Plancherel theorem. Alternatively use Fubini's theorem and the fact $\lim\limits_{n\to\infty} \int_{-n}^{n} \frac{\sin y}{y} dy$ exists in $\mathbb{R}$.)

5.27 Find an example of a function $f \in L^2(\mathbb{R}) \cap \left(L^1(\mathbb{R})\right)^c$ such that its Plancherel transform $\hat{f} \in L^1(\mathbb{R})$.

(**Hint:** Examine Problem 5.26.)

6

Probability Spaces

6.1 Kolmogorov's probability model

Probability theory provides a mathematical model for random phenomena, i.e., those involving uncertainty. First one identifies the set Ω of possible outcomes of (random) experiment associated with the phenomenon. This set Ω is called the *sample space* and an individual element ω of Ω is called a *sample point*. Even though the outcome is not predictable ahead of time, one is interested in the "chances" of some particular statement to be valid for the resulting outcome. The set of ω's for which a given statement is valid is called an *event*. Thus, an event is a subset of Ω. One then identifies a class $\mathcal{F}$ of events, i.e., a class $\mathcal{F}$ of subsets of Ω (not necessarily all of $\mathcal{P}(\Omega)$, the power set of Ω), and then a set function P on $\mathcal{F}$ such that for A in $\mathcal{F}$, $P(A)$ represents the "chance" of the event A happening. Thus, it is reasonable to impose the following conditions on $\mathcal{F}$ and P:

(i) $A \in \mathcal{F} \Rightarrow A^c \in \mathcal{F}$
(i.e., if one can define the probability of an event A, then the probability of A not happening is also well defined).

(ii) $A_1, A_2 \in \mathcal{F} \Rightarrow A_1 \cup A_2 \in \mathcal{F}$
(i.e., if one can define the probabilities of A_1 and A_2, then the probability of at least one of A_1 or A_2 happening is also well defined).

(iii) for all A in $\mathcal{F}$, $0 \leq P(A) \leq 1$, $P(\emptyset) = 0$, and $P(\Omega) = 1$.

(iv) $A_1, A_2 \in \mathcal{F}$, $A_1 \cap A_2 = \emptyset \Rightarrow P(A_1 \cup A_2) = P(A_1) + P(A_2)$ (i.e., if A_1 and A_2 are mutually exclusive events, then the probability of at least one of the two happening is simply the sum of the probabilities).

The above conditions imply that $\mathcal{F}$ is an *algebra* and P is a *finitely additive* set function. Next, as explained in Section 1.2, it is natural to require that $\mathcal{F}$ be closed under monotone increasing unions and P be monotone continuous from below. That is, if $\{A_n\}_{n\geq 1}$ is a sequence of events in $\mathcal{F}$ such that A_n implies A_{n+1} (i.e., $A_n \subset A_{n+1}$) for all $n \geq 1$, then the probability of at least one of the A_n's happening is well defined and is the limit of the corresponding probabilities. In other words, the following conditions on $\mathcal{F}$ and P must hold in addition to (i)–(iv) above:

(v) $A_n \in \mathcal{F}$, $A_n \subset A_{n+1}$ for all $n = 1, 2, \ldots \Rightarrow \cup_{n\geq 1} A_n, \in \mathcal{F}$ and $P(A_n) \uparrow P(\cup_{n\geq 1} A_n)$.

As noted in Section 1.2, conditions (i)–(v) imply that $(\Omega, \mathcal{F}, P)$ is a *measure space*, i.e., $\mathcal{F}$ is a σ-algebra and P is a measure on $\mathcal{F}$ with $P(\Omega) = 1$. That is, $(\Omega, \mathcal{F}, P)$ is a *probability space*. This is known as *Kolmogorov's probability model* for random phenomena (see Kolmogorov (1956), Parthasarathy (2005)). Here are some examples.

Example 6.1.1: (*Finite sample spaces*). Let $\Omega \equiv \{\omega_1, \omega_2, \ldots, \omega_k\}, 1 \leq k < \infty$, $\mathcal{F} \equiv \mathcal{P}(\Omega)$, the power set of Ω and $P(A) = \sum_{i=1}^k p_i I_A(\omega_i)$ where $\{p_i\}_{i=1}^k$ are such that $p_i \geq 0$ and $\sum_{i=1}^k p_i = 1$. This is a probability model for random experiments with finitely many possible outcomes.

An important application of this probability model is finite population sampling. Let $\{U_1, U_2, \ldots, U_N\}$ be a finite population of N units or objects. These could be individuals in a city, counties in a state, etc. In a typical sample survey procedure, one chooses a subset of size n $(1 \leq n \leq N)$ from this population. Let Ω denote the collection of all possible subsets of size n. Here $k = \binom{N}{n}$, each ω_i is a sample of size n and p_i is the selection probability of ω_i. The assignment of $\{p_i\}_{i=1}^k$ is determined by a given *sampling scheme*. For example, in simple random sampling without replacement, $p_i = \frac{1}{k}$ for $i = 1, 2, \ldots, k$.

Other examples include coin tossing, rolling of dice, bridge hands, and acceptance sampling in statistical quality control (Feller (1968), Volume 1).

Example 6.1.2: (*Countably infinite sample spaces*). Let $\Omega \equiv \{\omega_1, \omega_2, \ldots\}$ be a countable set, $\mathcal{F} = \mathcal{P}(\Omega)$, and $P(A) \equiv \sum_{i=1}^\infty p_i I_A(\omega_i)$ where $\{p_i\}_{i=1}^\infty$ satisfy $p_i \geq 0$ and $\sum_{i=1}^\infty p_i = 1$. It is easy to verify that $(\Omega, \mathcal{F}, P)$ is a probability space. This is a probability model for random experiments with countably infinite number of outcomes. For example, the experiment of tossing a coin until a "head" is produced leads to such a probability space.

Example 6.1.3: (*Uncountable sample spaces*).

(a) (*Random variables*). Let $\Omega = \mathbb{R}$, $\mathcal{F} = \mathcal{B}(\mathbb{R})$, $P = \mu_F$, the Lebesgue-Stieltjes measure corresponding to a cdf F, i.e., corresponding to a function $F : \mathbb{R} \to \mathbb{R}$ that is nondecreasing, right continuous, and satisfies $F(-\infty) = 0$, $F(+\infty) = 1$. See Section 1.3. This serves as a model for a single random variable X.

(b) (*Random vectors*). Let $\Omega = \mathbb{R}^k$, $\mathcal{F} = \mathcal{B}(\mathbb{R}^k)$, $P = \mu_F$, the Lebesgue-Stieltjes measure corresponding to a (multidimensional) cdf F on $\mathbb{R}^k$ where $k \in \mathbb{N}$. See Section 1.3. This is a model for a random vector $(X_1, X_2, \ldots, X_k)$.

(c) (*Random sequences*). Let $\Omega = \mathbb{R}^\infty \equiv \mathbb{R} \times \mathbb{R} \times \ldots$ be the set of all sequences $\{x_n\}_{n \geq 1}$ of real numbers. Let $\mathcal{C}$ be the class of all finite dimensional sets of the form $A \times \mathbb{R} \times \mathbb{R} \times \ldots$, where $A \in \mathcal{B}(\mathbb{R}^k)$ for some $1 \leq k < \infty$. Let $\mathcal{F}$ be the σ-algebra generated by $\mathcal{C}$. For each $1 \leq k < \infty$, let μ_k be a probability measure on $\mathcal{B}(\mathbb{R}^k)$ such that $\mu_{k+1}(A \times \mathbb{R}) = \mu_k(A)$ for all $A \in \mathcal{B}(\mathbb{R}^k)$. Then there exists a probability measure μ on $\mathcal{F}$ such that $\mu(A \times \mathbb{R} \times \mathbb{R} \times \ldots) = \mu_k(A)$ if $A \in \mathcal{B}(\mathbb{R}^k)$. (This will be shown later as a special case of the Kolmogorov's consistency theorem in Section 6.3.) This will be a model for a sequence $\{X_n\}_{n \geq 1}$ of random variables such that for each k, $1 \leq k < \infty$, the distribution of $(X_1, X_2, \ldots, X_k)$ is μ_k.

6.2 Random variables and random vectors

Recall the following definitions introduced earlier in Sections 2.1 and 2.2.

Definition 6.2.1: Let $(\Omega, \mathcal{F}, P)$ be a probability space and $X : \Omega \to \mathbb{R}$ be $\langle \mathcal{F}, \mathcal{B}(\mathbb{R}) \rangle$-measurable, that is, $X^{-1}(A) \in \mathcal{F}$ for all $A \in \mathcal{B}(\mathbb{R})$. Then, X is called a *random variable* on $(\Omega, \mathcal{F}, P)$.

Recall that $X : \Omega \to \mathbb{R}$ is $\langle \mathcal{F}, \mathcal{B}(\mathbb{R}) \rangle$-measurable iff for all $x \in \mathbb{R}$, $\{\omega : X(\omega) \leq x\} \in \mathcal{F}$.

Definition 6.2.2: Let X be a random variable on $(\Omega, \mathcal{F}, P)$. Let

$$F_X(x) \equiv P(\{\omega : X(\omega) \leq x\}), \quad x \in \mathbb{R}. \tag{2.1}$$

Then $F_X(\cdot)$ is called the *cumulative distribution function* (cdf) of X.

Definition 6.2.3: Let X be a random variable on $(\Omega, \mathcal{F}, P)$. Let

$$P_X(A) \equiv P(X^{-1}(A)) \quad \text{for all} \quad A \in \mathcal{B}(\mathbb{R}). \tag{2.2}$$

Then the probability measure P_X is called the *probability distribution* of X.

Note that P_X is the measure induced by X on $(\mathbb{R}, \mathcal{B}(\mathbb{R}))$ under P and that the Lebesgue-Stieltjes measure μ_{F_X} on $\mathcal{B}(\mathbb{R})$ corresponding to the cdf F_X of X is the same as P_X.

Definition 6.2.4: Let $(\Omega, \mathcal{F}, P)$ be a probability space, $k \in \mathbb{N}$ and $X : \Omega \to \mathbb{R}^k$ be $\langle \mathcal{F}, \mathcal{B}(\mathbb{R}^k) \rangle$-measurable, i.e., $X^{-1}(A) \in \mathcal{F}$ for all $A \in \mathcal{B}(\mathbb{R}^k)$. Then X is called a (k-dimensional) *random vector* on $(\Omega, \mathcal{F}, P)$.

Let $X = (X_1, X_2, \ldots, X_k)$ be a random vector with components X_i, $i = 1, 2, \ldots, k$. Then each X_i is a random variable on $(\Omega, \mathcal{F}, P)$. This follows from the fact that the coordinate projection maps from $\mathbb{R}^k$ to $\mathbb{R}$, given by

$$\pi_i(x_1, x_2, \ldots, x_k) \equiv x_i, \ 1 \leq i \leq k$$

are continuous and hence, are Borel measurable. Conversely, if for $1 \leq i \leq k$, X_i is a random variable on $(\Omega, \mathcal{F}, P)$, then $X = (X_1, X_2, \ldots, X_k)$ is a random vector (cf. Proposition 2.1.3).

Definition 6.2.5: Let X be a k-dimensional random vector on $(\Omega, \mathcal{F}, P)$ for some $k \in \mathbb{N}$. Let

$$F_X(x) \equiv P(\{\omega : X_1(\omega) \leq x_1, X_2(\omega) \leq x_2, \ldots, X_k(\omega) \leq x_k\}) \qquad (2.3)$$

for $x = (x_1, x_2, \ldots, x_k) \in \mathbb{R}^k$. Then $F_X(\cdot)$ is called the *joint cumulative distribution function* (joint cdf) of the random vector X.

Definition 6.2.6: Let X be a k-dimensional random vector on $(\Omega, \mathcal{F}, P)$ for some $k \in \mathbb{N}$. Let

$$P_X(A) = P(X^{-1}(A)) \quad \text{for all} \quad A \in \mathcal{B}(\mathbb{R}^k). \qquad (2.4)$$

The probability measure P_X is called the *(joint) probability distribution* of X.

As in the case $k = 1$, the Lebesgue-Stieltjes measure μ_{F_X} on $\mathcal{B}(\mathbb{R}^k)$ corresponding to the joint cdf F_X is the same as P_X.

Next, let $X = (X_1, X_2, \ldots, X_k)$ be a random vector. Let $Y = (X_{i_1}, X_{i_2}, \ldots, X_{i_r})$ for some $1 \leq i_1 < i_2 < \ldots < i_r \leq k$ and some $1 \leq r \leq k$. Then, Y is also a random vector. Further, the joint cdf of Y can be obtained from F_X by setting the components x_j, $j \notin \{i_1, i_2, \ldots, i_r\}$ equal to ∞. Similarly, the probability distribution P_Y can be obtained from P_X as an induced measure from the projection map $\pi(x) = (x_{i_1}, x_{i_2}, \ldots, x_{i_r})$, $x \in \mathbb{R}^k$. For example, if $(i_1, i_2, \ldots, i_r) = (1, 2, \ldots, r)$, $r \leq k$, then

$$F_Y(y_1, \ldots, y_r) = F_X(y_1, \ldots, y_r, \infty, \ldots, \infty), \ (y_1, \ldots, y_r) \in \mathbb{R}^r$$

and

$$P_Y(A) = P_X(A \times \mathbb{R}^{(k-r)}), \ A \in \mathcal{B}(\mathbb{R}^r).$$

Definition 6.2.7: Let $X = (X_1, X_2, \ldots, X_k)$ be a random vector on $(\Omega, \mathcal{F}, P)$. Then, for each $i = 1, \ldots, k$, the cdf F_{X_i} and the probability distribution P_{X_i} of the random variable X_i are called the *marginal* cdf and the *marginal probability distribution* of X_i, respectively.

It is clear that the distribution of X determines the marginal distribution P_{X_i} of X_i for all $i = 1, 2, \ldots, k$. However, the marginal distributions $\{P_{X_i} : i = 1, 2, \ldots, k\}$ do not uniquely determine the joint distribution P_X, without additional conditions, such as independence (see Problem 6.1).

Definition 6.2.8: Let X be a random variable on $(\Omega, \mathcal{F}, P)$. The *expected value* of X, denoted by EX or $E(X)$, is defined as

$$EX = \int_\Omega X dP, \tag{2.5}$$

provided the integral is well defined. That is, at least one of the two quantities $\int X^+ dP$ and $\int X^- dP$ is finite.

If X is a random variable on $(\Omega, \mathcal{F}, P)$ and $h : \mathbb{R} \to \mathbb{R}$ is Borel measurable, then $Y = h(X)$ is also a random variable on $(\Omega, \mathcal{F}, P)$. The expected value of Y may be computed as follows.

Proposition 6.2.1: (*The change of variable formula*). *Let X be a random variable on $(\Omega, \mathcal{F}, P)$ and $h : \mathbb{R} \to \mathbb{R}$ be Borel measurable. Let $Y = h(X)$. Then*

(i) $\displaystyle\int_\Omega |Y| dP = \int_\mathbb{R} |h(x)| P_X(dx) = \int_\mathbb{R} |y| P_Y(dy).$

(ii) If $\int_\Omega |Y| dP < \infty$, *then*

$$\int_\Omega Y dP = \int_\mathbb{R} h(x) P_X(dx) = \int_\mathbb{R} y P_Y(dy). \tag{2.6}$$

Proof: If $h = I_A$ for A in $\mathcal{B}(\mathbb{R})$, the proposition follows from the definition of P_X. By linearity, this extends to a nonnegative and simple function h and by the MCT, to any nonnegative measurable h, and hence to any measurable h. □

Remark 6.2.1: Proposition 6.2.1 shows that the expectation of Y can be computed in three different ways, i.e., by integrating Y with respect to P on Ω or by integrating $h(x)$ on $\mathbb{R}$ with respect to the probability distribution P_X of the random variable X or by integrating y on $\mathbb{R}$ with respect to the probability distribution P_Y of the random variable Y.

Remark 6.2.2: If the function h is nonnegative, then the relation $EY = \int_\mathbb{R} h(x) P_X(dx)$ is valid even if $EY = \infty$.

Definition 6.2.9: For any positive integer n, the nth *moment* μ_n of a random variable X is defined by

$$\mu_n \equiv EX^n, \tag{2.7}$$

provided the expectation is well defined.

Definition 6.2.10: The variance of a random variable X is defined as $\text{Var}(X) = E(X - EX)^2$, provided $EX^2 < \infty$.

Definition 6.2.11: The *moment generating function (mgf)* of a random variable X is defined by

$$M_X(t) \equiv E(e^{tX}) \quad \text{for all} \quad t \in \mathbb{R}. \tag{2.8}$$

Since e^{tX} is always nonnegative, $E(e^{tX})$ is well defined but could be infinity. Proposition 6.2.1 gives a way of computing the moments and the mgf of X without explicitly computing the distribution of X^k or e^{tX}. As an illustration, consider the case of a random variable X defined on the probability space $(\Omega, \mathcal{F}, P)$ with $\Omega = \{H, T\}^n$, $n \in \mathbb{N}$, $\mathcal{F}$ = the power set of Ω and P = the probability distribution defined by

$$P(\{\omega\}) = p^{X(\omega)} q^{n-X(\omega)}$$

where $0 < p < 1$, $q = 1 - p$, and $X(\omega)$ = the number of H's in ω. By the change of variable formula, the mgf of X is given by

$$\begin{aligned} M_X(t) &\equiv \int e^{tx} P_X(dx) \\ &= \sum_{r=0}^{n} e^{tr} \binom{n}{r} p^r q^{n-r} = (pe^t + q)^n, \end{aligned}$$

since P_X, the probability distribution of X, is supported on $\{0, 1, 2, \ldots, n\}$ with $P_X(\{r\}) = \binom{n}{r} p^r q^{n-r}$. Note that P_X is the Binomial (n, p) distribution. Here, $M_X(t)$ is computed using the distribution of X, i.e., using the middle term in (2.6) only.

The connection between the mgf $M_X(\cdot)$ and the moments of a random variable X is given in the following propositions.

Proposition 6.2.2: *Let X be a nonnegative random variable and $t \geq 0$. Then*

$$M_X(t) \equiv E(e^{tX}) = \sum_{n=0}^{\infty} \frac{t^n \mu_n}{n!} \tag{2.9}$$

where μ_n is as in (2.7).

Proof: Since $e^{tX} = \sum_{n=0}^{\infty} t^n \frac{X^n}{n!}$ and X is nonnegative, (2.9) follows from the MCT. □

Proposition 6.2.3: *Let X be a random variable and let $M_X(t)$ be finite for all $|t| < \epsilon$, for some $\epsilon > 0$. Then*

(i) $E|X|^n < \infty$ *for all* $n \geq 1$,

(ii) $M_X(t) = \sum_{n=0}^{\infty} t^n \frac{\mu_n}{n!}$ *for all* $|t| < \epsilon$,

(iii) $M_X(\cdot)$ *is infinitely differentiable on* $(-\epsilon, +\epsilon)$ *and for* $r \in \mathbb{N}$, *the* r*th derivative of* $M_X(\cdot)$ *is*

$$M_X^{(r)}(t) = \sum_{n=0}^{\infty} \frac{t^n}{n!} \mu_{n+r} = E(e^{tX} X^r) \quad \text{for} \quad |t| < \epsilon. \tag{2.10}$$

In particular,

$$M_X^{(r)}(0) = \mu_r = EX^r. \tag{2.11}$$

Proof: Since $M_X(t) < \infty$ for all $|t| < \epsilon$,

$$E(e^{|tX|}) \leq E(e^{tX}) + E(e^{-tX}) < \infty \quad \text{for} \quad |t| < \epsilon. \tag{2.12}$$

Also, $e^{|tX|} \geq \frac{|t|^n |X|^n}{n!}$ for all $n \in \mathbb{N}$ and hence, (i) follows by choosing a t in $(0, \epsilon)$. Next note that $\left|\sum_{j=0}^{n} \frac{(tx)^j}{j!}\right| \leq e^{|tx|}$ for all x in $\mathbb{R}$ and all $n \in \mathbb{N}$. Hence, by (2.12) and the DCT, (ii) follows.

Turning to (iii), since $M_X(\cdot)$ admits a power series expansion convergent in $|t| < \epsilon$, it is infinitely differentiable in $|t| < \epsilon$ and the derivatives of $M_X(\cdot)$ can be found by term-by-term differentiation of the power series (see Rudin (1976), Chapter 9). Hence,

$$\begin{aligned} M_X^{(r)}(t) &= \frac{d^r}{dt^r} \left(\sum_{n=0}^{\infty} \frac{t^n \mu_n}{n!} \right) \\ &= \sum_{n=0}^{\infty} \frac{\mu_n}{n!} \frac{d^r(t^n)}{dt^r} \\ &= \sum_{n=r}^{\infty} \mu_n \frac{t^{n-r}}{(n-r)!} \\ &= \sum_{n=0}^{\infty} \frac{t^n}{n!} \mu_{n+r} . \end{aligned}$$

The verification of the second equality in (2.10) is left in an exercise (see Problem 6.4). □

Remark 6.2.3: If the mgf $M_X(\cdot)$ is finite for $|t| < \epsilon$ for some $\epsilon > 0$, then by part (ii) of the above proposition, $M_X(t)$ has a power series expansion in t around 0 and $\frac{\mu_n}{n!}$ is simply the coefficient of t^n. For example, if X has a $N(0,1)$ distribution, then for all $t \in \mathbb{R}$,

$$\begin{aligned} M_X(t) &= \int_{-\infty}^{\infty} e^{tx} \frac{1}{\sqrt{2\pi}} e^{-x^2/2} dx = e^{t^2/2} \\ &= \sum_{k=0}^{\infty} \frac{(t^2)^k}{k!} \frac{1}{2^k} . \end{aligned} \tag{2.13}$$

Thus, $\mu_n = \begin{cases} 0 & \text{if } n \text{ is odd} \\ \frac{(2k)!}{k!2^k} & \text{if } n = 2k, k = 1, 2, \ldots \end{cases}$

Remark 6.2.4: If $M_X(t)$ is finite for $|t| < \epsilon$ for some $\epsilon > 0$, then all the moments $\{\mu_n\}_{n\geq 1}$ of X are determined and also its probability distribution. However, in general, the sequence $\{\mu_n\}_{n\geq 1}$ of moments of X need not determine the distribution of X uniquely.

Table 6.2.1 gives the mean, variance, and the mgf of a number of standard probability distributions on the real line.

For future reference, some of the inequalities established in Section 3.1 are specialized for random variables and collected below without proofs.

Proposition 6.2.4: (*Markov's inequality*). *Let X be a random variable on $(\Omega, \mathcal{F}, P)$. Then for any $\phi : \mathbb{R}_+ \to \mathbb{R}_+$ nondecreasing and any $t > 0$ with $\phi(t) > 0$,*

$$P(|X| \geq t) \leq \frac{E(\phi(|X|))}{\phi(t)} . \tag{2.14}$$

In particular,

(i) for $r > 0$, $t > 0$,

$$P(X \geq t) \leq P(|X| \geq t) \leq \frac{E|X|^r}{t^r}, \tag{2.15}$$

(ii) for any $t \geq 0$,

$$P(|X| \geq t) \leq \frac{E(e^{\theta|X|})}{e^{\theta t}} ,$$

for any $\theta > 0$ and hence

$$P(|X| \geq t) \leq \inf_{\theta > 0} \frac{E(e^{\theta|X|})}{e^{\theta t}} . \tag{2.16}$$

Proposition 6.2.5: (*Chebychev's inequality*). *Let X be a random variable with $EX^2 < \infty$, $EX = \mu$, $Var(X) = \sigma^2$. Then for any $k > 0$,*

$$P(|X - \mu| \geq k\sigma) \leq \frac{1}{k^2} . \tag{2.17}$$

TABLE 6.2.1. Mean, variance, mgf of the distributions listed in Tables 4.5.1 and 4.5.2.

Distribution	Mean	Variance	mgf M(t)
Bernoulli (p), $0 < p < 1$	p	$p(1-p)$	$(1-p)+pe^t$, $t \in \mathbb{R}$
Binomial (n, p), $p \in (0,1)$, $n \in \mathbb{N}$	np	$np(1-p)$	$\big((1-p)+pe^t\big)^n$, $t \in \mathbb{R}$
Geometric (p), $p \in (0,1)$	$\frac{1}{p}$	$\frac{1-p}{p}$	$\frac{pe^t}{1-(1-p)e^t}$, $t \in \big(-\infty, -\log(1-p)\big)$
Poisson (λ) $\lambda \in (0,\infty)$	λ	λ	$\exp\big(\lambda(e^t-1)\big)$, $t \in \mathbb{R}$
Uniform (a, b), $-\infty < a < b < \infty$	$\frac{a+b}{2}$	$\frac{(b-a)^2}{12}$	$\frac{e^{bt}-e^{at}}{(b-a)t}$, $t \in \mathbb{R} \setminus \{0\}$; $M(0) = 1$
Exponential (β), $\beta \in (0,\infty)$	β	β^2	$\frac{1}{1-\beta t}$, $t \in (-\infty, \frac{1}{\beta})$
Gamma (α, β), $\alpha, \beta \in (0,\infty)$	$\alpha\beta$	$\alpha\beta^2$	$(1-\beta t)^{-\alpha}$, $t \in (-\infty, \frac{1}{\beta})$
Beta (α, β), $\alpha, \beta \in (0,\infty)$	$\frac{\alpha}{\alpha+\beta}$	$\frac{\alpha\beta}{(\alpha+\beta)^2(\alpha+\beta+1)}$	$\Big[1+\sum_{k=1}^{\infty}\Big(\prod_{r=0}^{k-1}\frac{\alpha+r}{\alpha+\beta+r}\Big)\frac{t^k}{k!}\Big]$, $t \in \mathbb{R}$
Cauchy (γ, σ), $\gamma \in \mathbb{R}$, $\sigma \in (0,\infty)$	not defined since $E\lvert X\rvert = \infty$	not defined since $E\lvert X\rvert^2 = \infty$	∞ for all $t \neq 0$
Normal (γ, σ^2), $\gamma \in \mathbb{R}$, $\sigma \in (0,\infty)$	γ	σ^2	$\exp(\gamma t + \frac{t^2\sigma^2}{2})$, $t \in \mathbb{R}$
Lognormal (γ, σ^2), $\gamma \in \mathbb{R}$, $\sigma \in (0,\infty)$	$e^{\gamma+\frac{\sigma^2}{2}}$	$[e^{2(\gamma+\sigma^2)} - e^{2\gamma+\sigma^2}]$	∞ for all $t > 0$

Proposition 6.2.6: *(Jensen's inequality). Let X be a random variable with $P(a < X < b) = 1$ for $-\infty \leq a < b \leq \infty$. Let $\phi : (a, b) \to \mathbb{R}$ be convex on (a, b). Then*

$$E\phi(X) \geq \phi(EX), \tag{2.18}$$

provided $E|X| < \infty$ and $E|\phi(X)| < \infty$.

Proposition 6.2.7: *(Hölder's inequality). Let X and Y be random variables on $(\Omega, \mathcal{F}, P)$ with $E|X|^p < \infty$, $E|Y|^q < \infty$, $1 < p < \infty$, $1 < q < \infty$, $\frac{1}{p} + \frac{1}{q} = 1$. Then*

$$E|XY| \leq (E|X|^p)^{1/p}(E|Y|^q)^{1/q}, \tag{2.19}$$

with equality holding iff $P(c_1|X|^p = c_2|Y|^q) = 1$ for some $0 \leq c_1, c_2 < \infty$.

Proposition 6.2.8: *(Cauchy-Schwarz inequality). Let X and Y be random variables on $(\Omega, \mathcal{F}, P)$ with $E|X|^2 < \infty, E|Y|^2 < \infty$. Then*

$$|Cov(X, Y)| \leq \sqrt{Var(X)}\sqrt{Var(Y)}\ , \tag{2.20}$$

where $Cov(X, Y) = EXY - EXEY$. If $Var(X) > 0$, then equality holds in (2.20) iff $P(Y = aX + b) = 1$ for some constants a, b in $\mathbb{R}$ (Problem 6.6).

Proposition 6.2.9: *(Minkowski's inequality). Let X and Y be random variables on $(\Omega, \mathcal{F}, P)$ such that $E|X|^p < \infty, E|Y|^p < \infty$ for some $1 \leq p < \infty$. Then*

$$(E|X + Y|^p)^{1/p} \leq (E|X|^p)^{1/p} + (E|Y|^p)^{1/p}. \tag{2.21}$$

Product Moments of Random Vectors

Definition 6.2.12: Let $X = (X_1, X_2, \ldots, X_k)$ be a random vector. The *product moment of order* $r = (r_1, r_2, \ldots, r_k)$, with r_i being a nonnegative integer for each i, is defined as

$$\mu_r \equiv \mu_{r_1, r_2, \ldots, r_k} \equiv E(X_1^{r_1} X_2^{r_2} \cdots X_k^{r_k}), \tag{2.22}$$

provided $E|X_1^{r_1} \cdots X_k^{r_k}| < \infty$. The *joint moment generating function (joint mgf)* of a random vector $X = (X_1, X_2, \ldots, X_k)$ is defined by

$$M_{X_1, \ldots, X_k}(t_1, t_2, \ldots, t_k) \equiv E(e^{t_1 X_1 + t_2 X_2 + \cdots + t_k X_k}), \tag{2.23}$$

for all $t_1, t_2, \ldots, t_k$ in $\mathbb{R}$.

As in the case of a random variable, if the joint mgf $M_{X_1, X_2, \ldots, X_k}(t_1, \ldots, t_k)$ is finite for all $(t_1, t_2, \ldots, t_k)$ with $|t_i| < \epsilon$ for all $i = 1, 2, \ldots, k$ for some $\epsilon > 0$, then an analog of Proposition 6.2.3 holds. For example, the following assertions are valid (cf. Problem 6.4):

(i)

$$E|X_i|^n < \infty \quad \text{for all} \quad i = 1, 2, \ldots, k \quad \text{and} \quad n \geq 1. \tag{2.24}$$

(ii) For $t = (t_1, \ldots, t_k) \in \mathbb{R}^k$ and $r = (r_1, r_2, \ldots, r_k) \in \mathbb{Z}_+^k$, let

$$\begin{aligned} t^r &= t_1^{r_1} t_2^{r_2} \cdots t_k^{r_k}, \\ r! &= r_1! r_2! \cdots r_k!, \quad \text{and} \\ \mu_r &= EX_1^{r_1} X_2^{r_2} \cdots X_k^{r_k}. \end{aligned}$$

Then,

$$M_X(t_1, \ldots, t_k) = \sum_{r \in \mathbb{Z}_+^k} \frac{t^r}{r!} \mu_r \tag{2.25}$$

for all $t = (t_1, t_2, \ldots, t_k) \in (-\epsilon, +\epsilon)^k$.

(iii) For any $r = (r_1, \ldots, r_k) \in \mathbb{Z}_+^k$,

$$\frac{d^r}{dt^r} M_X(t)\Big|_{t=0} = \mu_r, \tag{2.26}$$

where $\frac{d^r}{dt^r} = \frac{\partial^{r_1}}{\partial t_1^{r_1}} \frac{\partial^{r_2}}{\partial t_2^{r_2}} \cdots \frac{\partial^{r_k}}{\partial t_k^{r_k}}$.

6.3 Kolmogorov's consistency theorem

In the previous section, the case of a single random variable and that of a finite dimensional random vector were discussed. The goal of this section is to discuss infinite families of random variables such as a random sequence $\{X_n\}_{n\geq 1}$ or a random function $\{X(t) : 0 \leq t < T\}, 0 \leq T \leq \infty$. For example, X_n could be the population size of the nth generation of a randomly evolving biological population and $X(t)$ could be the temperature at time t in a chemical reaction over a period $[0, T]$. An example from modeling of spatial random phenomenon is a collection $\{X(s) : s \in S\}$ of random variables $X(s)$ where S is a specified region such as the U.S. and $X(s)$ is the amount of rainfall at location $s \in S$, during a specified month.

Let $(\Omega, \mathcal{F}, P)$ be a probability space and $\{X_\alpha : \alpha \in A\}$ be a collection of random variables defined on $(\Omega, \mathcal{F}, P)$, where A is a nonempty set. Then for any $(\alpha_1, \alpha_2, \ldots, \alpha_k) \in A^k, 1 \leq k < \infty$, the random vector $(X_{\alpha_1}, X_{\alpha_2}, \ldots, X_{\alpha_k})$ has a joint probability distribution $\mu_{(\alpha_1, \alpha_2, \ldots, \alpha_k)}$ over $(\mathbb{R}^k, \mathcal{B}(\mathbb{R}^k))$.

Definition 6.3.1: A (real valued) *stochastic process* with index set A is a family $\{X_\alpha : \alpha \in A\}$ of random variables defined on a probability space $(\Omega, \mathcal{F}, P)$.

Example 6.3.1: (*Examples of stochastic processes*). Let $\Omega = [0, 1]$, $\mathcal{F} = \mathcal{B}([0, 1])$, $P =$ the Lebesgue measure on $[0, 1]$. Let $A_1 = \{1, 2, 3, \ldots\}$, $A_2 = [0, T]$, $0 < T < \infty$. For $\omega \in \Omega$, $n \in A_1$, $t \in A_2$, let

$$\begin{aligned} X_n(\omega) &= \sin 2\pi n\omega \\ Y_t(\omega) &= \sin 2\pi t\omega \\ Z_n(\omega) &= n\text{th digit in the decimal expansion of } \omega \\ V_{n,t}(\omega) &= X_n^2(\omega) + Y_t^2(\omega). \end{aligned}$$

Then $\{X_n : n \in A_1\}$, $\{Z_n : n \in A_1\}$, $\{V_{n,t} : (n, t) \in A_1 \times A_2\}$, $\{Y_t : t \in A_2\}$ are all stochastic processes.

Note that a real valued stochastic process $\{X_\alpha : \alpha \in A\}$ may also be viewed as a random real valued function on the set A by the identification $\omega \to f(\omega, \cdot)$, where $f(\omega, \alpha) = X_\alpha(\omega)$ for α in A.

Definition 6.3.2: The family $\{\mu_{(\alpha_1,\alpha_2,\ldots,\alpha_k)}(\cdot) \equiv P((X_{\alpha_1}, \ldots, X_{\alpha_k}) \in \cdot)$: $(\alpha_1, \alpha_2, \ldots, \alpha_k) \in A^k, 1 \leq k < \infty\}$ of probability distributions is called the *family of finite dimensional distributions (fdds)* associated with the stochastic process $\{X_\alpha : \alpha \in A\}$.

This family of finite dimensional distributions satisfies the following *consistency* conditions: For any $(\alpha_1, \alpha_2, \ldots, \alpha_k) \in A^k$, $2 \leq k < \infty$, and any $B_1, B_2, \ldots, B_k$ in $\mathcal{B}(\mathbb{R})$,

C1: $\mu_{(\alpha_1,\alpha_2,\ldots,\alpha_k)}(B_1 \times \cdots \times B_{k-1} \times \mathbb{R}) = \mu_{(\alpha_1,\alpha_2,\ldots,\alpha_{k-1})}(B_1 \times \cdots \times B_{k-1})$;

C2: For any permutation $(i_1, i_2, \ldots, i_k)$ of $(1, 2, \ldots, k)$,

$$\mu_{(\alpha_{i_1},\alpha_{i_2},\ldots,\alpha_{i_k})}(B_{i_1} \times B_{i_2} \times \cdots \times B_{i_k}) = \mu_{(\alpha_1,\ldots,\alpha_k)}(B_1 \times B_2 \times \cdots \times B_k)\,.$$

To verify C1, note that

$$\begin{aligned}
&\mu_{(\alpha_1,\alpha_2,\ldots,,\alpha_k)}(B_1 \times B_2 \times \cdots \times B_{k-1} \times \mathbb{R}) \\
&= P(X_{\alpha_1} \in B_1, X_{\alpha_2} \in B_2, \ldots, X_{\alpha_{k-1}} \in B_{k-1}, X_{\alpha_k} \in \mathbb{R}) \\
&= P(X_{\alpha_1} \in B_1, X_{\alpha_2} \in B_2, \ldots, X_{\alpha_{k-1}} \in B_{k-1}) \\
&= \mu_{(\alpha_1,\alpha_2,\ldots,\alpha_{k-1})}(B_1 \times B_2 \times \cdots \times B_{k-1}).
\end{aligned}$$

Similarly, to verify C2, note that

$$\begin{aligned}
&\mu_{(\alpha_{i_1},\alpha_{i_2},\ldots,\alpha_{i_k})}(B_{i_1} \times B_{i_2} \cdots \times B_{i_k}) \\
&= P(X_{\alpha_{i_1}} \in B_{i_1}, X_{\alpha_{i_2}} \in B_{i_2}, \ldots, X_{\alpha_{i_k}} \in B_{i_k}) \\
&= P(X_{\alpha_1} \in B_1, X_{\alpha_2} \in B_2, \ldots, X_{\alpha_k} \in B_k) \\
&= \mu_{(\alpha_1,\alpha_2,\ldots,\alpha_k)}(B_1 \times B_2 \times \cdots \times B_k).
\end{aligned}$$

A natural question is that given a family of probability distributions $Q_A \equiv \{\nu_{(\alpha_1,\alpha_2,\ldots,\alpha_k)} : (\alpha_1, \alpha_2, \ldots, \alpha_k) \in A^k,\ 1 \leq k < \infty\}$ on finite dimensional Euclidean spaces, does there exist a real valued stochastic process $\{X_\alpha : \alpha \in A\}$ such that its family of finite dimensional distributions coincides with Q_A?

Kolmogorov (1956) showed that if Q_A satisfies C1 and C2, then such a stochastic process does exist. This is known as Kolmogorov's consistency theorem (also known as Kolmogorov's existence theorem).

Theorem 6.3.1: (*Kolmogorov's consistency theorem*). *Let A be a nonempty set. Let $Q_A \equiv \{\nu_{(\alpha_1,\alpha_2,\ldots,\alpha_k)} : (\alpha_1, \alpha_2, \ldots, \alpha_k) \in A^k, 1 \leq k < \infty\}$ be a family of probability distributions such that for each $(\alpha_1, \alpha_2, \ldots, \alpha_k) \in A^k$, $1 \leq k < \infty$,* *(i)* $\nu_{(\alpha_1,\alpha_2,\ldots,\alpha_k)}$ *is a probability distribution on* $(\mathbb{R}^k, \mathcal{B}(\mathbb{R}^k))$,

(ii) C1 and C2 hold, i.e., for all $B_1, B_2, \ldots, B_k \in \mathcal{B}(\mathbb{R})$, $2 \le k < \infty$,

$$\begin{aligned}\nu_{(\alpha_1,\alpha_2,\ldots,\alpha_k)}&(B_1 \times B_2 \times \cdots \times B_{k-1} \times \mathbb{R}) \\ &= \nu_{(\alpha_1,\alpha_2,\ldots,\alpha_{k-1})}(B_1 \times B_2 \times \cdots \times B_{k-1}) \end{aligned} \tag{3.1}$$

and for any permutation $(i_1, i_2, \ldots, i_k)$ *of* $(1, 2, \ldots, k)$,

$$\begin{aligned}\mu_{(\alpha_{i_1},\alpha_{i_2},\ldots,\alpha_{i_k})}&(B_{i_1} \times B_{i_2} \times \cdots \times B_{i_k}) \\ &= \mu_{(\alpha_1,\alpha_2,\ldots,\alpha_k)}(B_1 \times B_2 \times \cdots \times B_k). \end{aligned} \tag{3.2}$$

Then, there exists a probability space $(\Omega, \mathcal{F}, P)$ *and a stochastic process* $X_A \equiv \{X_\alpha : \alpha \in A\}$ *on* $(\Omega, \mathcal{F}, P)$ *such that* Q_A *is the family of finite dimensional distributions associated with* X_A.

Remark 6.3.1: Thus the above theorem says that given the family Q_A satisfying conditions (i) and (ii), there exists a real valued function on $A \times \Omega$ such that for each ω, $f(\cdot, \omega)$ is a function on A and for each $(\alpha_1, \alpha_2, \ldots, \alpha_k) \in A^k$, the vector $\big(f(\alpha_1, \omega), f(\alpha_2, \omega), \ldots, f(\alpha_k, \omega)\big)$ is a random vector with probability distribution $\nu_{(\alpha_1,\alpha_2,\ldots,\alpha_k)}$. This random function point of view is useful in dealing with functionals of the form $M(\omega) \equiv \{\sup f(\alpha, \omega) : \alpha \in A\}$. For example, if $A_1 = \{1, 2, \ldots\}$, then one might consider functionals such as $\lim_{n\to\infty} f(n, \omega)$, $\lim_{n\to\infty} \frac{1}{n} \sum_{j=1}^{n} f(j, \omega)$, $\sum_{j=1}^{\infty} f(j, \omega)$, etc. Since the random functionals are not fully determined by $f(\alpha, \omega)$ for finitely many α's, it is not possible to compute probabilities of events defined in terms of these functionals from the knowledge of the finite dimensional distribution of $(f(\alpha_1, \omega), \ldots, f(\alpha_k, \omega))$ for a given $(\alpha_1, \ldots, \alpha_k)$, no matter how large k is. Kolmogorov's consistency theorem allows one to compute these probabilities given *all* finite dimensional distributions (provided that the functionals satisfy appropriate measurability conditions).

Given a probability measure μ on $(\mathbb{R}, \mathcal{B}(\mathbb{R}))$, now consider the problem of constructing a probability space $(\Omega, \mathcal{F}, P)$ and a random variable X on it with distribution μ. A natural solution is to set the sample space Ω to be $\mathbb{R}$, the σ-algebra $\mathcal{F}$ to be $\mathcal{B}(\mathbb{R})$, and the probability measure P to be μ and the random variable X to be the identity map $X(\omega) \equiv \omega$. Similarly, given a probability measure μ on $(\mathbb{R}^k, \mathcal{B}(\mathbb{R})^k)$, one can set the sample space Ω to be $\mathbb{R}^k$ and the σ-algebra $\mathcal{F}$ to be $\mathcal{B}(\mathbb{R}^k)$ and the probability measure P to be μ and the random vector X to be the identity map.

Arguing in the same fashion, given a family Q_A of finite dimensional distributions with index set A, to construct a stochastic process $\{X_\alpha : \alpha \in A\}$ with index set A on some probability space $(\Omega, \mathcal{F}, P)$, it is natural to set the sample space Ω to be $\mathbb{R}^A$, the collection of all real valued functions on A, $\mathcal{F}$ to be a suitable σ-algebra that includes all finite dimensional events,

P to be an appropriate probability measure that yields Q_A, and X to be the identity map.

These considerations lead to the following definitions.

Definition 6.3.3: Let A be a nonempty set. Then $\mathbb{R}^A \equiv \{f \mid f : A \to \mathbb{R}\}$, the collection of all real valued functions on A.

If A is a finite set $\{a_1, a_2, \ldots, a_k\}$, then $\mathbb{R}^A$ can be identified with $\mathbb{R}^k$ by associating each $f \in \mathbb{R}^A$ with the vector $(f(a_1), f(a_2), \ldots, f(a_k))$ in $\mathbb{R}^k$. If A is a countably infinite set $\{a_1, a_2, a_3, \ldots\}$, then $\mathbb{R}^A$ can be similarly identified with $\mathbb{R}^\infty$, the set of all sequences $\{x_1, x_2, x_3, \ldots\}$ of real numbers. If A is the interval $[0, 1]$, then $\mathbb{R}^A$ is the collection of all real valued functions on $[0, 1]$.

Definition 6.3.4: Let A be a nonempty set. A subset $C \subset \mathbb{R}^A$ is called a *finite dimensional cylinder set* (*fdcs*) if there exists a finite subset $A_1 \subset A$, say, $A_1 \equiv \{\alpha_1, \alpha_2, \ldots, \alpha_k\}, 1 \le k < \infty$ and a Borel set B in $\mathcal{B}(\mathbb{R}^k)$ such that $C = \{f : f \in \mathbb{R}^A$ and $(f(\alpha_1), f(\alpha_2), \ldots, f(\alpha_k)) \in B\}$. The set B is called a *base* for C.

The collection of all finite dimensional cylinder sets will be denoted by $\mathcal{C}$.

The name *cylinder* is motivated by the following example:

Example 6.3.2: Let $A = \{1, 2, 3\}$ and $C = \{(x_1, x_2, x_3) : x_1^2 + x_2^2 \le 1\}$. Then C is a cylinder (in the usual sense of the English word), but with infinite height and depth. According to Definition 6.3.4, C is also a cylinder in $\mathbb{R}^3$ with the unit circle in $\mathbb{R}^2$ as its base.

Examples 6.3.3 and 6.3.4 below are examples of fdcs while Example 6.3.5 is an example of a set that is not a fdcs.

Example 6.3.3: Let $A = \{1, 2\}$ and $C = \{(x_1, x_2) : |\sin 2\pi x_1| \le \frac{1}{\sqrt{2}}\}$.

Example 6.3.4: Let $A = \{1, 2, 3, \ldots\}$ and $C = \{(x_1, x_2, x_3, \ldots) : \frac{x_{17}^2}{4} + \frac{x_{30}^2}{10} - \frac{x_{42}^2}{5} \le 10\}$.

Example 6.3.5: Let $A = \{1, 2, 3, \ldots\}$ and $D = \{(x_1, x_2, x_3, \ldots) : x_j \in \mathbb{R}$ for all $j \ge 1$ and $\lim_{n\to\infty} \frac{1}{n} \sum_{j=1}^n x_j$ exists$\}$ is *not* a finite dimensional cylinder set (Problem 6.8).

Proposition 6.3.2: *Let A be a nonempty set and $\mathcal{C}$ be the collection of all finite dimensional cylinder sets in $\mathbb{R}^A$. Then $\mathcal{C}$ is an algebra.*

Proof: Let $C_1, C_2 \in \mathcal{C}$ and let

$$\begin{aligned} C_1 &= \{f : f \in \mathbb{R}^A \text{ and } \big(f(\alpha_1), f(\alpha_2), \ldots, f(\alpha_k)\big) \in B_1\} \\ C_2 &= \{f : f \in \mathbb{R}^A \text{ and } \big(f(\beta_1), f(\beta_2), \ldots, f(\beta_j)\big) \in B_2\} \end{aligned}$$

for some $A_1 = \{\alpha_1, \alpha_2, \ldots, \alpha_k\} \subset A, A_2 = \{\beta_1, \beta_2, \ldots, \beta_j\} \subset A, B_1 \in \mathcal{B}(\mathbb{R}^k), B_2 \in \mathcal{B}(\mathbb{R}^j), 1 \leq k < \infty, 1 \leq j < \infty$. Let $A_3 = A_1 \cup A_2 = \{\gamma_1, \gamma_2, \ldots, \gamma_\ell\}$, where without loss of generality $(\gamma_1, \gamma_2, \ldots, \gamma_k) = (\alpha_1, \alpha_2, \ldots, \alpha_k)$ and $(\gamma_{\ell-j+1}, \ldots, \gamma_{\ell-1}, \gamma_\ell) = (\beta_1, \beta_2, \ldots, \beta_j)$. Then C_1 and C_2 may be expressed as

$$\begin{aligned} C_1 &= \{f : f \in \mathbb{R}^A \text{ and } \big(f(\gamma_1), f(\gamma_2), \ldots, f(\gamma_\ell)\big) \in \tilde{B}_1\} \\ C_2 &= \{f : f \in \mathbb{R}^A \text{ and } \big(f(\gamma_1), \ldots, f(\gamma_\ell)\big) \in \tilde{B}_2\} \end{aligned}$$

where $\tilde{B}_1 = B_1 \times \mathbb{R}^{\ell-k}$ and $\tilde{B}_2 = \mathbb{R}^{\ell-j} \times B_2$. Thus, $C_1 \cup C_2 = \{f : f \in \mathbb{R}^A$ and $(f(\gamma_1), \ldots, f(\gamma_\ell)) \in \tilde{B}_1 \cup \tilde{B}_2\}$. Since both $\tilde{B}_1$ and $\tilde{B}_2$ lie in $\mathcal{B}(\mathbb{R}^\ell)$, $C_1 \cup C_2 \in \mathcal{C}$.

Next note that, $C_1^c = \{f : f \in \mathbb{R}^A$ and $\big(f(\alpha_1), \ldots, f(\alpha_k)\big) \in B_1^c\}$. Since $B_1^c \in \mathcal{B}(\mathbb{R}^k)$, it follows that $C_1^c \in \mathcal{C}$. Thus, $\mathcal{C}$ is an algebra. □

Remark 6.3.2: If A is a *finite* nonempty set, the collection $\mathcal{C}$ is also a σ-algebra.

Definition 6.3.5: Let A be a nonempty set. Let $\mathcal{R}^A$ be the σ-algebra generated by the collection $\mathcal{C}$. Then $\mathcal{R}^A$ is called the *product σ-algebra* on $\mathbb{R}^A$.

Remark 6.3.3: If $A = \{1, 2, 3, \ldots\} \equiv \mathbb{N}$ and $\mathbb{R}^\mathbb{N}$ is identified with the set $\mathbb{R}^\infty$ of all sequences of real numbers, then the product σ-algebra $\mathbb{R}^\mathbb{N}$ coincides with the Borel σ-algebra $\mathcal{B}(\mathbb{R}^\infty)$ on $\mathbb{R}^\infty$ under the metric

$$d(x, y) = \sum_{j=1}^{\infty} \frac{1}{2^j} \left(\frac{|x_j - y_j|}{1 + |x_j - y_j|} \right) \tag{3.3}$$

for $x = (x_1, x_2, \ldots)$, $y = (y_1, y_2, \ldots)$ in $\mathbb{R}^\infty$ (Problem 6.9).

Definition 6.3.6: Let A be a nonempty set. For any $(\alpha_1, \alpha_2, \ldots, \alpha_k) \in A^k$, $1 \leq k < \infty$, the projection map $\pi_{(\alpha_1, \ldots, \alpha_k)}$ from $\mathbb{R}^A$ to $\mathbb{R}^k$ is defined by

$$\pi_{(\alpha_1, \alpha_2, \ldots, \alpha_k)}(f) = (f(\alpha_1), f(\alpha_2), \ldots, f(\alpha_k)). \tag{3.4}$$

In particular, for $\alpha \in A$,

$$\pi_\alpha(f) = f(\alpha) \tag{3.5}$$

is called a *co-ordinate* map.

The projection map π_{A_1} for any arbitrary subset $A_1 \subset A$ may be similarly defined. The next proposition follows from the definition of $\mathcal{R}^A$.

Proposition 6.3.3:

(i) For each $\alpha \in A$, the map π_α from $\mathbb{R}^A$ to $\mathbb{R}$ is $\langle \mathcal{R}^A, \mathcal{B}(\mathbb{R}) \rangle$-measurable.

(ii) For any $(\alpha_1, \alpha_2, \ldots, \alpha_k) \in A^k$, $1 \leq k < \infty$, the map $\pi_{(\alpha_1, \alpha_2, \ldots, \alpha_k)}$ from $\mathbb{R}^A$ to $\mathbb{R}^k$ is $\langle \mathcal{R}^A, \mathcal{B}(\mathbb{R}^k) \rangle$-measurable.

Proof of Theorem 6.3.1: Let $\Omega = \mathbb{R}^A$ and $\mathcal{F} \equiv \mathcal{R}^A$. Define a set function P on $\mathcal{C}$ by

$$P(C) = \mu_{(\alpha_1,\alpha_2,\ldots,\alpha_k)}(B) \tag{3.6}$$

for a C in $\mathcal{C}$ with representation

$$C = \{\omega : \omega \in \mathbb{R}^A,\ \big(\omega(\alpha_1), \omega(\alpha_2), \ldots, \omega(\alpha_k)\big) \in B\}. \tag{3.7}$$

The main steps in the proof are:

(i) To show that $P(C)$ as defined in (3.6) is independent of the representation (3.7) of C, and

(ii) $P(\cdot)$ is countably additive on $\mathcal{C}$.

Next, by the Caratheodory extension theorem (Theorem 1.3.3), there exists a unique extension of P (also denoted by P) to $\mathcal{F}$ such that $(\Omega, \mathcal{F}, P)$ is a probability space. Defining $X_\alpha(\omega) \equiv \pi_\alpha(\omega) = \omega(\alpha)$ for α in A yields a stochastic process $\{X_\alpha : \alpha \in A\}$ on the probability space

$$(\mathbb{R}^A, \mathcal{R}^A, P) \equiv (\Omega, \mathcal{F}, P)$$

with the family Q_A as its set of finite dimensional distributions. Hence, it remains to establish (i) and (ii). Let $C \in \mathcal{C}$ admit two representations:

$$\begin{aligned} C &\equiv \{\omega : \big(\omega(\alpha_1), \omega(\alpha_2), \ldots, \omega(\alpha_k)\big) \in B_1\} \\ &\equiv \pi_{(\alpha_1,\alpha_2,\ldots,\alpha_k)}(B_1) \end{aligned}$$

and

$$\begin{aligned} C &\equiv \{\omega : \big(\omega(\beta_1), \omega(\beta_2), \ldots, \omega(\beta_j)\big) \in B_2\} \\ &\equiv \pi^{-1}_{(\beta_1,\beta_2,\ldots,\beta_j)}(B_2) \end{aligned}$$

for some $A_1 = \{\alpha_1, \alpha_2, \ldots, \alpha_k\} \subset A$, $1 \le k < \infty$, and some $A_2 = \{\beta_1, \beta_2, \ldots, \beta_j\} \subset A$, $1 \le j < \infty$, $B_1 \in \mathcal{B}(\mathbb{R}^k)$ and $B_2 \in \mathcal{B}(\mathbb{R}^j)$. Let $A_3 = A_1 \cup A_2 = \{\gamma_1, \gamma_2, \ldots, \gamma_\ell\}$ and w.l.o.g., let $(\gamma_1, \gamma_2, \ldots, \gamma_k) = (\alpha_1, \alpha_2, \ldots, \alpha_k)$ and $(\gamma_{\ell-j+1}, \gamma_{\ell-j+2}, \gamma_{\ell-1}, \gamma_\ell) = (\beta_1, \beta_2, \ldots, \beta_j)$. Then C may be represented as

$$\begin{aligned} C &= \pi^{-1}_{\gamma_1,\gamma_2,\ldots,\gamma_\ell}(\tilde{B}_1) \\ &= \pi^{-1}_{\gamma_1,\gamma_2,\ldots,\gamma_\ell}(\tilde{B}_2) \end{aligned}$$

where $\tilde{B}_1 = B_1 \times \mathbb{R}^{\ell-k}$ and $\tilde{B}_2 = \mathbb{R}^{\ell-j} \times B_2$. Note that $(\omega(\gamma_1), \ldots, \omega(\gamma_\ell)) \in \tilde{B}_1$ iff $\omega \in C$ iff $(\omega(\gamma_1), \ldots, \omega(\gamma_\ell)) \in \tilde{B}_2$ and thus

$$\tilde{B}_1 = \tilde{B}_2. \tag{3.8}$$

Next by the first consistency condition (3.1) and induction,

$$\nu_{(\gamma_1,\gamma_2,\ldots,\gamma_\ell)}(\tilde{B}_1) = \nu_{(\alpha_1,\alpha_2,\ldots,\alpha_k)}(B_1). \tag{3.9}$$

Also by (3.2), for B_2 of the form $B_{21} \times B_{22} \times \cdots \times B_{2j}$ with $B_{2i} \in \mathcal{B}(\mathbb{R})$ for all $1 \leq i \leq j$,

$$\nu_{(\gamma_1,\gamma_2,\ldots,\gamma_\ell)}(\mathbb{R}^{\ell-j} \times B_2) = \nu_{(\gamma_{\ell-j+1},\ldots,\gamma_\ell,\gamma_1,\gamma_2,\ldots,\gamma_{\ell-j})}(B_2 \times \mathbb{R}^{\ell-j}).$$

Now note that

(a) $\nu_{(\gamma_1,\gamma_2,\ldots,\gamma_\ell)}(\mathbb{R}^{\ell-j} \times B)$ and $\nu_{(\gamma_{\ell-j+1},\ldots,\gamma_\ell,\gamma_1,\gamma_2,\ldots,\gamma_{\ell-j})}(B \times \mathbb{R}^{\ell-j})$, considered as set functions defined for $B \in \mathcal{B}(\mathbb{R}^j)$, are probability measures on $\mathcal{B}(\mathbb{R}^j)$,

(b) they coincide on the class Γ of sets of the form $B = B_{21} \times B_{22} \times \cdots \times B_{2j}$ with $B_{2i} \in \mathcal{B}(\mathbb{R})$ for all i, and

(c) the class Γ is a π-class and it generates $\mathcal{B}(\mathbb{R}^j)$.

Hence, by the uniqueness theorem (Theorem 1.3.6),

$$\nu_{(\gamma_1,\gamma_2,\ldots,\gamma_\ell)}(\mathbb{R}^{\ell-j} \times B) = \nu_{(\gamma_{\ell-j+1},\ldots,\gamma_\ell,\gamma_1,\gamma_2,\ldots,\gamma_{\ell-j})}(B \times \mathbb{R}^{\ell-j}) \tag{3.10}$$

for all $B \in \mathcal{B}(\mathbb{R}^j)$.

Again by (3.1) and induction

$$\begin{aligned} &\nu_{(\gamma_{\ell-j+1},\ldots,\gamma_\ell,\gamma_1,\gamma_2,\ldots,\gamma_{\ell-j})}(B_2 \times \mathbb{R}^{\ell-j}) \\ &= \nu_{(\gamma_{\ell-j+1},\ldots,\gamma_\ell)}(B_2) = \nu_{(\beta_1,\beta_2,\ldots,\beta_j)}(B_2). \end{aligned} \tag{3.11}$$

Since $\tilde{B}_2 = \mathbb{R}^{\ell-j} \times B_2$, by (3.10) and (3.11)

$$\nu_{(\gamma_1,\gamma_2,\ldots,\gamma_\ell)}(\tilde{B}_2) = \nu_{(\beta_1,\beta_2,\ldots,\beta_j)}(B_2).$$

Now from (3.8) and (3.9) it follows that

$$\begin{aligned} \nu_{(\alpha_1,\ldots,\alpha_k)}(B_1) &= \nu_{(\gamma_1,\gamma_2,\ldots,\gamma_\ell)}(\tilde{B}_1) \\ &= \nu_{(\gamma_1,\gamma_2,\ldots,\gamma_\ell)}(\tilde{B}_2) \\ &= \nu_{(\beta_1,\beta_2,\ldots,\beta_j)}(B_2), \end{aligned}$$

thus establishing (i).

To establish (ii), it needs to be shown that

(ii)^a $P(C_1 \cup C_2) = P(C_1) + P(C_2)$ if $C_1, C_2 \in \mathcal{C}$ and $C_1 \cap C_2 = \emptyset$.

(ii)^b $C_n \in \mathcal{C}, C_n \supset C_{n+1}$ for all n, $\cap_{n\geq 1} C_n = \emptyset \Rightarrow P(C_n) \downarrow 0$.

Let $C_1 = \pi^{-1}_{(\alpha_1,\ldots,\alpha_k)}(B_1)$ and $C_2 = \pi^{-1}_{(\beta_1,\ldots,\beta_j)}(B_2)$ for $B_1 \in \mathcal{B}(\mathbb{R}^k)$, $B_2 \in \mathcal{B}(\mathbb{R}^j)$, $\{\alpha_1,\ldots,\alpha_k\} \subset A$ and $\{\beta_1,\ldots,\beta_j\} \subset A$, $1 \leq j, k < \infty$.

As in the proof of Proposition 6.3.2, C_1 and C_2 may be represented as

$$C_i = \pi^{-1}_{(\gamma_1,\gamma_2,\ldots,\gamma_\ell)}(\tilde{B}_i),\ i = 1,2,$$

where $\tilde{B}_i \in \mathcal{B}(\mathbb{R}^\ell)$. Since C_1 and C_2 are disjoint by hypothesis, it follows that $\tilde{B}_1$ and $\tilde{B}_2$ are disjoint. Also since

$$P(C_i) = \nu_{(\gamma_1,\gamma_2,\ldots,\gamma_\ell)}(\tilde{B}_i),\ i = 1,2,$$

and $\nu_{(\gamma_1,\ldots,\gamma_\ell)}(\cdot)$ is a measure on $\mathcal{B}(\mathbb{R}^\ell)$, it follows that

$$\begin{aligned} P(C_1 \cup C_2) &= \nu_{(\gamma_1,\ldots,\gamma_\ell)}(\tilde{B}_1 \cup \tilde{B}_2) \\ &= \nu_{(\gamma_1,\ldots,\gamma_\ell)}(\tilde{B}_1) + \nu_{(\gamma_1,\ldots,\gamma_\ell)}(\tilde{B}_2) \\ &= P(C_1) + P(C_2), \end{aligned}$$

thus proving (ii)^a.

To prove (ii)^b, note that for any sequence $\{C_n\}_{n\geq 1} \subset \mathcal{C}$, there exists a countable set $A_1 = \{\alpha_1, \alpha_2, \ldots, \alpha_n, \ldots\}$, an increasing sequence $\{k_n\}_{n\geq 1}$ of positive integers and a sequence of Borel sets $\{B_n\}_{n\geq 1}$ such that $B_n \in \mathcal{B}(\mathbb{R}^{k_n})$ and $C_n = \pi^{-1}_{(\alpha_1,\alpha_2,\ldots,\alpha_{k_n})}(B_n)$ for all $n \in \mathbb{N}$. Now suppose that $\{C_n\}_{n\geq 1}$ is decreasing. It will be shown that if $\lim_{n\to\infty} P(C_n) = \delta > 0$, then $\cap_{n\geq 1} C_n \neq \emptyset$. For each n, by the regularity of measures (Corollary 1.3.5), there exists a compact set $G_n \subset B_n$ such that

$$\nu_{(\alpha_1,\ldots,\alpha_{k_n})}(B_n \setminus G_n) < \frac{\delta}{2^{n+1}}\,.$$

Let $D_n = \pi^{-1}_{(\alpha_1,\alpha_2,\ldots,\alpha_{k_n})}(G_n)$. Then $P(C_n \setminus D_n) < \frac{\delta}{2^{n+1}}$. Let $H_n = \cap_{j=1}^n D_j$. Then $\{H_n\}_{n\geq 1}$ is decreasing and

$$\begin{aligned} P(C_n \setminus H_n) &= P\big(C_n \cap H_n^c\big) = P\Big(\bigcup_{j=1}^n (C_n \cap D_j^c)\Big) \\ &\leq \sum_{j=1}^n P(C_n \setminus D_j) \leq \sum_{j=1}^n P(C_j \setminus D_j) \\ &\qquad\qquad \text{(since } \{C_n\}_{n\geq 1} \text{ is decreasing)} \\ &\leq \sum_{j=1}^n \frac{\delta}{2^{j+1}} < \frac{\delta}{2}\,. \end{aligned}$$

Since $P(C_n) \downarrow \delta > 0$, $H_n \subset C_n$, and $P(C_n \setminus H_n) < \frac{\delta}{2}$, it follows that $P(H_n) > \frac{\delta}{2}$ for all $n \geq 1$. This implies $H_n \neq \emptyset$ for each n. It will now be shown that $\cap_{n\geq 1} H_n \neq \emptyset$. Let $\{\omega_n\}_{n\geq 1}$ be a sequence of elements from $\Omega = \mathbb{R}^A$ such that for each n, $\omega_n \in H_n$. Then since $\{H_n\}_{n\geq 1}$ is a decreasing sequence, for each $1 \leq j < \infty$, $\omega_n \in H_j$ for $n \geq j$. This implies that the vector $(\omega_n(\alpha_1), \omega_n(\alpha_2), \ldots, \omega_n(\alpha_{k_j})) \in G_j$ for all $n \geq j$. Since G_1 is compact, there exists a subsequence $\{n_{1i}\}_{i\geq 1}$ such that $\lim_{i\to\infty} \omega_{n_{1i}}(\alpha_1) = \omega(\alpha_1)$ exists. Next, since G_2 is compact, there exists a further sequence $\{n_{2i}\}_{i\geq 1}$ of $\{n_{1i}\}_{i\geq 1}$ such that $\lim_{i\to\infty} \omega_{n_{2i}}(\alpha_2) = \omega(\alpha_2)$ exists. Proceeding this way and applying the usual 'diagonal method', a subsequence $\{n_i\}_{i\geq 1}$ is obtained such that $\lim_{i\to\infty} \omega_{n_i}(\alpha_j) = \omega(\alpha_j)$ for all $1 \leq j < \infty$. Let $\omega(\alpha) = 0$ for $\alpha \notin \{\alpha_1, \alpha_2, \ldots\}$. Since for each j, G_j is compact, $(\omega(\alpha_1), \omega(\alpha_2), \ldots, \omega(\alpha_{k_j})) \in G_j$ and hence $\omega \in H_j$. Thus, $\omega \in \cap_{j\geq 1} H_j \subset \cap_{j\geq 1} C_j$ implying $\cap_{j\geq 1} C_j \neq \emptyset$. The proof of the theorem is now complete. □

When the index set A is countable and identified with the set $\mathbb{N} \equiv \{1, 2, 3, \ldots\}$, it is possible to give a simpler formulation of the consistency conditions.

Theorem 6.3.4: *Let $\{\mu_n\}_{n\geq 1}$ be a sequence of probability measures such that*

(i) for each $n \in \mathbb{N}$, μ_n is a probability measure on $(\mathbb{R}^n, \mathcal{B}(\mathbb{R}^n))$,

(ii) for each $n \in \mathbb{N}$, $\mu_{n+1}(B \times \mathbb{R}) = \mu_n(B)$ for all $B \in \mathcal{B}(\mathbb{R}^n)$.

Then there exists a stochastic process $\{X_n : n \geq 1\}$ on a probability space $(\Omega, \mathcal{F}, P)$ with $\Omega = \mathbb{R}^\infty, \mathcal{F} = \mathcal{B}(\mathbb{R}^\infty)$ such that for each $n \geq 1$, the probability distribution $P_{(X_1, X_2, \ldots, X_n)}$ of the random vector $(X_1, X_2, \ldots, X_n)$ is μ_n.

Proof: For any $\{i_1, i_2, \ldots, i_k\} \subset \mathbb{N}$, let $j_1 < j_2 < \cdots < j_k$ be the increasing rearrangement of $i_1, i_2, \ldots, i_k$. Then there exists a permutation $(r_1, r_2, \ldots, r_k)$ of $(1, 2, \ldots, k)$ such that $j_1 = i_{r_1}, j_2 = i_{r_2}, \ldots, j_k = i_{r_k}$.

Now define

$$\nu_{(j_1, j_2, \ldots, j_k)}(\cdot) \equiv \mu_{j_k} \pi^{-1}_{j_1, j_2, \ldots, j_k}(\cdot)$$

where $\pi_{j_1, j_2, \ldots, j_k}(x_1, \ldots, x_{j_k}) = (x_{j_1}, x_{j_2}, \ldots, x_{j_k})$ for all $(x_1, x_2, \ldots, x_{j_k}) \in \mathbb{R}^{j_k}$.

Next define

$$\nu_{(i_1, i_2, \ldots, i_k)}(B_1 \times B_2 \times \ldots \times B_k) \equiv \nu_{(j_1, j_2, \ldots, j_k)}(B_{r_1} \times B_{r_2} \times \ldots \times B_{r_k})$$

where $B_i \in \mathcal{B}(\mathbb{R})$ for all i, $1 \leq i \leq k$. It can be verified that this family of finite dimensional distributions

$$Q_{\mathbb{N}} \equiv \{\nu_{(i_1, i_2, \ldots, i_k)}(\cdot) : \{i_1, i_2, \ldots, i_k\} \subset \mathbb{N},\ 1 \leq k < \infty\} \tag{3.12}$$

satisfies the consistency conditions (3.1) and (3.2) of Theorem 6.3.1 and hence the assertion follows. □

Example 6.3.6: (*Sequence of independent random variables*). Let $\{F_n\}_{n\geq 1}$ be a sequence of cdfs on $\mathbb{R}$. Consider the problem of constructing a sequence $\{X_n\}_{n\geq 1}$ of random variables on a probability space $(\Omega, \mathcal{F}, P)$ such that (i) for each $n \in \mathbb{N}$, X_n has cdf F_n and (ii) for any $n \in \mathbb{N}$ and any $\{i_1, i_2, \ldots, i_n\} \subset \mathbb{N}$, the random variables $\{X_{i_1}, X_{i_2}, \ldots, X_{i_n}\}$ are *independent*, i.e.,

$$P(X_{i_1} \leq x_1, X_{i_2} \leq x_2, \ldots, X_{i_n} \leq x_n) = \prod_{j=1}^{n} F_{i_j}(x_j) \tag{3.13}$$

for all $x_1, x_2, \ldots, x_n$ in $\mathbb{R}$.

This problem can be solved by using Theorem 6.3.4. Let μ_n be the Lebesgue-Stieltjes probability measure on $(\mathbb{R}^n, \mathcal{B}(\mathbb{R}^n))$ corresponding to the distribution function

$$F_{1,2,\ldots,n}(x_1, x_2, \ldots, x_n) \equiv \prod_{j=1}^{n} F(x_j), \quad x_1, \ldots, x_n \in \mathbb{R}.$$

It is easy to verify that the family $\{\mu_n : n \geq 1\}$ satisfies (i) and (ii) of Theorem 6.3.4. Hence, there exist a probability measure P on the sequence space $\Omega \equiv \mathbb{R}^\infty$ equipped with σ-algebra $\mathcal{F} \equiv \mathcal{B}(\mathbb{R}^\infty)$ and random variables $X_n(\omega) \equiv \pi_n(\omega) \equiv \omega(n)$, for $\omega = (\omega(1), \omega(2), \ldots)$ in $\mathbb{R}^\infty$, $n \geq 1$, such that (3.13) holds.

Example 6.3.7: (*Family of independent random variables*). Given a family $\{F_\alpha : \alpha \in A\}$ of cdfs on $\mathbb{R}$ for some index set A, a construction similar to Example 6.3.6, but using Theorem 6.3.1 yields the existence of a real valued stochastic process $\{X_\alpha : \alpha \in A\}$ such that for any $\{\alpha_1, \alpha_2, \ldots, \alpha_n\} \subset A$, $1 \leq n < \infty$, the random variables $\{X_{\alpha_1}, X_{\alpha_2}, \ldots, X_{\alpha_n}\}$ are independent, i.e., (3.13) holds.

Example 6.3.8: (*Markov Chains*). Let $Q = ((q_{ij}))$ be a $k \times k$ *stochastic matrix* for some $1 < k < \infty$. That is,

(a) for all $1 \leq i, j \leq k$, $q_{ij} \geq 0$ and

(b) for each $1 \leq i \leq k$, $\sum_{j=1}^{k} q_{ij} = 1$.

Let $p = (p_1, p_2, \ldots, p_k)$ be a probability vector, i.e., for all i, $p_i \geq 0$, and $\sum_{i=1}^{k} p_i = 1$. Consider the problem of constructing a sequence $\{X_n\}_{n\geq 1}$ of random variables such that for each $n \in \mathbb{N}$,

$$P(X_1 = j_1, X_2 = j_2, \ldots, X_n = j_n) = p_{j_1} q_{j_1 j_2} \cdots q_{j_{n-1} j_n} \tag{3.14}$$

for $1 \leq j_i \leq k$, $i = 1, 2, \ldots, n$.

Let μ_n be the discrete probability distribution determined by the right side of (3.14), that is,

$$\mu_n(\{(j_1, j_2, \ldots, j_n)\}) = p_{j_1} q_{j_1 j_2} \cdots q_{j_{n-1} j_n}$$

for all $(j_1, \ldots, j_n)$ such that $1 \leq j_i \leq k$ for all $1 \leq i \leq n$. It is easy to verify that $\{\mu_n\}_{n \geq 1}$ satisfies the conditions of Theorem 6.3.4 and hence there exist a sequence $\{X_n\}_{n \geq 1}$ of random variables satisfying (3.14). It may be verified that (3.14) is equivalent to

$$\begin{aligned} & P(X_{n+1} = j_{n+1} | X_1 = j_1, \ldots, X_n = j_n) \\ & = q_{j_n j_{n+1}} = P(X_{n+1} = j_{n+1} | X_n = j_n) \end{aligned} \tag{3.15}$$

for all $n \geq 1$, $1 \leq j_i \leq k$, $i = 1, 2, \ldots, n+1$ provided $P(X_1 = j_1, \ldots, X_n = j_n) > 0$ and $P(X_1 = j) = p_j$ for $1 \leq j \leq k$. This says that the conditional distribution of X_{n+1} given $X_1, X_2, \ldots, X_n$ depends only on X_n. This property is known as the *Markov property* and the sequence $\{X_n\}_{n \geq 1}$ is called a *Markov Chain* with state space $\mathbb{S} \equiv \{1, 2, \ldots, k\}$ and *time homogeneous transition probability matrix* $((q_{ij}))$.

When the state space $\mathbb{S} = \{1, 2, \ldots\}$, the above construction goes over with minor notational modifications.

Next consider the case $\mathbb{S} = \mathbb{R}$. A function $Q : \mathbb{R} \times \mathcal{B}(\mathbb{R}) \to [0, 1]$ is called a *probability transition function* if

(i) for each x in $\mathbb{R}$, $Q(x, \cdot)$ is a probability measure on $(\mathbb{R}, \mathcal{B}(\mathbb{R}))$ and

(ii) for each B in $\mathcal{B}(\mathbb{R})$, $Q(\cdot, B)$ is a Borel measurable function on $\mathbb{R}$.

Let μ be a probability distribution on $(\mathbb{R}, \mathcal{B}(\mathbb{R}))$. Using Theorem 6.3.4, it can be shown that there exists a stochastic process $\{X_n\}_{n \geq 1}$ such that

$$\begin{aligned} & P(X_1 \in B_1, X_2 \in B_2, \ldots, X_n \in B_n) \\ & = \int_{B_1} \int_{B_2} \cdots \int_{B_{n-1}} Q(x_{n-1}, B_n) Q(x_{n-2}, dx_{n-1}) \cdots Q(x_1, dx_2) \mu(dx_1), \end{aligned} \tag{3.16}$$

where right side of (3.16) is a well-defined probability measure on $(\mathbb{R}^n, \mathcal{B}(\mathbb{R}^n))$ (Problem 6.18). Such a sequence $\{X_n\}_{n \geq}$ is called a Markov Chain with *state space* $\mathbb{R}$, *initial distribution* μ, and *transition probability function* Q. For more on Markov Chains see Chapter 14.

Example 6.3.9: (*Gaussian processes*). Let A be a nonempty set and $\{X_\alpha : \alpha \in A\}$ be a stochastic process. Such a process is called *Gaussian* if for $\{\alpha_1, \alpha_2, \ldots, \alpha_k\} \subset A$ and real numbers $t_1, t_2, \ldots, t_k$, the random variable $\sum_{i=1}^k t_i X_{\alpha_i}$ has a univariate normal distribution (with possibly zero variance). For such a process, the functions $\mu(\alpha) \equiv EX_\alpha$ and $\sigma(\alpha, \beta) \equiv$

$\text{Cov}(X_\alpha, X_\beta)$ are called the *mean and covariance functions*, respectively. Since $\text{Var}(\sum_{i=1}^k t_i X_{\alpha_i}) \geq 0$, it follows that for any $t_1, t_2, \ldots, t_k$,

$$\sum_{i=1}^k \sum_{j=1}^k t_i t_j \sigma(\alpha_i, \alpha_j) \geq 0. \tag{3.17}$$

This property of the covariance function $\sigma(\cdot, \cdot)$ is called *nonnegative definiteness.*

A natural question is: Given functions $\mu : A \to \mathbb{R}$ and $\sigma : A \times A \to \mathbb{R}$ such that σ is symmetric and satisfies (3.17), does there exist a Gaussian process $\{X_\alpha : \alpha \in A\}$ with $\mu(\cdot)$ and $\sigma(\cdot;)$ as its mean and covariance functions, respectively? The answer is yes and it follows from Theorem 6.3.1 by defining the family Q_A of finite dimensional distributions as follows. Let $\nu_{(\alpha_1,\alpha_2,\ldots,\alpha_k)}$ be the unique probability distribution on $(\mathbb{R}^k, \mathcal{B}(\mathbb{R}^k))$ with the moment generating function

$$\begin{aligned} & M_{(\alpha_1,\alpha_2,\ldots,\alpha_k)}(s_1, s_2, \ldots, s_k) \\ = \ & \exp\left(\sum_{i=1}^k s_i \mu(\alpha_i) + \frac{1}{2}\sum_{i=1}^k \sum_{j=1}^k s_i s_j \sigma(\alpha_i, \alpha_j)\right) \end{aligned} \tag{3.18}$$

for $s_1, s_2, \ldots, s_k$ in $\mathbb{R}$. If the matrix $\Sigma \equiv ((\sigma(\alpha_i, \alpha_j)))$, $1 \leq i, j \leq k$ is *positive definite*, i.e., it is such that in (3.17) equality holds iff $t_i = 0$ for all i, then $\nu_{(\alpha_1,\ldots,\alpha_k)}(\cdot)$ can be shown to be a probability measure that is absolutely continuous w.r.t. m^k, the Lebesgue measure on $\mathbb{R}^k$ with density $\frac{1}{(2\pi)^{k/2}}|\Sigma|^{-\frac{1}{2}} e^{-\sum_{i=1}^k \sum_{j=1}^k (x_i - \mu(\alpha_i))\tilde{\sigma}_{ij}(x_j - \mu(\alpha_j))/2}$ where $\tilde{\Sigma} \equiv ((\tilde{\sigma}_{ij})) = \Sigma^{-1}$, the inverse of Σ and $|\Sigma| =$ the determinant of Σ.

The verification of conditions (3.1) and (3.2) for this family is left as an exercise (Problem 6.12).

Remark 6.3.4: Kolmogorov's consistency theorem (Theorem 6.3.1) remains valid when the real line $\mathbb{R}$ is replaced by a complete separable metric space $\mathbb{S}$. More specifically, let A be a nonempty set and for $\{\alpha_1, \alpha_2, \ldots, \alpha_k\} \subset A$, $1 \leq k < \infty$, let $\nu_{(\alpha_1,\alpha_2,\ldots,\alpha_k)}(\cdot)$ be a probability measure on $(\mathbb{S}^k, \mathcal{B}(\mathbb{S}^k))$. If the family $Q_A \equiv \{\nu_{(\alpha_1,\alpha_2,\ldots,\alpha_k)} : \{\alpha_1, \alpha_2, \ldots, \alpha_k\} \subset A, \ 1 \leq k < \infty\}$ satisfies the natural analogs of (3.1) and (3.2), then there exists a probability measure P on $(\Omega \equiv \mathbb{S}^A, \mathcal{F} \equiv (\mathcal{B}(\mathbb{S}))^A)$ and an $\mathbb{S}$-valued stochastic process $\{X_\alpha : \alpha \in A\}$ on $(\Omega, \mathcal{F}, P)$ such that $\nu_{(\alpha_1,\alpha_2,\ldots,\alpha_k)}(\cdot) = P(X_{\alpha_1}, X_{\alpha_2}, \ldots X_{\alpha_K})^{-1}(\cdot)$. Here $\mathbb{S}^A$ is the set of all $\mathbb{S}$ valued functions on A, $(\mathcal{B}(\mathbb{S}))^A$ is the σ-algebra generated by the cylinder sets of the form

$$C = \{f : f : A \to \mathbb{S}, \ f(\alpha_i) \in B_i, \ i = 1, 2, \ldots, k\}$$

where $\{\alpha_1, \alpha_2, \ldots, \alpha_k\} \subset A$, $B_i \in \mathcal{B}(\mathbb{S})$, $1 \leq i \leq k$, $1 \leq k < \infty$, and also $X_\alpha(\omega)$ is the projection map $X_\alpha(\omega) \equiv \omega(\alpha)$. The main step in the proof of Theorem 6.3.1 was to establish the countable additivity of the set function P on the algebra $\mathcal{C}$ of finite dimensional cylinder sets. This in turn depended upon the fact that any probability measure μ on $(\mathbb{R}^k, \mathcal{B}(\mathbb{R}^k))$ for $1 \leq k < \infty$ is *regular*, i.e., for every Borel set B in $\mathcal{B}(\mathbb{R}^k)$ and for every $\epsilon > 0$, there exists a compact set $G \subset B$ such that $\mu(B \backslash G) < \epsilon$. If $\mathbb{S}$ is a Polish space, then any probability measure on $(\mathbb{S}^k, (\mathcal{B}(\mathbb{S}))^k)$, $1 \leq k < \infty$ is regular (see Billingsley (1968)), and hence, the main steps in the proof of Theorem 6.3.1 go through in this case.

Remark 6.3.5: (*Limitations of Theorem 6.3.1*). In this construction, $\Omega = \mathbb{R}^A$ is rather large and the σ-algebra $\mathcal{F} \equiv (\mathcal{B}(\mathbb{R}))^A$ is not large enough to include many events of interest when the index set A is uncountable. In fact, it can be shown that $\mathcal{F}$ coincides with the class of all sets $G \subset \Omega$ that depend only on a countable number of coordinates of ω. More precisely, the following holds.

Proposition 6.3.5: *The σ-algebra*

$$\begin{aligned} \mathcal{F} &= \{G : G = \pi_{A_1}^{-1}(B) \textit{ for some } B \textit{ in } \mathcal{B}(\mathbb{R}^\infty) \\ &\qquad \textit{and} \quad A_1 \subset A, \ A_1 \quad \textit{countable}\}. \end{aligned} \tag{3.19}$$

Proof: Verify that the right side of (3.19) is a σ-algebra containing the class $\mathcal{C}$ of cylinder sets and also that, it is contained in $\mathcal{F}$. □

For example, if $A = [0, 1]$, then the set $C[0, 1]$ of all continuous functions from $[0, 1] \to \mathbb{R}$ is not a member of $\mathcal{F} \equiv (\mathcal{B}(\mathbb{R}))^A$. Similarly, if $M(\omega) \equiv \sup\{|\omega(\alpha)| : \alpha \in [0, 1]\}$, then the set $\{M(\omega) \leq 1\}$ is not in $\mathcal{F} = (\mathcal{B}(\mathbb{R}))^{[0,1]}$. When A is an interval in $\mathbb{R}$, this difficulty can be overcome in several ways. One approach pioneered by J.L. Doob is the notion of separable stochastic processes (Doob (1953)). Another approach pioneered by Kolmogorov and Skorohod is to restrict Ω to the class of all continuous functions or functions that are right continuous and have left limits (Billingsley (1968)). For more on stochastic processes see Chapter 15.

Independent Random Experiments

If $\mathcal{E}_1$ and $\mathcal{E}_2$ are two random experiments with associated probability spaces $(\Omega_1, \mathcal{F}_1, P_1)$ and $(\Omega_2, \mathcal{F}_2, P_2)$, it is possible to model the experiment of performing both $\mathcal{E}_1$ and $\mathcal{E}_2$ independently by the product probability space $(\Omega_1 \times \Omega_2, \mathcal{F}_1 \times \mathcal{F}_2, P_1 \times P_2)$ (see Chapter 5). The same idea carries over to an arbitrary collection $\{\mathcal{E}_\alpha : \alpha \in A\}$ of random experiments. It is possible to think of a grand experiment $\mathcal{E}$ in which all the $\mathcal{E}_\alpha$'s are independent components by considering the product probability space

$$(\times_{\alpha \in A} \Omega_\alpha, \times_{\alpha \in A} \mathcal{F}_\alpha, \times_{\alpha \in A} P_\alpha) \equiv (\Omega, \mathcal{F}, P) \tag{3.20}$$

where $(\Omega_\alpha, \mathcal{F}_\alpha, P_\alpha)$ is the probability space corresponding to $\mathcal{E}_\alpha$. Here $\Omega \equiv \times_{\alpha \in A} \Omega_\alpha$ is the collection of all functions ω on A such that $\omega(\alpha) \in \Omega_\alpha$,

$\mathcal{F} \equiv \times_{\alpha \in A} \mathcal{F}_\alpha$ is the σ-algebra generated by finite dimensional cylinder sets of the form

$$C = \{\omega : \omega(\alpha_i) \in B_{\alpha_i},\ i = 1, 2, \ldots, k\}, \tag{3.21}$$

$1 \leq k < \infty, \{\alpha_1, \alpha_2, \ldots, \alpha_k\} \subset A$, $B_{\alpha_i} \in \mathcal{F}_{\alpha_i}$ and $P \equiv \times_{\alpha \in A} P_\alpha$ is the probability measure on $\mathcal{F}$ such that for C of (3.21),

$$P(C) = \prod_{i=1}^{k} P_{\alpha_i}(B_{\alpha_i}). \tag{3.22}$$

The proof of the existence of such a P on $\mathcal{F}$ is an application of the extension theorem (Theorem 1.3.3). The verification of countable additivity on the class $\mathcal{C}$ of cylinder sets is not difficult. See Kolmogorov (1956).

6.4 Problems

6.1 Let $\mu_1 = \mu_2$ be the probability distribution on $\Omega = \{1, 2\}$ with $\mu_1(\{1\}) = 1/2$. Find two distinct probability distributions on $\Omega \times \Omega$ with μ_1 and μ_2 as the set of marginals.

6.2 Let $\Omega = (0, 1)$, $\mathcal{F} = \mathcal{B}((0, 1))$ and P be the Lebesgue measure on (0,1). Let $X(\omega) = -\log \omega$, $h(x) = x^2$ and $Y = h(X)$. Find P_X and P_Y and evaluate EY by applying the change of variables formula (Proposition 6.2.1).

6.3 In the change of variables formula, one of the three integrals is usually easier to evaluate than the other two. In this problem, in part (a), the first integral is easier to evaluate than the other two while in part (b), the second one is easier.

(a) Let $Z \sim N(0, 1)$, $X = Z^2$, and $Y = e^{-X}$.

(i) Find the distributions P_X and P_Y on $(\mathbb{R}, \mathcal{B}(\mathbb{R}))$.

(ii) Compute the integrals

$$\int_{\mathbb{R}} e^{-z^2} \phi(z) dz, \quad \int_{\mathbb{R}} e^{-x} P_X(dx) \quad \text{and} \quad \int_{\mathbb{R}} y P_Y(dy),$$

where $\phi(z) = \frac{1}{\sqrt{2\pi}} e^{-z^2/2}$, $-\infty < z < \infty$. Verify that all three integrals agree.

(b) Let $X_1, X_2, \ldots, X_n$ be iid $N(0, 1)$ random variables. Let $Y = (X_1 + \cdots + X_k)$ and $Z = Y^2$.

(i) Find the distributions of Y and Z.

(ii) Evaluate $\int_{\mathbb{R}^k} (x_1 + \cdots + x_k)^2 dP_{X_1, \ldots, X_k}(x_1, \ldots, x_k)$, $\int_{\mathbb{R}} y^2 P_Y(dy)$, and $\int_{\mathbb{R}_+} z P_Z(dz)$.

(c) Let $X_1, X_2, \ldots, X_k$ be independent Binomial (n_i, p), $i = 1, 2, \ldots, k$ random variables. Let $Y = (X_1 + \cdots + X_k)$.

(i) Find the distribution P_Y of Y.

(ii) Evaluate $\int_{\mathbb{R}^k}(x_1 + \cdots + x_k) dP_{X_1,\ldots,X_k}(x_1, \ldots, x_k)$ and $\int_{\mathbb{R}} y P_Y(dy)$.

6.4 Let X be a random variable such that $M_X(t) \equiv E(e^{tX}) < \infty$ for $|t| < \epsilon$ for some $\epsilon > 0$.

(a) Show that $E(e^{tX}|X|^r) < \infty$ for all $r > 0$ and $|t| < \epsilon$.

(b) Show that $M_X^{(r)}(t)$, the rth derivative of $M_X(t)$ for $r \in \mathbb{N}$, satisfies

$$M_X^{(r)}(t) = E(e^{tX} X^r) \quad \text{for} \quad |t| < \epsilon.$$

(c) Verify (2.25).

(**Hint:** (a) First show that for $t_1 \in (-\epsilon, \epsilon)$, there exist a $t_2 \in (-\epsilon, \epsilon)$ such that $|t_1| < |t_2| < \epsilon$ and for some $C < \infty$, $e^{t_1 x}|x|^r \leq Ce^{|t_2 x|}$ for all x in $\mathbb{R}$.
(b) Verify that for all $x \in \mathbb{R}$, $|e^x - 1| \leq |x|e^{|x|}$. Now use (a) and the DCT to show that $M_X(t)$ is differentiable and $M_X^{(1)}(t) = E(e^{tX}X)$ for all $|t| < \epsilon$. Now complete the proof by induction.)

6.5 Let X be a random variable.

(a) Show that $\phi(r) \equiv (E|X|^r)^{1/r}$ is nondecreasing on $(0, \infty)$.

(b) Show that $\phi(r) \equiv \log E|X|^r$ is convex in $(0, r_0)$ if $E|X|^{r_0} < \infty$.

(c) Let $M = \sup\{x : P(|X| > x) > 0\}$. Show that

(i) $\lim_{r \uparrow \infty} \phi(r) = M$.

(ii) $\lim_{n \to \infty} \frac{E|X|^{n+1}}{E|X|^n} = M$.

(**Hint:** For $M < \infty$, note that $E|X|^r \geq (M - \epsilon)^r P(|X| > M - \epsilon)$ for any $\epsilon > 0$.)

6.6 Show that if equality holds in (2.20), then there exist constants a and b such that $P(Y = aX + b) = 1$.

(**Hint:** Show that there exist a constant a such that $\text{Var}(Y - aX) = 0$.)

6.7 Determine C and its base B explicitly in Examples 6.3.3 and 6.3.4.

6.8 (a) Show that D in Example 6.3.5 is not a finite dimensional cylinder set.

(**Hint:** Note that $\lim_{n \to \infty} \frac{1}{n} \sum_{j=1}^n x_j$ is not determined by the values of finitely many x_i's.)

(b) Find three other such examples of sets D in $\mathbb{R}^\infty$ that are not finite dimensional cylinder sets.

6.9 Establish the assertion in Remark 6.3.3 by completing the following steps:

(a) Show that the coordinate map $f_n(x) \equiv x_n$ from $\mathbb{R}^\infty$ to $\mathbb{R}$ is continuous under the metric d of (3.3). (Conclude, using Example 1.1.6, that $\mathcal{R}^{\mathbb{N}} \subset \mathcal{B}(\mathbb{R}^\infty)$).

(b) Let $\mathcal{C}_1 \equiv \{A : A = (a_1, b_1) \times \cdots \times (a_k, b_k) \times \mathbb{R}^\infty,\ -\infty \le a_i < b_i \le \infty,\ 1 \le i \le k$, for some $k < \infty\}$ and $\mathcal{C}_2 \equiv \{A : A$ is an open ball in $(\mathbb{R}^\infty, d)\}$. Show that $\sigma\langle \mathcal{C}_2 \rangle \subset \sigma\langle \mathcal{C}_1 \rangle$.

(c) Show that $\sigma\langle \mathcal{C}_2 \rangle = \mathcal{B}(\mathbb{R}^\infty)$ by showing that every open set in $(\mathbb{R}^\infty, d)$ is a countable union of open balls.

6.10 Show that the family $Q_{\mathbb{N}}$ defined in (3.12) satisfies the consistency conditions (3.1) and (3.2) of Theorem 6.3.1.

6.11 Verify that the family of finite dimensional distributions defined by the right side of (3.14) satisfies the conditions of Theorem 6.3.4.

6.12 Verify that the family of distributions defined in (3.18) satisfies conditions (3.1) and (3.2) of Theorem 6.3.1.

(**Hint:** Use the fact that for any $k \ge 1$, any $\mu = (\mu_1, \mu_2, \ldots, \mu_k) \in \mathbb{R}^k$, and any nonnegative definite $k \times k$ matrix $\Sigma \equiv ((\sigma_{ij}))_{k \times k}$, there is a unique probability distribution ν such that for any $s = (s_1, s_2, \ldots, s_k)$ in $\mathbb{R}^k$,

$$\begin{aligned} &\int_{\mathbb{R}^k} \exp\Big(\sum_{i=1}^k s_i x_k\Big) \nu(dx) \\ &= \exp\Big(\sum_{i=1}^k s_i \mu_i + \frac{1}{2} \sum_{i=1}^k \sum_{j=1}^k s_i s_j \sigma_{ij}\Big). \end{aligned}$$

Observe that this implies that for $s = (s_1, s_2, \ldots, s_k)$ in $\mathbb{R}^k$, the induced distribution (under ν) on $\mathbb{R}$ by the map $g(x) = \sum_{i=1}^k s_i x_i$ from $\mathbb{R}^k \to \mathbb{R}$ is univariate normal with mean $\sum_{i=1}^k s_i \mu_i$ and variance $\sum_{i=1}^k \sum_{j=1}^k s_i s_j \sigma_{ij}$.)

6.13 Show that the set $D \equiv C[0,1]$ of continuous functions from $[0,1]$ to $\mathbb{R}$ is not a member of the σ-algebra $\mathcal{F} \equiv (\mathcal{B}(\mathbb{R}))^{[0,1]}$.

(**Hint:** If $D \in \mathcal{F}$, then by Proposition 6.3.5, D is of the form $\pi_{A_1}^{-1}(B)$ for some B in $\mathcal{B}(\mathbb{R}^\infty)$, where $A_1 \subset [0,1]$ is countable. Show that for any such A_1 and B, there exist functions $f : [0,1] \to \mathbb{R}$ such that $f \in \pi_{A_1}^{-1}(B)$ but f is not continuous on $[0,1]$.)

6.14 Show that $K \equiv \{\omega : \omega \in \mathbb{R}^{[0,1]}, \sup_{0 \le \alpha \le 1} |\omega(\alpha)| < 1\}$ is not in $\mathcal{F} \equiv (\mathcal{B}(\mathbb{R}))^{[0,1]}$.

(**Hint:** Observe that $\sup_{0 \le \alpha \le 1} |\omega(\alpha)|$ is not determined by the values of $\omega(\alpha)$ for countably many α's.)

6.15 Let $\{\mu_i\}_{i \ge 1}$ be a sequence of probability distributions on $(\mathbb{R}, \mathcal{B}(\mathbb{R}))$ and let μ be a probability distribution on $\mathbb{N}$ with $p_i \equiv \mu(\{i\})$, $i \ge 1$.

(a) Verify that $\nu(\cdot) = \sum_{i \ge 1} p_i \mu_i(\cdot)$ is a probability distribution on $(\mathbb{R}, \mathcal{B}(\mathbb{R}))$.

(b) (i) Show that there exists a probability space $(\Omega, \mathcal{F}, P)$ and a collection of independent random variables $\{J, X_1, X_2, \ldots\}$ on $(\Omega, \mathcal{F}, P)$ such that for each $i \ge 1$, X_i has distribution μ_i and $J \sim \mu$.

(ii) Let $Y = X_J$, i.e., $Y(\omega) \equiv X_{J(\omega)}(\omega)$. Show that Y is a random variable on $(\Omega, \mathcal{F}, P)$ and $Y \sim \nu$.

6.16 Let F be a cdf on $\mathbb{R}$ and let F be decomposed as

$$F = \alpha F_d + \beta F_{ac} + \gamma F_{sc}$$

where $\alpha, \beta, \gamma \in [0,1]$ and $\alpha + \beta + \gamma = 1$ and F_d, F_{ac}, F_{sc} are discrete, absolutely continuous, and singular continuous cdfs on $\mathbb{R}$ (cf. (4.5.3)). Show that there exist independent random variables X_1, X_2, X_3 and J on some probability space such that $X_1 \sim F_d$, $X_2 \sim F_{ac}$, $X_3 \sim F_{sc}$, $P(J = 1) = \alpha$, $P(J = 2) = \beta$, $P(J = 3) = \gamma$ and $X_J \sim F$, where $\sim$ means "has cdf".

6.17 Let μ be a probability measure on $(\mathbb{R}, \mathcal{B}(\mathbb{R}))$. Let for each x in $\mathbb{R}$, $F(x, \cdot)$ be a cdf on $(\mathbb{R}, \mathcal{B}(\mathbb{R}))$. Let $\psi(x,t) \equiv \inf\{y : F(x,y) \ge t\}$, for x in $\mathbb{R}$, $0 < t < 1$. Assume that $\psi(\cdot,\cdot) : \mathbb{R} \times (0,1) \to \mathbb{R}$ is measurable. Let X and U be independent random variables on some probability space $(\Omega, \mathcal{F}, P)$ such that $X \sim \mu$ and $U \sim$ uniform (0,1).

(a) Show that $Y = \psi(X, U)$ is a random variable.

(b) Show that $P(Y \le y) = \int_{\mathbb{R}} F(x,y)\mu(dx)$. (The distribution of Y is called a mixture of distributions with μ as the mixing distribution. This is of relevance in Bayesian statistical inference.)

6.18 (a) Let $(\mathbb{S}_i, \mathcal{S}_i)$, $i = 1, 2$ be two measurable spaces. Let μ be a probability measure on $(\mathbb{S}_1, \mathcal{S}_1)$ and let $Q : \mathbb{S}_1 \times \mathcal{S}_2 \to [0,1]$ be such that for each x in $\mathbb{S}_1$, $Q(x, \cdot)$ is a probability measure on $(\mathbb{S}_2, \mathcal{S}_2)$ and for each B in $\mathcal{S}_2$, $Q(\cdot, B)$ is $\mathcal{S}_1$-measurable. Define

$$\nu(B_1 \times B_2) \equiv \int_{B_1} Q(x, B_2)\mu(dx)$$

on $\mathcal{C} \equiv \{B_1 \times B_2 : B_i \in s_i, i = 1, 2\}$. Show that ν can be extended to be a probability measure on $\sigma\langle\mathcal{C}\rangle \equiv \mathcal{S}_1 \times \mathcal{S}_2$.

(b) Let μ and Q be as in Example 6.3.8 (cf. (3.16)). For each $n \geq 1$ let ν_n be a set function defined by the recursive scheme

$$\begin{aligned} \nu_1(\cdot) &= \mu(\cdot), \\ \nu_{n+1}(A \times B) &= \int_A Q(x, B)\nu_n(dx), \ A \in \mathcal{B}(\mathbb{R}^n), \ B \in \mathcal{B}(\mathbb{R}). \end{aligned}$$

Show that for each n, ν_n can be extended to be a probability measure on $(\mathbb{R}^n, \mathcal{B}(\mathbb{R}^n))$. (Thus the right side of (3.16) is defined to be $\nu_n(B_1 \times B_2 \times \cdots \times B_n)$.)

6.19 (*Bayesian paradigm*). Consider the setup in Problem 6.18 (a). Let $\lambda(B) \equiv \nu(\mathbb{S}_1 \times B) = \int_{\mathbb{S}_1} Q(x, B)\mu(dx)$ for all B in $\mathcal{S}_2$.

(a) Verify that λ is a probability measure on $(\mathbb{S}_2, \mathcal{S}_2)$.

(b) Now fix B_1 in $\mathcal{S}_1$. Show that there exists a function $\tilde{Q}(x, B_1)$, $\mathbb{S}_2 \to [0, 1]$ that is $\langle \mathcal{S}_2, \mathcal{B}(\mathbb{R})\rangle$-measurable such that

$$\nu(B_1 \times B_2) = \int_{B_2} \tilde{Q}(x, B_1)\lambda(dx).$$

(**Hint:** Apply the Radon-Nikodym theorem to the pair $\nu(B_1 \times \cdot)$ and $\lambda(\cdot)$.)

(c) Let $\Omega = \mathbb{S}_1 \times \mathbb{S}_2$, $\mathcal{F} = \sigma\langle\mathcal{C}\rangle$. For $\omega = (s_1, s_2)$, let $\theta(\omega) = s_1$ and $X(\omega) = s_2$. Think of θ as the *parameter*, X as the *data*, $Q(\theta, \cdot)$ as the *distribution of X given θ*, $\mu(\cdot)$ as the *prior distribution of* θ and $\tilde{Q}(x, B_1)$ as the *posterior* probability that θ is in B_1 given the data $X = x$. Compute $\tilde{Q}(x, B_1)$ when $(\mathbb{S}_i, \mathcal{S}_i) = (\mathbb{R}, \mathcal{B}(\mathbb{R}))$, $i = 1, 2$, $\mu(\cdot) \sim N(0, 1)$, $Q(\theta, \cdot) \sim N(\theta, 1)$.

6.20 Let X be a random variable on some probability space $(\Omega, \mathcal{F}, P)$. Recall that a random variable X is

(a) *discrete* if there is a finite or countable set $D \equiv \{a_j : 1 \leq j \leq k \leq \infty\}$ such that $P(X \in D) = 1$,

(b) *continuous* if for every $x \in \mathbb{R}$, $P(X = x) = 0$ or equivalently the cdf $F_X(\cdot)$ is continuous on all of $\mathbb{R}$,

(c) *absolutely continuous* if there exists a nonnegative Borel measurable function $f_X(\cdot)$ on $\mathbb{R}$ such that for any $-\infty < a < b < \infty$,

$$P(a < X \leq b) = \int_{(a,b]} f_X(\cdot)dm$$

or equivalently the induced measure PX^{-1} is $\ll m$,

(d) *singular* if $PX^{-1} \perp m$ or equivalently $F_X(\cdot) = 0$ a.e. m,

(e) *singular continuous* if it is singular and continuous.

Let $g : \mathbb{R} \to \mathbb{R}$ be Borel measurable and $Y = g(X)$.

(a) Show that if X is discrete then so is Y but not conversely.

(b) Show that if X is continuous and g is (1–1) on the range of X, then Y is continuous.

(c) Show if X is absolutely continuous with p.d.f. $f_X(\cdot)$ and g is absolutely continuous on bounded intervals such that $g'(\cdot) > 0$ a.e. (m), then Y is also absolutely continuous with p.d.f.

$$f_Y(y) = \frac{f_X\left(g^{-1}(y)\right)}{g'\left(g^{-1}(y)\right)}.$$

(d) Let X be as in (c) above. Suppose g is absolutely continuous on bounded intervals and there exist disjoint intervals $\{I_j\}_{1 \le j \le k}$, $1 \le k \le \infty$, such that $\cup_{1 \le j \le k} I_j = \mathbb{R}$ and for each j, either $g'(\cdot) > 0$ a.e. (m) on I_j or $g'(\cdot) < 0$ a.e. (m) on I_j. Show that Y is also absolutely continuous with p.d.f.

$$f_Y(y) = \sum_{x_j \in D(y)} \frac{f_X(x_j)}{|g'(x_j)|}$$

where $D\{y\} \equiv \{x_j : x_j \in I_j,\ g(x_j) = y\}$.

(e) Use (c) to compute the p.d.f. of Y when

(i) $X \sim N(0,1)$, $g(x) = e^x$.

(ii) $X \sim N(0,1)$, $g(x) = x^2$.

(iii) $X \sim N(0,1)$, $g(x) = \sin 2\pi x$.

(iv) $X \sim \exp(1)$, $g(x) = e^{-x}$.

6.21 (*Simple random sampling without replacement*). Let $S \equiv \{1, 2, \ldots, m\}$, $1 < m < \infty$. Fix $1 \le n \le m$. Choose an element X_1 from S such that the probability that $X_1 = j$ is $\frac{1}{m}$ for all $j \in S$. Next, choose an element X_2 from $S - \{X_1\}$ such that the probability that $X_2 = j$ is $\frac{1}{(m-1)}$ for $j \in S - \{X_1\}$. Continue this procedure for n steps. Write the outcome as the ordered vector $\omega \equiv (X_1, X_2, \ldots, X_n)$.

(a) Identify the sample space Ω, the σ-algebra $\mathcal{F}$ and the probability measure P for this experiment.

(b) Show that for any permutation σ of $\{1, 2, \ldots, n\}$, the random vector $Y_\sigma = (X_{\sigma(1)}, X_{\sigma(2)}, \ldots, X_{\sigma(n)})$ has the same distribution as $(X_1, X_2, \ldots, X_n)$.

(c) Conclude that $\{X_i\}_{1 \le i \le n}$ are identically distributed and that EX_i, $\text{Cov}(X_i, X_j)$, $i \neq j$ are independent of i and j and compute them.

(d) Answer the same questions (a)–(c) if the sampling is changed to *with replacement*, i.e., at each stage i, the probability that $P(X_i = j) = \frac{1}{m}$ for all $j \in S$.

(e) In (d), let D be the number of *distinct units* in the sample. Find $E(D)$ and $\text{Var}(D)$.

6.22 Let X be a nonnegative random variable. Show that

$$\sqrt{1 + (EX)^2} \le E\sqrt{1 + X^2} \le 1 + EX.$$

(Note that $f(x) \equiv \sqrt{1 + x^2}$ is convex on $[0, \infty)$ and bounded by $1+x$.)

6.23 Let X and Y be nonnegative random variables defined on a probability space $(\Omega, \mathcal{F}, P)$. Suppose $X \cdot Y \ge 1$ w.p. 1. Show that

$$EX \cdot EY \ge 1.$$

(**Hint:** Use Cauchy-Schwarz on $\sqrt{X}\sqrt{Y}$.)

6.24 Let μ be a probability measure on $\big(\mathbb{R}, \mathcal{B}(\mathbb{R})\big)$. Show that there is a random variable X on the Lebesgue space $([0,1], \mathcal{B}([0,1]), m)$ such that $m X^{-1} \equiv \mu$ where m is the Lebesgue measure. Extend this to $\big(\mathbb{R}^k, \mathcal{B}(\mathbb{R}^k)\big)$, where k is an integer > 1.

(Note: This is true for any Polish space, i.e., a complete separable metric space, see Billingsley (1968).)

Appendix A
Advanced Calculus: A Review

This Appendix is a brief review of elementary set theory, real numbers, limits, sequences and series, continuity, differentiability, Riemann integration, complex numbers, exponential and trigonometric functions and metric spaces. For proofs and further details see Rudin (1976) and Royden (1988).

A.1 Elementary set theory

This section reviews the following: sets, set operations, product sets (finite and infinite), equivalence relation, axiom of choice, countability, and uncountability.

Definition A.1.1: A *set* is a collection of objects.

It is typically defined as a collection of objects with a *common defining property*. For example, the collection of even integers can be written as $\mathbb{E} \equiv \{n : n \text{ is an even integer}\}$. In general a set Ω with defining property p is written as

$$\Omega = \{\omega : \omega \text{ has property } p\}.$$

The individual elements are denoted by the small letters ω, a, x, s, t, etc. and the sets by capital letters Ω, A, X, S, T, etc.

Example A.1.1: The closed interval $[0, 1] \equiv \{x : x \text{ a real number}, 0 \leq x \leq 1\}$.

Example A.1.2: The set of positive rationals $\equiv \{x : x = \frac{m}{n}$, m and n positive integers$\}$.

Example A.1.3: The set of polynomials in x of degree 10 $\equiv \{P(x) : P(x) = \sum_{j=0}^{10} a_j x^j$, a_j real, $j = 0, \ldots, 10$, $a_{10} \neq 0\}$.

Example A.1.4: The set of all polynomials in $x \equiv \{P(x) : P(x) = \sum_{j=0}^{n} a_j x^j$, n a nonnegative integer, a_j real, $j = 0, 1, 2, \ldots, n\}$.

A.1.1 Set operations

Definition A.1.2: Let A be a set. A set B is called a *subset of* A and written as $B \subset A$ if every element of B is also an element of A. Two sets A and B are the *same* and written as $A = B$, if each is a subset of the other. A subset $A \subset \Omega$ is called *empty* and denoted by $\emptyset$ if there exists no ω in Ω such that $\omega \in A$.

Using the mathematical notation $\in$ and $\Rightarrow$, one writes

$$B \subset A \quad \text{if} \quad x \in B \Rightarrow x \in A.$$

Here '$\in$' means "belongs to" and $\Rightarrow$ means "implies".

Example A.1.5: Let $\mathbb{N}$ be the set of natural numbers, i.e., $\mathbb{N} = \{1, 2, 3, \ldots\}$. Let $\mathbb{E}$ be the set of even natural numbers, i.e., $\mathbb{E} = \{n : n \in \mathbb{N}$, $n = 2k$ for some $k \in \mathbb{N}\}$. Then $\mathbb{E} \subset \mathbb{N}$.

Example A.1.6: Let $A = [0, 1]$ and B be the set of x in A such that $x^2 < \frac{1}{4}$. Then $B = \{x : 0 \leq x < \frac{1}{2}\} \subset A$.

Definition A.1.3: (*Intersection and union*). Let A_1, A_2 be subsets of a set Ω. Then A_1 *union* A_2, written as $A_1 \cup A_2$ is the set defined by

$$A_1 \cup A_2 = \{\omega : \omega \in A_1 \text{ or } \omega \in A_2 \text{ or both}\}.$$

Similarly, A_1 *intersection* A_2, written as $A_1 \cap A_2$, is the set defined by

$$A_1 \cap A_2 = \{\omega : \omega \in A_1 \text{ and } \omega \in A_2\}.$$

Example A.1.7: Let $\Omega \equiv \mathbb{N} \equiv \{1, 2, 3, \ldots\}$,

$$\begin{aligned} A_1 &= \{\omega : \omega = 3k \text{ for some } k \in \mathbb{N}\} \quad \text{and} \\ A_2 &= \{\omega : \omega \equiv 5k \text{ for some } k \in \mathbb{N}\}. \end{aligned}$$

Then $A_1 \cup A_2 = \{\omega : \omega$ is divisible by at least one of the two integers 3 and 5$\}$, $A_1 \cap A_2 = \{\omega : \omega$ is divisible by both 3 and 5$\}$.

Definition A.1.4: Let Ω and I be nonempty sets. Let $\{A_\alpha : \alpha \in I\}$ be a collection of subsets of Ω. Then I is called the *index set.*

The *union* of $\{A_\alpha : \alpha \in I\}$ is defined as

$$\bigcup_{\alpha \in I} A_\alpha \equiv \{\omega : \omega \in A_\alpha \text{ for some } \alpha \in I\}.$$

The *intersection* of $\{A_\alpha : \alpha \in I\}$ is defined as

$$\bigcap_{\alpha \in I} A_\alpha \equiv \{\omega : \omega \in A_\alpha \text{ for every } \alpha \subset I\}.$$

Definition A.1.5: (*Complement of a set*). Let $A \subset \Omega$. Then the *complement* of the set A, written as A^c (or $\tilde{A}$) is defined by $A^c \equiv \{\omega : \omega \notin A\}$.

Example A.1.8: If $\Omega = \mathbb{N}$ and A is the set of all integers that are divisible by 2, then A^c is the set of all odd integers, i.e., $A^c = \{1, 3, 5, 7, \ldots\}$.

Proposition A.1.1: (*DeMorgan's law*). *For any* $\{A_\alpha : a \in I\}$ *of subsets of* Ω, $(\cup_{\alpha \in I} A_\alpha)^c = \cap_{\alpha \in I} A_\alpha^c$, $(\cap_{\alpha \in I} A_\alpha)^c = \cup_{\alpha \in I} A_\alpha^c$.

Proof: To show that two sets A and B are the same, it suffices to show that

$$\omega \in A \Rightarrow \omega \in B \quad \text{and} \quad \omega \in B \Rightarrow \omega \in A.$$

Let $\omega \in (\cup_{\alpha \in I} A_\alpha)^c$. Then $\omega \notin \cup_{\alpha \in I} A_\alpha$

$$\begin{aligned} &\Rightarrow \quad \omega \notin A_\alpha \text{ for any } \alpha \in I \\ &\Rightarrow \quad \omega \in A_\alpha^c \text{ for each } \alpha \in I \\ &\Rightarrow \quad \omega \in \bigcap_{\alpha \in I} A_\alpha^c. \end{aligned}$$

Thus $(\cup_{\alpha \in I} A_\alpha)^c \subset \cap_{\alpha \in I} A_\alpha^c$. The opposite inclusion and the second identity are similarly proved. □

Definition A.1.6: (*Product sets*). Let Ω_1 and Ω_2 be two nonempty sets. Then the *product set* of Ω_1 and Ω_2 denoted by $\Omega \equiv \Omega_1 \times \Omega_2$, consists of all ordered pairs (ω_1, ω_2) such that $\omega_1 \in \Omega_1, \omega_2 \in \Omega_2$.

Note that if $\Omega_1 = \Omega_2$ and $\omega_1 \neq \omega_2$, then the pair (ω_1, ω_2) is not the same as (ω_2, ω_1), i.e., the order is important.

Example A.1.9: $\Omega_1 = [0, 1]$, $\Omega_2 = [2, 3]$. Then $\Omega_1 \times \Omega_2 = \{(x, y) : 0 \leq x \leq 1,\ 2 \leq y \leq 3\}$.

Definition A.1.7: (*Finite products*). If $\Omega_i, i = 1, 2 \ldots, k$ are nonempty sets, then

$$\Omega = \Omega_1 \times \Omega_2 \times \ldots \times \Omega_k$$

is the set of all ordered k vectors $(\omega_1, \omega_2, \ldots, \omega_k)$ where $\omega_i \in \Omega_i$. If $\Omega_i = \Omega_1$ for all $1 \leq i \leq k$, then $\Omega_1 \times \Omega_2 \times \ldots \times \Omega_k$ is written as $\Omega_1^{(k)}$ or Ω_1^k.

Definition A.1.8: (*Infinite products*). Let $\{\Omega_\alpha : \alpha \in I\}$ be an infinite collection of nonempty sets. Then $\times_{\alpha \in I} \Omega_\alpha$, *the product set*, is defined as $\{f : f$ is a function defined on I such that for each α, $f(\alpha) \in \Omega_\alpha\}$. If $\Omega_\alpha = \Omega$ for all $\alpha \in I$, then $\times_{\alpha \in I} \Omega_\alpha$ is also written as Ω^I.

It is a *basic axiom of set theory*, known as the *axiom of choice* (A.C.), that this space is nonempty. That is, given an arbitrary collection of nonempty sets, it is possible to form a *parliament* with one representative from each set.

For a long time it was thought this should follow from the other axioms of set theory. But it is shown in Cohen (1966) that it is an independent axiom. That is, both the A.C. and its negation are consistent with the rest of the axioms of set theory.

There are several equivalent versions of A.C. These are: Zorn's Lemma, Hausdorff's Maximality Principle, The 'Principle of Well Ordering' and Tukey's Lemma. For a proof of these equivalences, see Hewitt and Stromberg (1965).

Definition A.1.9: (*Functions, countability and uncountability*). A *function* f is a correspondence between the elements of a set X and another set Y and is written as $f : X \to Y$. It satisfies the condition that for each x, there is a unique y in Y that corresponds to it and is denoted as

$$y = f(x).$$

The set X is called the *domain of* f and the set $f(X)$, defined as, $f(X) \equiv \{y :$ there exists x in X such that $f(x) = y\}$ is called the *range of* f. It is possible that many x's may correspond to the same y and also there may exist y in Y for which there is no x such that $f(x) = y$. If $f(X)$ is all of Y, then the map is called *onto*. If for each y in $f(X)$, there is a unique x in X such that $f(x) = y$, then f is called (1–1) or *one-to-one*. If f is one-to-one and onto, then X and Y are said to have the same *cardinality*.

Definition A.1.10: Let $f : X \to Y$ be (1–1) and onto. Then, for each y in Y, there is a unique element x in X such that $f(x) = y$. This x is denoted as $f^{-1}(y)$. Note that in this case $g(y) \equiv f^{-1}(y)$ is a (1–1) onto map from Y to X and is called the *inverse* of f.

Example A.1.10: Let $X = \mathbb{N} \equiv \{1, 2, 3, \ldots\}$. Let $Y = \{n : n = 2k, k \in \mathbb{N}\}$ be the set of even integers. Then the map $f(x) = 2x$ is a (1–1) onto map from X to Y.

Example A.1.11: Let X be $\mathbb{N}$ and let $\mathbb{P}$ be the set of all prime numbers. Then X and $\mathbb{P}$ have the same cardinality.

Definition A.1.11: A set X is *finite* if there exists $n \in \mathbb{N}$ such that X and $Y \equiv \{1, 2, \ldots, n\}$ have the same *cardinality*, i.e., there exists a (1–1) onto map from Y to X. A set X is *countable* if X and $\mathbb{N}$ have the *same cardinality*, i.e., there exists a (1–1) onto map from $\mathbb{N}$ to X. A set X is *uncountable* if it is not *finite* or *countable*.

Example A.1.12: The set $\{0, 1, 2, \ldots, 9\}$ is finite, the set $\mathbb{N}^k$ $(k \in \mathbb{N})$ is countable and $\mathbb{N}^{\mathbb{N}}$ is uncountable (Problem A.6).

Definition A.1.12: Let Ω be a nonempty set. Then the power set of Ω, denoted by $\mathcal{P}(\Omega)$, is the collection of all subsets of Ω, i.e.,

$$\mathcal{P}(\Omega) \equiv \{A : A \subset \Omega\}.$$

Remark A.1.1: $\mathcal{P}(\mathbb{N})$ is an uncountable set (Problem A.5).

A.1.2 The principle of induction

The set $\mathbb{N}$ of natural numbers has the *well ordering property* that every nonempty subset A of $\mathbb{N}$ has a smallest element s such that (i) $s \in A$ and (ii) $a \in A \Rightarrow a \geq s$. This property is one of the basic postulates in the definition of $\mathbb{N}$. The *principle of induction* is a consequence of the well ordering property. It says the following:

Let $\{P(n) : n \in \mathbb{N}\}$ be a collection of propositions (or statements). Suppose that

(i) $P(1)$ is true.

(ii) For each $n \in \mathbb{N}$, $P(n)$ true $\Rightarrow P(n+1)$ true.

Then, $P(n)$ is true for all $n \in \mathbb{N}$.

See Problem A.9 for some examples.

A.1.3 Equivalence relations

Definition A.1.13:

(a) Let Ω be a nonempty set. Let G be a nonempty subset of $\Omega \times \Omega$. Write $x \sim y$ if $(x, y) \in G$ and call it a *relation* defined by G.

(b) A relation defined by G is an *equivalence relation* if

  (i) (*reflexive*) for all x in Ω, $x \sim x$, i.e., $(x, x) \in G$;

  (ii) (*symmetric*) $x \sim y \Rightarrow y \sim x$, i.e., $(x, y) \in G \Rightarrow (y, x) \in G$;

  (iii) (*transitive*) $x \sim y$, $y \sim z \Rightarrow x \sim z$, i.e., $(x, y) \in G$, $(y, z) \in G \Rightarrow (x, z) \in G$.

Example A.1.13: Let $\Omega = \mathbb{Z}$, the set of all integers, $G = \{(m, n) : m - n$ is divisible by $3\}$. Thus, $m \sim n$ if $(m - n)$ is a multiple of 3. It is easy to verify that this is an equivalence relation.

Definition A.1.14: (*Equivalence classes*). Let Ω be a nonempty set. Let G define an equivalence relation on Ω. For each x in Ω, the set $[x] \equiv \{y : x \sim y\}$ is called the *equivalence class generated by* x.

Proposition A.1.2: *Let $\mathcal{C}$ be the set of all equivalence classes in Ω generated by an equivalence relation defined by G. Then*

(i) $C_1, C_2 \in \mathcal{C} \Rightarrow C_1 = C_2$ *or* $C_1 \cap C_2 = \emptyset$.

(ii) $\bigcup_{C \in \mathcal{C}} C = \Omega$.

Proof:

(i) Suppose $C_1 \cap C_2 \neq \emptyset$. Then there exist x_1, x_2, y such that $C_1 = [x_1]$, $C_2 = [x_2]$ and $y \in C_1 \cap C_2$. This implies $x_1 \sim y$, $x_2 \sim y$. But by symmetry $y \sim x_2$ and this implies by transitivity that $x_1 \sim x_2$, i.e., $x_2 \in C_1$ implying $C_2 \subset C_1$. Similarly, $C_1 \subset C_2$, i.e., $C_1 = C_2$.

(ii) For each x in Ω, $(x, x) \in G$ and so $[x]$ is not empty and $x \in [x]$. □

The above proposition says that every equivalence relation on Ω leads to a decomposition of Ω into equivalence classes that are disjoint and whose union is all of Ω. In the example given above, the set $\mathbb{Z}$ of all integers can be decomposed to three equivalence classes $C_j \equiv \{n : n = 3m + j$ for some $m \in \mathbb{Z}\}$, $j = 0, 1, 2$.

A.2 Real numbers, continuity, differentiability and integration

A.2.1 Real numbers

This section reviews the following: integers, rationals, real numbers; algebraic, order and completeness axioms; Archimedean property, denseness of rationals.

There are at least two approaches to defining the real number system.

APPROACH 1. Start with the natural numbers $\mathbb{N}$, construct the set $\mathbb{Z}$ of all integers $(\mathbb{N} \cup \{0\} \cup (-\mathbb{N}))$ and next, the set $\mathbb{Q}$ of rationals and then the set $\mathbb{R}$ of real numbers either as the set of all Cauchy sequences of rationals or as Dedekind cuts. The step going from $\mathbb{Q}$ to $\mathbb{R}$ via Cauchy sequences is also available for completing any incomplete metric space (see Section A.4).

APPROACH 2. Define the set of real numbers $\mathbb{R}$ as a set that satisfies three sets of axioms. The first set is *algebraic* involving addition and multiplication. The second set is on *ordering* that, with the first, makes $\mathbb{R}$ an ordered field (see Royden (1988) for a definition). The third set is a single axiom known as the *completeness* axiom. Thus $\mathbb{R}$ is defined as a complete ordered field.

The *algebraic axioms* say that there are two binary operations known as addition (+) and multiplication ($\cdot$) that render $\mathbb{R}$ a field. See Royden (1988) for the nine axioms for this set.

The *order axiom* says that there is a set $\mathbb{P} \subset \mathbb{R}$, to be called positive numbers such that

(i) $x, y \in \mathbb{P} \Rightarrow x \cdot y \in \mathbb{P}$, $x + y \in \mathbb{P}$

(ii) $x \in \mathbb{P} \Rightarrow -x \notin \mathbb{P}$

(iii) $x \in \mathbb{R} \Rightarrow x = 0$ or $x \in \mathbb{P}$ or $-x \in \mathbb{P}$.

The set $\mathbb{Q}$ of rational numbers is an ordered field (i.e., it satisfies the algebraic and order axioms). But $\mathbb{Q}$ does not satisfy the completeness axiom (see below).

Given $\mathbb{P}$, one can define an order on $\mathbb{R}$ by defining $x < y$ (read x less than y) to mean $y - x \in \mathbb{P}$. Since for all x, y in $\mathbb{R}$, $(x - y)$ is either 0 or $(x - y) \in \mathbb{P}$ or $(y - x) \in \mathbb{P}$, it follows that for all x, y in $\mathbb{R}$, either $x = y$ or $x < y$ or $x > y$. This is called *total* or *linear order*.

Definition A.2.1: (*Upper and lower bounds*).

(a) Let $A \subset \mathbb{R}$. A real number M is an *upper bound* for A if $a \in A \Rightarrow a \leq M$ and m is a *lower bound* for A if $a \in A \Rightarrow a \geq m$.

(b) The *supremum* of a set A, denoted by *sup A* or the *least upper bound (l.u.b.)* of A, is defined by the following conditions:

 (i) $x \in A \Rightarrow x \leq \sup A$,

 (ii) $K < \sup A \Rightarrow$ there exists $x \in A$ such that $K < x$.

The completeness axiom says that if $A \subset \mathbb{R}$ has an upper bound M in $\mathbb{R}$, then there exists a $\tilde{M}$ in $\mathbb{R}$ such that $\tilde{M} = \sup A$.

That is, every set A that is bounded above in $\mathbb{R}$ has a l.u.b. in $\mathbb{R}$. The ordered field of rationals $\mathbb{Q}$ does not possess this property. One well-known example is the set

$$A = \{r : r \in \mathbb{Q}, r^2 < 2\}.$$

Then A is bounded above in $\mathbb{Q}$ but has no l.u.b. in $\mathbb{Q}$ (Problem A.11).

Next some consequences of the completeness axiom are discussed.

Proposition A.2.1: (*Axiom of Eudoxus and Archimedes (AOE)*). *For all x in $\mathbb{R}$, there exists a natural number n such that $n > x$.*

Proof: If $x \leq 1$, take $n = 2$. If $x > 1$, let $S_x \equiv \{k : k \in \mathbb{N}, k \leq x\}$. Then S_x is not empty and is bounded above. By the completeness axiom, there is a real number y which is the l.u.b. of S_x. Thus $y - \frac{1}{2}$ is not an upper bound for S_x and so there exists $k_0 \in S_x$ such that $y - \frac{1}{2} < k_0$. This implies that $(k_0 + 1) > y - \frac{1}{2} + 1 = y + \frac{1}{2} > y$ and so $(k_0 + 1) \notin S_x$. By the linear order in $\mathbb{R}$, $(k_0 + 1) > x$ and so $(k_0 + 1)$ is the desired integer. □

Corollary A.2.2: *For any x, $y \in \mathbb{R}$ with $x < y$, there is a r in $\mathbb{Q}$ such that $x < r < y$.*

Proof: Let $z = (y - x)^{-1}$. Then there is an integer k such that $0 < z < k$ (by AOE.) Again by AOE, there is a positive integer n such that $n > yk$. Let $S = \{n : n \in \mathbb{N}, n > yk\}$. Since $S \neq \emptyset$, it has a smallest element (by the well ordering property of $\mathbb{N}$) say, p. Then $p - 1 < yk < p$, i.e., $\frac{p-1}{k} < y < \frac{p}{k}$. Since $\frac{1}{k} < \frac{1}{z} = (y - x)$ and $\frac{p}{k} > y$, it follows that $\frac{p-1}{k} > x$. Now take $r = \frac{p-1}{k}$. □

Remark A.2.1: This property is often stated as: *The set $\mathbb{Q}$ of rationals is dense in the set $\mathbb{R}$ of real numbers.*

Definition A.2.2: The set $\bar{\mathbb{R}}$ of *extended real numbers* is the set consisting of $\mathbb{R}$ and two elements identified as $+\infty$ (plus infinity) and $-\infty$ (negative infinity) with the following definition of addition $(+)$ and multiplication $(\cdot)$. For any x in $\mathbb{R}$, $x + \infty = \infty$, $x - \infty = -\infty$, $x \cdot \infty = \infty$ if $x > 0$, $x \cdot \infty = -\infty$ if $x < 0$, $0 \cdot \infty = 0$, $\infty + \infty = \infty$, $-\infty - \infty = -\infty$, $\infty \cdot (\pm\infty) = \pm\infty$, $(-\infty) \cdot (\pm\infty) = \mp\infty$. But $\infty - \infty$ is not defined. The order property on $\bar{\mathbb{R}}$ is defined by extending that on $\mathbb{R}$ with the additional condition $x \in \mathbb{R} \Rightarrow -\infty < x < +\infty$. Finally, if $A \subset \mathbb{R}$ does not have an upper bound in $\mathbb{R}$, then $\sup A$ is defined as $+\infty$ and if $A \subset \mathbb{R}$ does not have a lower bound in $\mathbb{R}$, then $\inf A$ is defined as $-\infty$.

A.2.2 Sequences, series, limits, limsup, liminf

Definition A.2.3: Let $\{x_n\}_{n \geq 1}$ be a sequence of real numbers.

(i) For a real number a, $\lim_{n \to \infty} x_n = a$ if for every $\epsilon > 0$, there exists a positive integer N_ϵ such that $n \geq N_\epsilon \Rightarrow |x_n - a| < \epsilon$.

(ii) $\lim_{n \to \infty} x_n = \infty$ if for any K in $\mathbb{R}$, there exists an integer N_K such that $n \geq N_K \Rightarrow x_n > K$.

(iii) $\lim_{n \to \infty} x_n = -\infty$ if $\lim_{n \to \infty} (-x_n) = \infty$.

(iv) $\limsup_{n \to \infty} x_n \equiv \overline{\lim_{n \to \infty}} x_n = \inf_{n \geq 1} (\sup_{j \geq n} x_j)$.

(v) $\liminf_{n \to \infty} x_n \equiv \underline{\lim_{n \to \infty}} x_n = \sup_{n \geq 1} (\inf_{j \geq n} x_j)$.

Definition A.2.4: (*Cauchy sequences*). A sequence $\{x_n\}_{n\geq 1} \subset \mathbb{R}$ is called a *Cauchy sequence* if for every $\varepsilon > 0$, there is N_ε such that $n, m \geq N_\varepsilon \Rightarrow |x_m - x_n| < \varepsilon$.

Proposition A.2.3: *If $\{x_n\}_{n\geq 1} \subset \mathbb{R}$ is convergent in $\mathbb{R}$ (i.e., $\lim_{n\to\infty} x_n = a$ exists in $\mathbb{R}$), then $\{x_n\}_{n\geq 1}$ is Cauchy. Conversely, if $\{x_n\}_{n\geq 1} \subset \mathbb{R}$ is Cauchy, then there exists an $a \in \mathbb{R}$ such that $\lim_{n\to\infty} x_n = a$.*

The proof is based on the use of the l.u.b. axiom (Problem A.14).

Definition A.2.5: Let $\{x_n\}_{n\geq 1}$ be a sequence of real numbers. For $n \geq 1$, $s_n \equiv \sum_{j=1}^{n} x_j$ is called the *nth partial sum* of the series $\sum_{j=1}^{\infty} x_j$. The series $\sum_{j=1}^{\infty} x_j$ is said to *converge* to s in $\mathbb{R}$ if $\lim_{n\to\infty} s_n = s$. If $\lim_{n\to\infty} s_n = \pm\infty$, then the series $\sum_{j=1}^{\infty} x_j$ is said to *diverge* to $\pm\infty$.

Note that if $x_j \geq 0$ for all j, then either $\lim_{n\to\infty} s_n = s \in \mathbb{R}$, or $\lim_{n\to\infty} s_n = \infty$.

Example A.2.1: (*Geometric series*). Fix $0 < r < 1$. Let $x_n = r^n$, $n \geq 0$. Then $s_n = 1 + r + \ldots + r^n = \frac{1-r^{n+1}}{1-r}$ and $\sum_{j=1}^{\infty} r^j$ converges to $s = \frac{1}{1-r}$.

Example A.2.2: Consider the series $\sum_{j=1}^{\infty} \frac{1}{j^p}$, $0 < p < \infty$. It can be shown that this converges for $p > 1$ and diverges to ∞ for $0 < p \leq 1$.

Definition A.2.6: The series $\sum_{j=1}^{\infty} x_j$ *converges absolutely* if the series $\sum_{j=1}^{\infty} |x_j|$ converges in $\mathbb{R}$.

There exist series $\sum_{j=1}^{\infty} x_j$ that converge but not absolutely. For example, $\sum_{j=1}^{\infty} \frac{(-1)^j}{j}$. For further material on convergence properties of series, such as tests for convergence, rates of convergence, etc., see Rudin (1976).

Definition A.2.7: (*Power series*). Let $\{a_n\}_{n\geq 0}$ be a sequence of real numbers. For $x \in \mathbb{R}$, the series $\sum_{n=0}^{\infty} a_n x^n$ is called a *power series*. If the series $\sum_{n=0}^{\infty} a_n x^n$ converges for all x in $B \subset \mathbb{R}$, the *power series* $\sum_{n=0}^{\infty} a_n x^n$ is said to be *convergent on B*.

Proposition A.2.4: *Let $\{a_n\}_{n\geq 0}$ be a sequence of real numbers. Let $\rho = (\limsup_{n\to\infty} |a_n|^{\frac{1}{n}})^{-1}$. Then*

(i) $|x| < \rho \Rightarrow \sum_{n=0}^{\infty} |a_n x^n|$ *converges.*

(ii) $|x| > \rho \Rightarrow \sum_{n=0}^{\infty} |a_n x^n|$ *diverges to* $+\infty$.

Proof of this is left as an exercise (Problem A.15).

Definition A.2.8: $\rho \equiv (\limsup_{n\to\infty} |a_n|^{\frac{1}{n}})^{-1}$ is called the *radius of convergence* of the power series $\sum_{n=0}^{\infty} a_n x^n$.

A.2.3 Continuity and differentiability

Definition A.2.9: Let $f : A \to \mathbb{R}$, $A \subset \mathbb{R}$. Then

(a) f is *continuous at* x_0 in A if for every $\epsilon > 0$, there exists a $\delta > 0$ such that $x \in A$, $|x - x_0| < \delta$, implies $|f(x) - f(x_0)| < \epsilon$. (Here, δ may depend on ϵ and x_0.)

(b) f is *continuous on* $B \subset A$ if it is continuous at every x_0 in B.

(c) f is *uniformly continuous* on $B \subset A$ if for every $\epsilon > 0$, there exists a $\delta_\epsilon > 0$ such that $\sup\{|f(x) - f(y)| : x, y \in B, |x - y| < \delta_\epsilon\} < \epsilon$.

Some properties of continuous functions are listed below.

Proposition A.2.5:

(i) (Sums, products, and ratios of continuous functions). Let $f, g : A \to \mathbb{R}$, $A \subset \mathbb{R}$. Let f and g be continuous on $B \subset A$. Then

(a) $f + g$, $f - g$, $\alpha \cdot f$ for any $\alpha \in \mathbb{R}$ are all continuous on B.

(b) $f(x)/g(x)$ is continuous at x_0 in B, provided $g(x_0) \neq 0$.

(ii) (Continuous functions on a closed bounded interval). Let f be continuous on a closed and bounded interval $[a, b]$. Then

(a) f is bounded, i.e., $\sup\{|f(x)| : a \leq x \leq b\} < \infty$,

(b) it achieves its maximum and minimum, i.e., there exist x_0, y_0 in $[a, b]$ such that $f(x_0) \geq f(x) \geq f(y_0)$ for all x in $[a, b]$ and f attains all values in $[f(y_0), f(x_0)]$, i.e., for all $\ell \in [f(y_0), f(x_0)]$, there exists $z \in [a, b]$ such that $f(z) = \ell$. Thus, f maps bounded closed intervals onto bounded closed intervals.

(c) f is uniformly continuous on $[a, b]$.

(iii) (Composition of functions). Let $f : A \to \mathbb{R}$, $g : B \to \mathbb{R}$ be continuous on A and B, respectively. Let $f(A) \subset B$, i.e., for any x in A, $f(x) \in B$. Let $h(x) = g(f(x))$ for x in A. Then $h : A \to \mathbb{R}$ is continuous.

(iv) (Uniform limits of continuous functions). Let $\{f_n\}_{n \geq 1}$, be a sequence of functions continuous on A to $\mathbb{R}$, $A \subset \mathbb{R}$. If $\sup\{|f_n(x) - f(x)| : x \in A\} \to 0$ as $n \to \infty$ for some $f : A \to \mathbb{R}$, i.e., f_n converges to f uniformly on A, then f is continuous on A.

Remark A.2.2: The function $f(x) \equiv x$ is clearly continuous on $\mathbb{R}$. Now by Proposition A.2.5 (i) and (iv), it follows that all polynomials are continuous on $\mathbb{R}$ and hence, so are their uniform limits. *Weierstrass' approximation theorem* is a sort of converse to this. That is, every continuous function on a closed and bounded interval is the uniform limit of polynomials. More precisely, one has the following:

Theorem A.2.6: *Let $f : [a, b] \to \mathbb{R}$ be continuous. Then for any $\epsilon > 0$ there is a polynomial $p(x) = \sum_0^n a_j x^j$, $a_j \in \mathbb{R}$, $j = 0, 1, 2, \ldots, n$ such that* $\sup\{|f(x) - p(x)| : x \in [a, b]\} < \epsilon$.

It should be noted that a power series $A(x) \equiv \sum_0^\infty a_n x^n$ is the uniform limits of polynomials on $[-\lambda, \lambda]$ for any $0 < \lambda < \rho \equiv \left(\overline{\lim}_{n\to\infty} |a_n|^{1/n}\right)^{-1}$ and hence is continuous on $(-\rho, \rho)$.

Definition A.2.10: Let $f : (a, b) \to \mathbb{R}$, $(a, b) \subset \mathbb{R}$. The function f is said to be *differentiable at* $x_0 \in (a, b)$ if

$$\lim_{h\to 0} \frac{f(x_0 + h) - f(x_0)}{h} \equiv f'(x_0) \quad \text{exists in} \quad \mathbb{R}.$$

A function is differentiable in (a, b) if it is differentiable at each x in (a, b).

Some important consequences of differentiability are listed below.

Proposition A.2.7: *Let f, $g : (a, b) \to \mathbb{R}$, $(a, b) \subset \mathbb{R}$. Then*

(i) f differentiable at x_0 in (a, b) implies f is continuous at x_0.

(ii) (Mean value theorem). f differentiable on (a, b), f continuous on $[a, b]$ implies that for some $a < c < b$, $f(b) - f(a) = (b - a)f'(c)$.

(iii) (Maxima and minima). f differentiable at x_0 and for some $\delta > 0$, $f(x) \le f(x_0)$ for all $x \in (x_0 - \delta, x_0 + \delta)$ implies that $f'(x_0) = 0$.

(iv) (Sums, products and ratios). f, g differentiable at x_0 implies that for any α, β in $\mathbb{R}$, $(\alpha f + \beta g)$, $f - g$ are differentiable at x_0 with

$$\begin{aligned} (\alpha f + \beta g)'(x_0) &= \alpha f'(x_0) + \beta g'(x_0), \\ (fg)'(x_0) &= f'(x_0)g(x_0) + f(x_0)g'(x_0), \end{aligned}$$

and if $g'(x_0) \neq 0$, then f/g is differentiable at x_0 with

$$(f/g)'(x_0) = \frac{f'(x_0)g(x_0) - f(x_0)g'(x_0)}{(g(x_0))^2}.$$

(v) (Chain rule). If f is differentiable at x_0 and g is differentiable at $f(x_0)$, then $h(x) \equiv g(f(x))$ is differentiable at x_0 with $h'(x_0) = g'(f(x_0))f'(x_0)$.

(vi) (Differentiability of power series). Let $A(x) \equiv \sum_{n=0}^\infty a_n x^n$ be a power series with radius of convergence $\rho \equiv \left(\overline{\lim}_{n\to\infty} |a_n|^{1/n}\right)^{-1} > 0$. Then $A(\cdot)$ is differentiable infinitely many times on $(-\rho, \rho)$ and for x in $(-\rho, \rho)$,

$$\frac{d^k A(x)}{dx^k} = \sum_{n=k}^\infty n(n-1)\cdots(n-k+1)x^{n-k}, \ k \ge 1.$$

Remark A.2.3: It should be noted that the converse to (a) in the above proposition does not hold. For example, the function $f(x) = |x|$ is continuous at $x_0 = 0$ but is not differentiable at x_0. Indeed, Weierstrass showed that there exists a function $f : [0,1] \to \mathbb{R}$ such that it is continuous on $[0,1]$ but is not differentiable at any x in $(0,1)$.

Also note that the mean value theorem implies that if $f'(\cdot) \geq 0$ on (a,b), then f is nondecreasing on (a,b).

Definition A.2.11: (*Taylor series*). Let f be a map from $I \equiv (a-\eta, a+\eta)$ to $\mathbb{R}$ for some $a \in \mathbb{R}$, $\eta > 0$. Suppose f is n times differentiable in I, for each $n \geq 1$. Let $a_n = \frac{f^{(n)}(a)}{n!}$. Then power series $\sum_{n=0}^{\infty} a_n(x-a)^n = \sum_{n=0}^{\infty} \frac{f^{(n)}(a)}{n!}(x-a)^n$ is called the *Taylor series* of f at a.

Remark A.2.4: Let f be as in Definition A.2.11. *Taylor's remainder theorem* says that for any x in I and any $n \geq 1$, if f is $(n+1)$ times differentiable in I, then

$$\left| f(x) - \sum_{j=0}^{n} a_j x^j \right| \leq \frac{|f^{(n+1)}(y_n)|}{(n+1)!}$$

for some y_n in I.

Thus, if for some $\epsilon > 0$, $\sup_{|y-a|<\varepsilon} |f^{(k)}(y)| \equiv \lambda_k$ satisfies $\frac{\lambda_k}{k!} \to 0$ as $k \to \infty$, then the Taylor series satisfies

$$\sup_{|x-a|<\varepsilon} \left| \sum_{j=0}^{n} (x-a)^j \frac{f^{(j)}(a)}{j!} - f(x) \right| \to 0 \quad \text{as} \quad n \to \infty.$$

A.2.4 Riemann integration

Let $f : [a,b] \to \mathbb{R}$, $[a,b] \subset \mathbb{R}$. Let $\sup\{|f(x)| : a \leq x \leq b\} < \infty$. For any partition $P \equiv \{a = x_0 < x_1 < x_2 < x_k = b\}$, let

$$\begin{aligned} U(P,f) &\equiv \sum_{i=1}^{n} M_i(P)\Delta_i(P) \quad \text{and} \\ L(P,f) &\equiv \sum_{i=1}^{n} m_i(P)\Delta_i(P) \end{aligned}$$

where

$$\begin{aligned} M_i(P) &= \sup\{f(x) : x_i \leq x \leq x_i\}, \\ m_i(P) &= \inf\{f(x) : x_{i-1} \leq x \leq x_i\}, \\ \Delta_i(P) &= (x_i - x_{i-1}) \end{aligned}$$

for $i = 1, 2, \ldots, k$.

Definition A.2.12: The upper and lower Riemann integrals of f over $[a,b]$, denoted by $\overline{\int}_a^b f(x)dx$ and $\underline{\int}_a^b f(x)dx$, respectively defined as

$$\overline{\int}_a^b f(x)dx \equiv \inf\{U(P,f) : P \text{ a partition of } [a,b]\}$$
$$\underline{\int}_a^b f(x)dx \equiv \sup\{L(P,f) : P \text{ a partition of } [a,b]\}.$$

Definition A.2.13: Let $f : [a,b] \to \mathbb{R}$, $[a,b] \subset \mathbb{R}$ and let $\sup\{|f(x)| : a \leq x \leq b\} < \infty$. The function f is Riemann integrable on $[a,b]$ if

$$\overline{\int}_a^b f(x)dx = \underline{\int}_a^b f(x)dx$$

and the common value is denoted as $\int_a^b f(x)dx$.

The following are some important results on Riemann integration.

Proposition A.2.8:

(i) *Let $f : [a,b] \to \mathbb{R}$, $[a,b] \subset \mathbb{R}$ be continuous. Then f is Riemann integrable and $\int_a^b f(x)dx = \lim_{n\to\infty} \frac{1}{k_n} \sum_{i=0}^{k_n} f(x_{ni})\Delta_{ni}$ where $\{P_n \equiv \{x_{ni} : 1 \leq i \leq k_n\}\}$ is a sequence of partitions of $[a,b]$ such that $\Delta_n \equiv \{\max(x_{ni} - x_{n,i-1}) : 1 \leq i \leq k_n\} \to 0$ as $n \to \infty$.*

(ii) *If f, g are Riemann integrable on $[a,b]$, then $\alpha f + \beta g$, $\alpha, \beta \in \mathbb{R}$ and fg are Riemann integrable on $[a,b]$ and*

$$\int_a^b (\alpha f + \beta g)(x)dx = \alpha \int_a^b f(x)dx + \beta \int_a^b g(x)dx.$$

(iii) *Let f be Riemann integrable on $[a,b]$ and $[b,c]$, $-\infty < a < b < c < \infty$. Then f is Riemann integrable on $[a,c]$ and*

$$\int_a^c f(x)dx = \int_a^b f(x)dx + \int_b^c f(x)dx.$$

(iv) *(Fundamental theorem of Riemann integration: Part I). Let f be Riemann integrable on $[a,b]$. Then it is Riemann integrable on $[a,c]$ for all $a \leq c \leq b$. Let $F(x) \equiv \int_a^x f(u)du$, $a \leq x \leq b$. Then*

- *(a)* *$F(\cdot)$ is continuous on $[a,b]$.*
- *(b)* *If $f(\cdot)$ is continuous at $x_0 \in (a,b)$, then $F(\cdot)$ is differentiable at x_0 and $F'(x_0) = f(x_0)$.*

(v) *(Fundamental theorem: Part II). If $F : [a,b] \to \mathbb{R}$ is differentiable on (a,b), $f : [a,b] \to \mathbb{R}$ is continuous on $[a,b]$ and $F'(x) = f(x)$ for all $a < x < b$, then $F(x) = F(a) + \int_a^c f(x)dx$, $a \le c \le b$.*

(vi) *(Integration by parts). If f and g are continuous and differentiable on (a,b) with f' and g' continuous on (a,b), then*

$$\begin{aligned}\int_c^d f(x)g'(x)dx + \int_c^d f'(x)g(x)dx \\ = f(d)g(d) - f(c)g(c) \quad \text{for all} \quad a < c < d < b.\end{aligned}$$

(vii) *(Interchange of limits and integration). Let f_n, f be continuous on $[a,b]$ to $\mathbb{R}$. Let f_n converge to f uniformly on $[a,b]$. Then*

$$F_n(c) \equiv \int_a^c f_n(x)dx \to F(c) \equiv \int_a^c f(x)dx$$

uniformly on $[a,b]$.

A.3 Complex numbers, exponential and trigonometric functions

Definition A.3.1: The set $\mathbb{C}$ of complex numbers is defined as the set $\mathbb{R} \times \mathbb{R}$ of all ordered pairs of real numbers endowed with addition and multiplication as follows:

$$\begin{aligned}(a_1,b_1) + (a_2,b_2) &= (a_1 + a_2, b_1 + b_2) \\ (a_1,b_1) \cdot (a_2,b_2) &= (a_1a_2 - b_1b_2, a_1b_2 + a_2b_1).\end{aligned} \quad (3.1)$$

It can be verified that $\mathbb{C}$ satisfies the field axioms Royden (1988). By defining ι to be the element (0,1) one can write the elements of $\mathbb{C}$ in the form $(a,b) = a + \iota b$ and do addition and multiplication with the rule of replacing ι^2 by -1. Clearly the set $\mathbb{R}$ of real numbers can be identified with the set $\{(a,0) : a \in \mathbb{R}\}$. The set $\{(0,b) : b \in \mathbb{R}\}$ is called the set of *purely imaginary numbers.*

Definition A.3.2: For any complex number $z = (a,b)$ in $\mathbb{C}$,

$$\begin{aligned}\text{Re}(z), &\quad \text{called the } \textit{real part of } z \text{ is } a, \\ \text{Im}(z), &\quad \text{called the } \textit{imaginary part of } z \text{ is } b, \\ \bar{z}, &\quad \text{called the } \textit{complex conjugate of } z \text{ is } \bar{z} = a - \iota b \\ |z|, &\quad \text{called the } \textit{absolute value of } z \text{ is } |z| = \sqrt{a^2 + b^2}\,.\end{aligned} \quad (3.2)$$

Clearly, $z\bar{z} = \bar{z}z = |z|^2$ and any $z \neq 0$ can be written as $z = |z|\omega$ where $|\omega| = 1$. A set $A \subset \mathbb{C}$ is *bounded* if $\sup\{|z| : z \in A\} < \infty$. If $d(z_1, z_2) \equiv |z_1 - z_2|$, then $(\mathbb{C}, d)$ is a *complete separable metric space* (see Section A.4). Clearly, $z_n = (a_n, b_n)$ converges to $z = (a, b)$ in this metric iff $\text{Re}(z_n) = a_n \to \text{Re}(z) = a$ and $\text{Im}(z_n) = b_n \to \text{Im}(z) = b$.

Definition A.3.3: The *exponential function* is a map from $\mathbb{C}$ to $\mathbb{C}$ defined by

$$\exp(z) \equiv \sum_{n=0}^{\infty} \frac{z^n}{n!} \tag{3.3}$$

where the right side is defined as the limit of the partial sum sequence $\{\sum_{n=0}^{m} \frac{z^n}{n!}\}_{m\geq 0}$ which exists since $\sum_{n=0}^{\infty} \frac{|z|^n}{n!} < \infty$ for each $z \in \mathbb{C}$. In fact, the convergence of the partial sum sequence is uniform on every bounded subset A of $\mathbb{C}$. This in turn implies that the exponential function is continuous on $\mathbb{C}$.

Notation: $\exp(z)$ will also be written as e^z.

Definition A.3.4: The number $e^1 \equiv \sum_{n=0}^{\infty} \frac{1}{n!}$ will be called e. It is not difficult to show that $e = \lim_{n\to\infty}(1 + \frac{1}{n})^n$. By definition of e^z, $e^0 = 1$.

Definition A.3.5: The *cosine* and *sine* functions from $\mathbb{R} \to \mathbb{R}$ are defined by

$$\begin{aligned} \cos t &\equiv \text{Re}(e^{\iota t}) = \sum_{k=0}^{\infty} (-1)^k \frac{t^{2k}}{(2k)!} \\ \sin t &\equiv \text{Im}(e^{\iota t}) = \sum_{k=0}^{\infty} (-1)^k \frac{t^{2k+1}}{(2k+1)!}. \end{aligned}$$

Thus, one gets Euler's formula $e^{\iota t} = \cos t + \iota \sin t$, for all $t \in \mathbb{R}$.

The following are some important and useful properties of the exponential and the cosine and sine (also called *trigonometric*) functions.

Theorem A.3.1:

(i) For all z_1, z_2 in $\mathbb{C}$

$$e^{z_1}e^{z_2} = e^{z_1+z_2} \quad and \quad e^z \neq 0 \quad for\ any \quad z.$$

(ii) e^z is differentiable for all z (i.e. $(e^z)' \equiv \lim_{h\to 0} \frac{e^{z+h}-e^z}{h}$ exists) and $(e^z)' = e^z$.

(iii) The function $e^x : \mathbb{R} \to \mathbb{R}_+$ is strictly increasing with $e^x \uparrow \infty$ as $x \uparrow \infty$ and $\downarrow 0$ as $x \downarrow -\infty$.

(iv) For t in $\mathbb{R}$, $(\cos t)^2 + (\sin t)^2 = 1$, $(\cos t)' = -\sin t$, $(\sin t)' = \cos t$.

(v) There is a smallest positive number, called π, such that $e^{\iota\pi/2} = \iota$.

(vi) e^z is a periodic function such that $e^z = e^{z+\iota 2\pi}$.

Proof:

(i) Since $\sum_{n=0}^{\infty} \frac{|z|^n}{n!}$ converges for all $z \in \mathbb{C}$,

$$\begin{aligned} e^{z_1} \cdot e^{z_2} &= \lim_{m\to\infty} \left(\sum_{n=0}^{m} \frac{z_1^n}{n!}\right)\left(\sum_{n=0}^{m} \frac{z_2^n}{n!}\right) \\ &= \lim_{m\to\infty} \sum_{0\le r,s\le m} \frac{z_1^r z_2^s}{r!s!} \\ &= \lim_{m\to\infty} \sum_{k=0}^{2m} \frac{1}{k!} \sum_{r=0}^{k} \frac{k!}{r!(k-r)!} z_1^r z_2^{k-r} \\ &= \lim_{m\to\infty} \sum_{k=0}^{2m} \frac{(z_1+z_2)^k}{k!} = e^{z_1+z_2}. \end{aligned}$$

Since $e^z \cdot e^{-z} = e^0 = 1$, $e^z \neq 0$ for any $z \in \mathbb{C}$.

(ii) Fix $z \in \mathbb{C}$. For any $h \in \mathbb{C}$, $h \neq 0$, by (i),

$$\frac{e^{z+h} - e^z}{h} = e^z \frac{e^h - 1}{h}.$$

But

$$\begin{aligned} \left|\frac{e^h - 1}{h} - 1\right| &\le \left(\sum_{k=2}^{\infty} \frac{|h|^k}{k!}\right) \frac{1}{|h|} \\ &\le |h| \sum_{k=0}^{\infty} \frac{1}{k!} \quad \text{if} \quad |h| \le 1. \end{aligned}$$

Thus

$$\lim_{h\to 0} \left(\frac{e^h - 1}{h} - 1\right) = 0$$

and

$$\lim_{h\to 0} \left(\frac{e^{z+h} - e^z}{h}\right) = e^z, \quad \text{i.e. (ii) holds.}$$

(iii) That the map $t \to e^t$ is strictly increasing on $[0, \infty)$ and that $e^t \uparrow \infty$ as $t \uparrow \infty$ is clear from the definition of e^z. Since $e^{-t}e^t = e^0 = 1$, $e^{-t} = \frac{1}{e^t}$ for all $t \in \mathbb{R}$, so that e^t is strictly increasing on all of $\mathbb{R}$ and $e^t \downarrow 0$ as $t \downarrow -\infty$.

(iv) From the definition of e^z it follows that $\overline{e^z} = e^{\bar{z}}$ and in particular, for t real $\overline{e^{\iota t}} = e^{\overline{\iota t}} = e^{-\iota t}$ and hence

$$|e^{\iota t}|^2 = \overline{e^{\iota t}} e^{\iota t} = e^{-\iota t + \iota t} = e^0 = 1.$$

Thus, for all t real $|e^{\iota t}| = 1$ and since $e^{\iota t} = \cos t + \iota \sin t$ it follows that

$$|e^{\iota t}|^2 = (\cos t)^2 + (\sin t)^2 = 1.$$

Also from (ii), for $t \in \mathbb{R}$

$$(e^{\iota t})' = (\cos t)' + \iota(\sin t)' = \iota e^{\iota t} = -\sin t + \iota \cos t$$

yielding $(\cos t)' = -\sin t$, $(\sin t)' = \cos t$ proving (iv).

(v) By definition

$$\cos 2 = \sum_{k=0}^{\infty} (-1)^k \frac{2^{2k}}{(2k)!} .$$

Now $a_k \equiv \frac{2^{2k}}{(2k)!}$ satisfies $a_{k+1} < a_k$ for $k = 0, 1, 2, \ldots$ and hence $\sum_{k \geq 3} (-1)^k a_k = -\{(a_3 - a_4) + (a_5 - a_6) + \cdots\} < 0$. Thus,

$$\cos 2 < 1 - \frac{2^2}{2!} + \frac{2^4}{4!} = -\frac{1}{3} < 0.$$

Since $\cos t$ is a continuous function on $\mathbb{R}$ with $\cos 0 = 1$ and $\cos 2 < 0$ there exists a smallest $t_0 > 0$ such that $\cos t_0 = 0$, defined by $t_0 = \inf\{t : t > 0, \cos t = 0\}$. Set $\pi = 2t_0$. Since $\cos \frac{\pi}{2} = 0$, (iv) implies $\sin \frac{\pi}{2} = 1$ and hence that $e^{\iota \frac{\pi}{2}} = \iota$.

(vi) Clearly, $e^{\iota \frac{\pi}{2}} = \iota$ implies that $e^{\iota \pi} = -1$ and $e^{\iota 2\pi} = 1$ and $e^{\iota 2\pi k} = 1$ for all integers k. Since $e^{\iota 2\pi} = 1$, it follows that $e^z = e^{z + \iota 2\pi}$ for all $z \in \mathbb{C}$,

□

It is now possible to prove various results involving π that one learns in calculus from the above definition. For example, that the arc length of the *unit circle* $\{z : |z| = 1\}$ is 2π and that $\int_{-\infty}^{\infty} \frac{1}{1+x^2} dx = \pi$, etc. (Problems A.19 and A.20).

The following assertions about e^z can be proved with some more effort.

Theorem A.3.2:

(i) $e^z = 1$ *iff* $z = 2\pi\iota k$ *for some integer* k.

(ii) The map $t \to e^{\iota t}$ *from* $\mathbb{R}$ *is* onto *the unit circle.*

(iii) For any $\omega \in \mathbb{C}$, $\omega \neq 0$ *there is a* $z \in \mathbb{C}$ *such that* $\omega = e^z$.

For a proof of this theorem as well as more details on Theorem A.3.1 see Rudin (1987).

Theorem A.3.3: *(Orthogonality of $\{e^{\iota 2\pi nt}\}_{n\in\mathbb{Z}}$). For any $n \in \mathbb{Z}$, $\int_0^1 e^{\iota 2\pi nt}dt = 0$ if $n \neq 0$ and 1 if $n = 0$.*

Proof: Since $(e^{\iota t})' = \iota e^{\iota t}$

$$(e^{\iota 2\pi nt})' = \iota 2\pi n\, e^{\iota 2\pi nt}, \; n \in \mathbb{Z}$$

and so for $n \neq 0$,

$$\begin{aligned}\int_0^1 e^{\iota 2\pi nt}dt &= \frac{1}{\iota 2\pi n}\int_0^1 (e^{\iota 2\pi nt})'dt \\ &= \frac{1}{\iota 2\pi n}(e^{\iota 2\pi n} - 1) = 0.\end{aligned}$$

Corollary A.3.4: *The family $\{\cos 2\pi nt : n = 0, 1, 2, \ldots\} \cup \{\sin 2\pi nt : n = 1, 2, \ldots\}$ are orthogonal in $L^2[0,1]$ (Problem A.22), i.e., for any two f, g in this family, $\int_0^1 f(x)g(x)dx = 0$ for $f \neq g$.*

A.4 Metric spaces

A.4.1 Basic definitions

This section reviews the following: metric spaces, Cauchy sequences, completeness, functions, continuity, compactness, convergence of sequences functions, and uniform convergence.

Definition A.4.1: Let $\mathbb{S}$ be a nonempty set. Let $d : \mathbb{S} \times \mathbb{S} \to \mathbb{R}_+ = [0, \infty)$ be such that

(i) $d(x, y) = d(y, x)$ for any x, y in $\mathbb{S}$.

(ii) $d(x, z) \leq d(x, y) + d(y, z)$ for any x, y, z in $\mathbb{S}$.

(iii) $d(x, y) = 0$ iff $x = y$.

Such a d is called a *metric* on $\mathbb{S}$ and the pair $(\mathbb{S}, d)$ a *metric space*. Property (ii) is called the *triangle inequality.*

Example A.4.1: Let $\mathbb{R}^k \equiv \{(x_1, \ldots, x_k) : x_i \in \mathbb{R},\ 1 \leq i \leq k\}$ be the k-dimensional Euclidean space. For $1 \leq p < \infty$ and $x = (x_1, \ldots, x_k)$, $y = (y_1, y_2, \ldots, y_k) \in \mathbb{R}^k$, let

$$d_p(x, y) = \left(\sum_{i=1}^{k} |x_i - y_i|^p\right)^{\frac{1}{p}},$$

and $d_\infty(x, y) = \max\{|x_i - y_i| : 1 \le i \le k\}$.

It can be shown that $d_p(\cdot, \cdot)$ is a metric on $\mathbb{R}$ for all $1 \le p \le \infty$ (Problem A.24).

Definition A.4.2: A sequence $\{x_n\}_{n\ge1}$ in a metric space $(\mathbb{S}, d)$ *converges* to an x in $\mathbb{S}$ if for every $\varepsilon > 0$, there is a N_ε such that $n \ge N_\varepsilon \Rightarrow d(x_n, x) < \varepsilon$ and is written as $\lim_{n\to\infty} x_n = x$.

Definition A.4.3: A sequence $\{x_n\}_{n\ge1}$ in a metric space $(\mathbb{S}, d)$ is *Cauchy* if for all $\varepsilon > 0$, there exists N_ε such that $n, m \ge N_\varepsilon \Rightarrow d(x_n, x_m) < \varepsilon$.

Definition A.4.4: A metric space $(\mathbb{S}, d)$ is *complete* if every Cauchy sequence $\{x_n\}_{n\ge1}$ converges to some x in $\mathbb{S}$.

Example A.4.2:

(a) Let $\mathbb{S} = \mathbb{Q}$, the set of rationals and $d(x, y) \equiv |x - y|$. Then $(\mathbb{Q}, d)$ is a metric space that is *not* complete.

(b) Let $\mathbb{S} = \mathbb{R}$ and $d(x, y) = |x - y|$. Then $(\mathbb{R}, d)$ is *complete* (cf. Proposition A.2.3).

(c) Let $\mathbb{S} = \mathbb{R}^k$. Then $(\mathbb{R}^k, d_p)$ is complete for every $1 \le p \le \infty$, where d_p is as in Example A.4.1.

Remark A.4.1: (*Completion of an incomplete metric space*). Let $(\mathbb{S}, d)$ be a metric space. Let $\tilde{\mathbb{S}}$ be the set of all Cauchy sequences in $\mathbb{S}$. Identify each x in $\mathbb{S}$ with the Cauchy sequence $\{x_n = x\}_{n\ge1}$. Define a function from $\tilde{\mathbb{S}} \times \tilde{\mathbb{S}}$ to $\mathbb{R}_+$ by

$$\tilde{d}(\{x_n\}_{n\ge1}, \{y_n\}_{n\ge1}) = \limsup_{n\to\infty} d(x_n, y_n).$$

It is easy to verify that $\tilde{d}$ is symmetric and satisfies the triangle inequality. Define $s_1 = \{x_n\}_{n\ge1}$ and $s_2 = \{y_n\}_{n\ge1}$ to be *equivalent* (write $\{x_n\} \sim \{y_n\}$) if $\tilde{d}(s_1, s_2) = 0$. Let $\bar{\mathbb{S}}$ be the set of all equivalence classes in $\tilde{\mathbb{S}}$ and define $\bar{d}(c_1, c_2) \equiv \tilde{d}(s_1, s_2)$, where c_1, c_2 are equivalence classes and s_1, s_2 are arbitrary elements of c_1 and c_2, respectively.

It can now be verified that $(\bar{\mathbb{S}}, \bar{d})$ is a complete metric space and $(\mathbb{S}, d)$ is embedded in $(\bar{\mathbb{S}}, \bar{d})$ by identifying each x in $\mathbb{S}$ with the equivalence class containing the sequence $\{x_n = x\}_{n\ge1}$.

Definition A.4.5: A metric space $(\mathbb{S}, d)$ is *separable* if there exists a subset $D \subset \mathbb{S}$ that is countable and *dense in* $\mathbb{S}$, i.e., for each x in $\mathbb{S}$ and $\varepsilon > 0$, there is a y in D such that $d(x, y) < \varepsilon$.

Example A.4.3: By the Archimedean property, $\mathbb{Q}$ is dense in $\mathbb{R}$. Similarly $\mathbb{Q}^k$, the set of all k vectors with components from $\mathbb{Q}$ is dense in $\mathbb{R}^k$.

Definition A.4.6: A metric space $(\mathbb{S}, d)$ is called *Polish* if it is complete and separable.

Example A.4.4: $(\mathbb{R}^k, d_p)$ in Example A.4.2 is Polish.

A.4.2 Continuous functions

Let $(\mathbb{S}, d)$ and $(\mathbb{T}, \rho)$ be two metric spaces. Let $f : \mathbb{S} \to \mathbb{T}$ be a map from $\mathbb{S}$ to $\mathbb{T}$.

Definition A.4.7:

(a) f is *continuous at* p in $\mathbb{S}$ if for each $\varepsilon > 0$, there exists $\delta > 0$ such that $d(x, p) < \delta \Rightarrow \rho(f(x), f(p)) < \varepsilon$. (Here the δ may depend on ε and p.)

(b) f is *continuous on a set* $B \subset \mathbb{S}$ if it is continuous at every $p \in B$.

(c) f is *uniformly continuous* on B if for each $\varepsilon > 0$, there exists $\delta > 0$ such that for each pair x, y in $\mathbb{S}$, $d(x, y) < \delta \Rightarrow \rho(f(x), f(y)) < \varepsilon$.

Definition A.4.8: Let $(\mathbb{S}, d)$ be a metric space.

(a) A set $O \subset (\mathbb{S}, d)$ is *open* if $x \in O \Rightarrow$ there exists $\delta > 0$ such that $d(x, y) < \delta \Rightarrow y \in O$. That is, at every point x in O, an *open ball* $B_x(\delta) \equiv \{y : d(x, y) < \delta\}$ of positive radius δ is a subset of O.

(b) A set $C \subset (\mathbb{S}, d)$ is *closed* if C^c is open.

Theorem A.4.1: *Let $(\mathbb{S}, d)$ and $(\mathbb{T}, \rho)$ be metric spaces. A map $f : \mathbb{S} \to \mathbb{T}$ in continuous on $\mathbb{S}$ iff for each O open in $\mathbb{T}$, $f^{-1}(O)$ is open in $\mathbb{S}$.*

Proof is left as an exercise (Problem A.28).

A.4.3 Compactness

Definition A.4.9: A collection of open sets $\{O_\alpha : \alpha \in I\}$ is an *open cover* for a set $B \subset (\mathbb{S}, d)$ if for each $x \in B$, there exists $\alpha \in I$ such that $x \in O_\alpha$.

Example A.4.5: Let $B = (0, 1)$. Then the collection $\{(\alpha - \frac{\alpha}{2}, \alpha + \frac{(1-\alpha)}{2}) : \alpha \in \mathbb{Q} \cap (0, 1)\}$ is an open cover for B.

Definition A.4.10: Let $(\mathbb{S}, d)$ be a metric space. A set $K \subset \mathbb{S}$ is called *compact* if given any open cover $\{O_\alpha : \alpha \in I\}$ for K, there exists a finite subcollection $\{O_{\alpha_i} : \alpha_i \in I, i = 1, 2, \ldots, n, n < \infty\}$ that is an open cover for K.

Example A.4.6: The set $B = (0, 1)$ is not *compact* as the open cover in the above Example A.3.4 does not admit a finite subcover.

The next result is the well-known *Heine-Borel theorem.*

Theorem A.4.2:

(i) For any $-\infty < a < b < \infty$, the closed interval $[a, b]$ is compact in $\mathbb{R}$.

(ii) Any $K \subset \mathbb{R}$ is compact iff it is bounded and closed.

For a proof see Rudin (1976). From Proposition A.4.1, it is seen that the inverse image of an open set under a continuous function is open but the forward image may not have this property. But the following is true.

Theorem A.4.3: *Let $(\mathbb{S}, d)$ and $(\mathbb{T}, \rho)$ be two metric spaces and let $f : (\mathbb{S}, d) \to (\mathbb{T}, \rho)$ be continuous. Let $K \subset \mathbb{S}$ be compact. Then $f(K)$ is compact.*

The proof is left as an exercise (Problem A.35).

A.4.4 Sequences of functions and uniform convergence

Definition A.4.11: Let $(\mathbb{S}, d)$ and $(\mathbb{T}, \rho)$ be two metric spaces and let $\{f_n\}_{n\geq 1}$ be a sequence of functions from $(\mathbb{S}, d)$ to $(\mathbb{T}, \rho)$. The sequence $\{f_n\}_{n\geq 1}$ is said to:

(a) *converge pointwise to f on a set $A \subset \mathbb{S}$* if $\lim_{n\to\infty} f_n(x) = f(x)$ for each x in A;

(b) *converges uniformly to f on a set $A \subset \mathbb{S}$* if for each $\varepsilon > 0$, there exists $N_\varepsilon > 0$ (depending on ε and A) such that

$$n \geq N_\varepsilon \Rightarrow \rho\big(f_n(x), f(x)\big) < \varepsilon \quad \text{for all} \quad x \quad \text{in} \quad A.$$

A consequence of uniform convergence is the preservation of the continuity property.

Theorem A.4.4: *Let $(\mathbb{S}, d)$ and $(\mathbb{T}, \rho)$ be two metric spaces and let $\{f_n\}_{n\geq 1}$ be a sequence of functions from $(\mathbb{S}, d)$ to $(\mathbb{T}, \rho)$. Let for each $n \geq 1$, f_n be continuous on $A \subset \mathbb{S}$. Let $\{f_n\}_{n\geq 1}$ converge to f uniformly on A. Then f is continuous on A.*

Proof: The proof is based on the "break up into three parts" idea. By the triangle inequality,

$$\rho\big(f(x), f(y)\big) \leq \rho\big(f(x), f_n(x)\big) + \rho\big(f_n(x), f_n(y)\big) + \rho\big(f_n(y), f(y)\big).$$

Fix x in A. By the uniform convergence on A, $\sup\{\rho\big(f_n(u), f(u)\big) : u \in A\} \to 0$ as $n \to \infty$. So for each $\varepsilon > 0$, there exists $N_\varepsilon < \infty$ such that $n \geq N_\varepsilon \Rightarrow \rho\big(f_n(u), f(u)\big) < \frac{\varepsilon}{3}$ for all u in A. Now since $f_{N_\varepsilon}(\cdot)$ is continuous on A, there exists a $\delta > 0$ (depending on N_ε and x), such that $d(x, y) < \delta, y \in A \Rightarrow \rho\big(f_{N_\varepsilon}(y), f_{N_\varepsilon}(x)\big) < \frac{\varepsilon}{3}$. Thus, $y \in A$, $d(x, y) < \delta \Rightarrow \rho\big(f(x), f(y)\big) < \frac{2\varepsilon}{3} + \frac{\varepsilon}{3} = \varepsilon$. □

A.5 Problems

A.1 Express the following sets in the form $\{x : x \text{ has property } p\}$.

(a) The set A of all integers which when divided by 7 leave a remainder ≤ 3.

(b) The set B of all functions form $[0, 1]$ to $\mathbb{R}$ with at most two discontinuity points.

(c) The set C of all students at a given university who are graduate students with at least one course in mathematics at the graduate level.

(d) The set D of all algebraic numbers. (A number x is called an *algebraic number*, if it is the root of a polynomial with rational coefficients.)

(e) The set E of all possible sequences whose elements are either 0 or 1.

A.2 Give an example of sets A_1, A_2 such that $A_1 \cap A_2 \neq A_1 \cup A_2$.

A.3 Let $I = [0, 1]$, $\Omega = \mathbb{R}$ and for $\alpha \in \mathbb{R}$, $A_\alpha = (\alpha - 1, \alpha + 1)$, the open interval $\{x : \alpha - 1 < x < \alpha + 1\}$.

(a) Show that $\cup_{\alpha \in I} A_\alpha = (-1, 2)$ and $\cap_{\alpha \in I} A_\alpha = (0, 1)$.

(b) Suppose $J = \{x : x \in I, x \text{ is rational}\}$. Find $\cup_{x \in J} A_x$ and $\cap_{x \in J} A_x$.

A.4 With $\Omega \equiv \mathbb{N} \equiv \{1, 2, 3, \ldots\}$, find A^c in the following cases:

(a) $A = \{\omega : \omega \text{ is divisible by 2 or 3 or both}\}$. If $\omega \in A^c$, what can be said about its prime factors?

(b) $A = \{\omega : \omega \text{ is divisible by 15 and 16}\}$.

(c) $A = \{\omega : \omega \text{ is a perfect square}\}$.

A.5 Show that $X \equiv \{0, 1\}^{\mathbb{N}}$, the set of all sequences $\{\omega_i\}_{i \in \mathbb{N}}$ where each $\omega_i \in \{0, 1\}$, is *uncountable*. Conclude that $\mathcal{P}(\mathbb{N})$ is uncountable.

A.6 Show that if Ω_i is countable for each $i \in \mathbb{N}$, then for each $k \in \mathbb{N}$, $\times_{i=1}^{k} \Omega_i$ is countable and $\cup_{i \in \mathbb{N}} \Omega_i$ is also countable but $\times_{i \in \mathbb{N}} \Omega_i$ is *not countable*.

A.7 Show that the set of all polynomials in x with integer coefficients is countable.

A.8 Show that the well ordering property implies the principle of induction.

A.9 Apply the principle of induction to establish the following:

(a) For each $n \in \mathbb{N}$, $\sum_{j=1}^{n} j^2 = \frac{n(n+1)(2n+1)}{6}$.

(b) For each $n \in \mathbb{N}$, $x_1, x_2, \ldots, x_k \in \mathbb{R}$,

(i) (*The binomial formula*). $(x_1 + x_2)^n = \sum_{r=0}^{n} \binom{n}{r} x_1^r x_2^{n-r}$.

(ii) (*The multinomial formula*).

$$(x_1 + x_2 + \ldots + x_k)^n = \sum \frac{n!}{r_1! r_2! \ldots r_k!} x_1^{r_1} x_2^{r_2} \ldots x_k^{r_k},$$

where the summation extends over all $(r_1, r_2, \ldots, r_k)$ such that $r_i \in \mathbb{N}$, $0 \leq r_i \leq n$, $\sum_{r=1}^{k} r_i = n$.

A.10 Verify that on $\mathbb{R}$, the relation $x \sim y$ if $x - y$ is rational is an equivalence relation but the relation $x \sim y$ if $x - y$ is irrational is not.

A.11 Show that the set $A = \{r : r \in \mathbb{Q}, r^2 < 2\}$ is bounded above in $\mathbb{Q}$ but has no l.u.b. in $\mathbb{Q}$.

A.12 Show that for any two sequences $\{x_n\}_{n \geq 1}$, $\{y_n\}_{n \geq 1} \subset \mathbb{R}$,

$$\varliminf_{n\to\infty} x_n + \varliminf_{n\to\infty} y_n \leq \varliminf_{n\to\infty} (x_n + y_n) \leq \varlimsup_{n\to\infty} (x_n + y_n)$$
$$\leq \varlimsup_{n\to\infty} x_n + \varlimsup_{n\to\infty} y_n.$$

A.13 Verify that $\lim_{n\to\infty} x_n = a \in \mathbb{R}$ iff $\varliminf_{n\to\infty} x_n = \varlimsup_{n\to\infty} x_n = a$.

A.14 Establish Proposition A.2.3.

(**Hint:** First show that a Cauchy sequence is bounded and then show that $\varliminf_{n\to\infty} x_n = \varlimsup_{n\to\infty} x_n$.)

A.15 (a) Prove Proposition A.2.4 by comparison with the geometric series.

(b) Show that for integer $k \geq 1$, the power series $\sum_{n=k}^{\infty} n(n-1)(n-k+1)a_n x^{n-k}$ has the same radius of convergence as $\sum_{n=0}^{\infty} a_n x^n$.

A.16 Show that the series $\sum_{j=2}^{\infty} \frac{1}{j(\log j)^p}$ converges for $p > 1$ and diverges for $p \leq 1$.

A.17 Find the radius of convergence, ρ, for the powers series $A(x) \equiv \sum_{n=0}^{\infty} a_n x^n$ where

(a) $a_n = \frac{n}{(n+1)}$, $n \geq 0$.

(b) $a_n = n^p$, $n \geq 0$, $p \in \mathbb{R}$.

(c) $a_n = \frac{1}{n!}$, $n \geq 0$ (where $0! = 1$).

A.18 (a) Find the Taylor series at $a = 0$ for the function $f(x) = \frac{1}{1-x}$ in $I \equiv (-1, +1)$ and show that it converges to $f(x)$ on I.

(b) Find the Taylor series of $1 + x + x^2$ in $I = (1, 3)$, centered at 2.

(c) Let

$$f(x) = \begin{cases} e^{-\frac{1}{x^2}} & \text{if } \ |x| < 1, x \neq 0 \\ 0 & \text{if } \ x = 0 \ . \end{cases}$$

(i) Show that f is infinitely differentiable at 0 and compute $f^{(j)}(0)$ for all $j \geq 1$.

(ii) Show that the Taylor series at $a = 0$ converges but not to f on $(-1, 1)$.

A.19 Let $S = \{z : z \in \mathbb{C}, |z| = 1\}$ be the unit circle. Using the parameterization $t \to e^{\iota t} = (\cos t + \iota \sin t)$ from $[0, 2\pi]$ to S, show that the arc length of S (i.e., the circumference of the limit circle) is 2π.

A.20 Set $\phi(t) = \frac{\sin t}{\cos t}$ for $-\frac{\pi}{2} < t < \frac{\pi}{2}$. Verify that $\phi' = 1 + \phi^2$ and that $\phi : (-\frac{\pi}{2}, \frac{\pi}{2})$ to $(-\infty, \infty)$ is strictly monotone increasing and onto. Conclude that

$$\int_{-\infty}^{\infty} \frac{1}{1 + x^2} dx = \int_{-\pi/2}^{\pi/2} \frac{\phi'(t)}{1 + (\phi(t))^2} dt = \pi.$$

A.21 Using the property that $e^{\iota \frac{\pi}{2}} = \iota$ verify that for all t in $\mathbb{R}$

$$\begin{aligned} &\cos(\frac{\pi}{2} - t) = \sin t, \ \sin(\frac{\pi}{2} - t) = \cos t, \\ &\cos(\pi + t) = -\cos t, \ \sin(\pi + t) = -\sin t, \\ &\cos(2\pi + t) = \cos t, \ \sin(2\pi + t) = \sin t. \end{aligned}$$

Also show that $\cos t$ is a strictly decreasing map from $[0, \pi]$ onto $[-1, 1]$ and that $\sin t$ is a strictly increasing map from $[-\frac{\pi}{2}, \frac{\pi}{2}]$ onto $[-1, 1]$.

A.22 Using (i) of Theorem A.3.1, express $\cos(t_1 + t_2)$, $\sin(t_1 + t_2)$ in terms $\cos t_i$, $\sin t_i$, $i = 1, 2$ and in turn use this to prove Corollary A.3.4 from Theorem A.3.3.

A.23 Verify that $p_n(z) \equiv (1 + \frac{z}{n})^n$ converges to e^z uniformly on bounded sets in $\mathbb{C}$.

A.24 (a) Verify that for $p = 1$, $p = 2$ and $p = \infty$, d_p is a metric on $\mathbb{R}^k$.

(b) Show that for fixed x and y, $\varphi(p) \equiv d_p(x, y)$ is continuous in p on $[1, \infty]$.

(c) Draw the open unit ball $B_p \equiv \{x : x \in \mathbb{R}^2,\ d_p(x,0) < 1\}$ in $\mathbb{R}^2$ for $p = 1, 2$ and ∞.

A.25 Let $\mathbb{S} = C[0,1]$ be the set of all real valued continuous functions on $[0,1]$. Now let

$$
\begin{aligned}
d_1(f,g) &= \int_0^1 |f(x) - g(x)|dx, \quad \text{(area metric)} \\
d_2(f,g) &= \left(\int_0^1 |f(x) - g(x)|^2 dx\right)^{\frac{1}{2}}, \quad \text{(least square metric)} \\
d_\infty(f,g) &= \sup\{|f(x) - g(x)| : 0 \le x \le 1\} \quad \text{(sup metric)}.
\end{aligned}
$$

Show that all these are metrics on $\mathbb{S}$.

A.26 Let $\mathbb{S} = \mathbb{R}^\infty \equiv \{\{x_n\}_{n\ge1} : x_n \in \mathbb{R}$ for all $n \ge 1\}$ be the space of all sequences of real numbers. Let $d(\{x_n\}_{n\ge1}, \{y_n\}_{n\ge1}) = \sum_{j=1}^\infty \left(\frac{|x_j - y_j|}{1+|x_j-y_j|}\right)\frac{1}{2^j}$. Show that $(\mathbb{S}, d)$ is a Polish space.

A.27 If $s_k = \{x_{kn}\}_{n\ge1}$ and $s = \{x_n\}_{n\ge1}$, are elements of $\mathbb{S} = \mathbb{R}^\infty$ as in Problem A.26, verify that as $k \to \infty$, $s_k \to s$ iff $x_{kn} \to x_n$ for all $n \ge 1$.

A.28 Establish Theorem A.4.1.

A.29 Let $\mathbb{S} = C[0,1]$ and $d_p(f,g) \equiv \left(\int_0^1 |f(t) - g(t)|^p dt\right)^{\frac{1}{p}}$ for $1 \le p < \infty$ and $d_\infty(f,g) = \sup\{|f(t) - g(t)| : t \in [0,1]\}$.

(a) Let $f(x) \equiv 1$. Let $f_n(t) \equiv 1$ for $0 \le t \le 1 - \frac{1}{n}$, and $f_n(t) = n(1-t)$ for $1 - \frac{1}{n} \le t \le 1$. Show that $d_p(f_n, f) \to 0$ for $1 \le p < \infty$ but $d_\infty(f_n, f) \not\to 0$.

(b) Fix $f \in C[0,1]$. Let $g_n(t) = f(t)$, $0 \le t \le 1 - \frac{1}{n}$, and $g_n(t) = f(1 - \frac{1}{n}) + (f(1) - f(1 - \frac{1}{n}))n(t + \frac{1}{n} - 1)$, $1 - \frac{1}{n} \le t \le 1$.

Show that $d_p(g_n, f) \to 0$ for all $1 \le p \le \infty$.

A.30 Show that if $\{x_n\}_{n\ge1}$ is a convergent sequence in a metric space $(\mathbb{S}, d)$, then it is *Cauchy*.

A.31 Verify (b) of Example A.4.2 from the axioms of real numbers (cf. Proposition A.2.3). Verify (c) of the same example from (b).

A.32 Let $\mathbb{S} = C[0,1]$ and d be the *supremum metric*, i.e.,

$$d(f,g) = \sup\{|f(x) - g(x)| : 0 \le x \le 1\}.$$

By approximating any continuous function with piecewise linear functions with rational end points and rational values, show that $(\mathbb{S}, d)$ is Polish, i.e., it is complete and separable.

A.33 Show that the function $f(x) = x^2$ is continuous on $\mathbb{R}$, uniformly so on any bounded set $B \subset \mathbb{R}$ but not uniformly on $\mathbb{R}$.

A.34 Show that unions of open sets are open and intersection of any two open sets is open. Give an example to show that the intersection of an infinite number of open sets need not be open.

A.35 Prove Theorem A.4.3.

A.36 Let $f_n(x) = x^n$ and $g(x) \equiv 0$ on $\mathbb{R}$. Then $\{f_n\}_{n\geq 1}$ converges pointwise to g on $(-1, 1)$, uniformly on $[a, b]$ for $-1 < a < b < 1$, but not uniformly on $(0, 1)$.

A.37 Let $\{f_n\}_{n\geq 1}$, $f \in C[0, 1]$. Let $\{f_n\}_{n\geq 1}$ converge to f uniformly on $[0, 1]$. Show that $\lim\limits_{n\to\infty} \int_0^1 |f_n(x) - f(x)|dx = 0$ and $\lim\limits_{n\to\infty} \int_0^1 f_n(x) = \int_0^1 f(x)dx$.

A.38 Give a proof of Proposition A.2.6 (vi) (term by term differentiability of a power series) using Proposition A.2.7 (iv) (the fundamental theorem of Riemann integration).

Appendix B
List of Symbols and Abbreviations

B.1 Symbols

$\ll$	$\mu \ll \nu$: absolute continuity of a measure
$(\cdot) * (\cdot)$	convolution of measures, functions, etc.
$\mathcal{B}(\mathbb{S})$	Borel σ-algebra on a metric space $\mathbb{S}$ such as $\mathbb{S} = \mathbb{R}$, $\mathbb{R}^k$, $\mathbb{R}^\infty$
$\mathbb{C}$	the set of all complex numbers
$C[a, b]$	$= \{f : f : [a, b] \to \mathbb{R},\ f \text{ continuous}\}$
$C_c(\mathbb{R})$	$\equiv \{f : f : \mathbb{R} \to \mathbb{R}$, continuous and $f \equiv 0$ outside a bounded interval$\}$
$C_0(\mathbb{R})$	$\equiv \{f : f : \mathbb{R} \to \mathbb{R}$, continuous and $\lim_{\lvert x \rvert \to \infty} f(x) = 0\}$
$\frac{d\mu}{d\nu}$	Radon-Nikodym derivative of μ w.r.t. ν
$H^\perp$	orthogonal complement of a subspace H of a Hilbert space
ι	$\sqrt{-1}$
$\lambda\langle\mathcal{A}\rangle$	λ-class generated by a class of sets $\mathcal{A}$
$L^p(\Omega, \mathcal{F}, \mu)$	$= \{f : f : \Omega \to \mathbb{F}$, $\mathcal{F}$-measurable, $\int \lvert f \rvert^p d\mu < \infty\}$, with $\mathbb{F} = \mathbb{R}$ or $\mathbb{F} = \mathbb{C}$ (in Sections 5.6, 5.7 only)
$L^p(\mathbb{R})$	$= L^p(\mathbb{R}, \mathcal{B}(\mathbb{R}), m)$
m	the Lebesgue measure
μ^*	extension of a measure μ
μ_F	Lebesgue-Stieltjes measure corresponding to F

$\mu \perp \nu$	singularity of measures μ and ν
$\mathbb{N}$	the set of natural numbers
$\emptyset$	the null set
$\mathcal{P}(\Omega)$	the power set of $\Omega = \{A : A \subset \Omega\}$
$\mathbb{Q}$	the set of all rationals
$\mathbb{R}$	the set of real numbers, $(-\infty, \infty)$
$\mathbb{R}_+$	the set of nonnegative real numbers, $[0, \infty)$
$\bar{\mathbb{R}}$	the set of all extended real numbers, $[-\infty, \infty]$
$\bar{\mathbb{R}}_+$	$= [0, \infty]$
$\sigma\langle \mathcal{A} \rangle$	σ-algebra generated by a class of sets $\mathcal{A}$
$\sigma\langle \{f_a : a \in A\} \rangle$	σ-algebra generated by a collection of mappings $\{f_a : a \in A\}$
$\lfloor x \rfloor$	the integer part of x, i.e., $\lfloor x \rfloor = k$ if $k \leq x < k+1$, $k \in \mathbb{Z}, x \in \mathbb{R}$
$\lceil x \rceil$	the smallest integer not less than x, i.e., $\lceil x \rceil = k+1$ if $k < x \leq k+1$, $k \in \mathbb{Z}$, $x \in \mathbb{R}$
$\lvert z \rvert$	$= \sqrt{a^2 + b^2}$, the absolute value of a complex number $z = a + \iota b, a, b \in \mathbb{R}$
$Re(z)$	$= a$, the real part of a complex number $z = a + \iota b, a$, $b \in \mathbb{R}$
$Im(z)$	$= b$, the imaginary part of a complex number $z = a + \iota b, a, b \in \mathbb{R}$
$\mathbb{Z}$	the set of all integers $= \{0, \pm 1, \pm 2, \ldots\}$
$\mathbb{Z}_+$	the set of all nonnegative integers $= \{0, 1, 2, \ldots\}$

B.2 Abbreviations

a.c.	absolutely continuous (functions)
a.e.	almost everywhere
BCT	bounded convergence theorem
cdf	cumulative distribution function
DCT	dominated convergence theorem
EDCT	extended dominated convergence theorem
fdds	finite dimensional distributions
i.o.	infinitely often
m.c.f.a.	monotone continuity from above
m.c.f.b.	monotone continuity from below
MCT	monotone convergence theorem
o.n.b.	orthonormal basis
UI	uniform integrability

w.p. 1	with probability one
w.r.t.	with respect to
w.l.o.g.	without loss of generality

References

Bhatia, R. (2003), *Fourier Series*, 2nd edn, Hindustan Book Agency, New Delhi, India.

Billingsley, P. (1968), *Convergence of Probability Measures*, John Wiley, New York.

Billingsley, P. (1995), *Probability and Measure*, 3rd edn, John Wiley, New York.

Chung, K. L. (1974), *A Course in Probability Theory*, 2nd edn, Academic Press, New York.

Cohen, P. (1966), *Set Theory and the Continuum Hypothesis*, Benjamin, New York.

Doob, J. L. (1953), *Stochastic Processes*, John Wiley, New York.

Feller, W. (1966), *An Introduction to Probability Theory and Its Applications*, Vol. II, John Wiley, New York.

Feller, W. (1968), *An Introduction to Probability Theory and Its Applications*, Vol. I, 3rd edn, John Wiley, New York.

Hewitt, E. and Stromberg, K. (1965), *Real and Abstract Analysis*, Springer-Verlag, New York.

Kolmogorov, A. N. (1956), *Foundations of Probability*, 2nd edn, Chelsea, New York.

Körner, T. W. (1989), *Fourier Analysis*, Cambridge University Press, New York.

Munkres, J. R. (1975), *Topology, A First Course*, Prentice Hall, Englewood Cliffs, NJ.

Parthasarathy, K. R. (2005), *Introduction to Probability and Measure*, Vol. 33 of *Texts and Readings in Mathematics*, Hindustan Book Agency, New Delhi, India.

Royden, H. L. (1988), *Real Analysis*, 3rd edn, Macmillan Publishing Co., New York.

Rudin, W. (1976), *Principles of Mathematical Analysis*, International Series in Pure and Applied Mathematics, 3rd edn, McGraw-Hill Book Co., New York.

Rudin, W. (1987), *Real and Complex Analysis*, 3rd edn, McGraw-Hill Book Co., New York.

Szego, G. (1939), *Orthogonal Polynomials*, Vol. 23 of *American Mathematical Society Colloquium Publications*, American Mathematical Society, Providence, RI.

Subject Index

λ-system, 13
π-λ theorem, 13
π-system, 13
σ-algebra, 9, 44
  Borel, 12
  product, 147, 157, 203
  trace, 32
  trivial, 10

absolute continuity, 53, 113, 128
algebra, 9

BCT (*See* theorem)
binary expansion, 135

cardinality, 222
Cauchy sequence, 91, 227
change of variables, 81, 132, 141, 193
complete
  measure space, 160
  metric space, 91, 96, 124, 237
  orthogonal basis, 171
  orthogonal set, 100
  orthogonal system, 169
complex
  conjugate, 232
  numbers, 232
convergence
  almost everywhere, 61
  in measure, 62
  pointwise, 61
  radius of, 164, 227
convex functions, 84
convolution of
  cdfs, 184
  functions, 163
  sequence, 162
  signed measures, 161
cumulative distribution function, 45, 46, 133, 191
  singular, 133

DCT (*See* theorem)
de Morgan's laws, 72, 221
differentiation, 118, 130, 229
  chain rule of, 229
distribution
  discrete univariate, 144
  finite dimensional, 200

dual space, 94

EDCT (*See* theorem)
equivalence relation, 223
essential supremum, 89

Fourier
 coefficients, 99, 166, 170
 series, 186
 transform (*See* transforms)
function
 absolutely continuous, 129
 Cantor, 114
 Cantor ternary, 136
 continuous, 41, 228, 238
 differentiable, 229
 exponential, 233
 generating, 164
 Haar, 109
 integrable, 54
 simple, 49
 trigonometric, 233

Gaussian process, 209

independence, 208, 211
inequalities
 Bessel's, 99
 Cauchy-Schwarz, 88, 98, 198
 Chebychev's, 83, 196
 Cramer's, 84
 Hölder's, 87, 88, 198
 Jensen's, 86, 197
 Markov's, 83, 196
 Minkowski's, 88, 198
inner-product, 98
integration
 by parts formula, 155, 232
 Lebesgue, 51, 54, 61
 Riemann, 51, 59, 60, 61, 231
inversion formula, 175
isometry, 94
isomorphism, 100

Jordan decomposition, 123

L^p-
 norm, 89
 space, 89
lemma
 Fatou's, 7, 54
 Riemann-Lebesgue, 171
liminf, 226, 241
limits, 226
limsup, 226, 241
linear operator, 96

map
 co-ordinate, 203
 projection, 203
Markov Chains, 208
MCT (*See* theorem)
mean
 arithmetic, 87
 geometric, 87
measure, 14, 19
 complete, 23, 29
 counting, 16
 finite, 15
 induced, 45
 Lebesgue, 25
 Lebesgue-Stieltjes, 16, 25, 27, 58, 59
 negative variation, 121
 outer, 21
 positive variation, 121
 probability, 15
 product, 151
 Radon, 29, 131
 regularity of, 29
 signed, 119
 singular, 114
 space, 39
 total variation, 121
 uniqueness of, 29
measurable
 rectangle, 147
 space, 39
metric space, 11, 236, 243
 discrete, 90
 Polish, 238

separable, 237
moment, 198
moment generating function, 194, 198

norm, 95
total variation, 123
normed
vector space, 95

orthogonal
basis, 100
polynomials, 107
orthogonality, 99
orthonormality, 99

Parseval identity, 170
partition, 31
polynomial
trigonometric, 169
power series, 227
differentiability of, 229
probabiilty
distribution, 45, 191
space, 39
product space, 147
projection, 108

Radon-Nikodym derivative, 118
random
variable, 39, 191
vector, 192

sample space, 189
semialgebra, 3, 19
σ-finite, 27
set, 219
Cantor, 37, 134
compact, 238
complement, 221
cylinder, 202
empty, 220
function, 14
intersection, 220
Lebesgue measurable, 25
power, 9
product, 147, 221
union, 220
stochastic processes, 199
subspace, 96
symmetric difference, 4

ternary expansion, 135
theorem
binomial, 241
bounded convergence, 57
dominated convergence, 7, 57, 77
Egorov's, 69, 71
extended dominated convergence, 57
extension, 23, 157, 204
Fubini's, 153
Hahn decomposition, 122
Heine-Borel, 239
Kolmogorov's consistency, 200, 210
Lebesgue decomposition, 115
Lusin's, 69
mean value, 229
monotone convergence, 6, 52, 77
multinomial, 241
Radon-Nikodym, 115
Riesz representation, 94, 97, 101, 144
Scheffe's, 64
Taylor's, 230
Tonelli's, 152
topological space, 11
transformation
linear, 96
transforms
Fourier, 173
Laplace, 165
Laplace-Stieltjes, 166
Plancherel, 180
transition probability
function, 209
matrix, 209

uniform integrability, 65

variation
- bounded, 127
- negative, 126
- positive, 126
- total, 126

vector space, 95

Texts and Readings in Mathematics

1. R. B. Bapat: Linear Algebra and Linear Models (Second Edition)
2. Rajendra Bhatia: Fourier Series (Second Edition)
3. C. Musili: Representations of Finite Groups
4. H. Helson: Linear Algebra (Second Edition)
5. D. Sarason: Notes on Complex Function Theory
6. M. G. Nadkarni: Basic Ergodic Theory (Second Edition)
7. H. Helson: Harmonic Analysis (Second Edition)
8. K. Chandrasekharan: A Course on Integration Theory
9. K. Chandrasekharan: A Course on Topological Groups
10. R. Bhatia (ed.): Analysis, Geometry and Probability
11. K. R. Davidson: C* – Algebras by Example
12. M. Bhattacharjee et al.: Notes on Infinite Permutation Groups
13. V. S. Sunder: Functional Analysis — Spectral Theory
14. V. S. Varadarajan: Algebra in Ancient and Modern Times
15. M. G. Nadkarni: Spectral Theory of Dynamical Systems
16. A. Borel: Semisimple Groups and Riemannian Symmetric Spaces
17. M. Marcolli: Seiberg – Witten Gauge Theory
18. A. Bottcher and S. M. Grudsky: Toeplitz Matrices, Asymptotic Linear Algebra and Functional Analysis
19. A. R. Rao and P. Bhimasankaram: Linear Algebra (Second Edition)
20. C. Musili: Algebraic Geometry for Beginners
21. A. R. Rajwade: Convex Polyhedra with Regularity Conditions and Hilbert's Third Problem
22. S. Kumaresan: A Course in Differential Geometry and Lie Groups
23. Stef Tijs: Introduction to Game Theory
24. B. Sury: The Congruence Subgroup Problem
25. R. Bhatia (ed.): Connected at Infinity
26. K. Mukherjea: Differential Calculus in Normed Linear Spaces
27. Satya Deo: Algebraic Topology: A Primer
28. S. Kesavan: Nonlinear Functional Analysis: A First Course
29. S. Szabó: Topics in Factorization of Abelian Groups
30. S. Kumaresan and G. Santhanam: An Expedition to Geometry
31. D. Mumford: Lectures on Curves on an Algebraic Surface (Reprint)
32. J. W. Milnor and J. D. Stasheff: Characteristic Classes (Reprint)
33. K. R. Parthasarathy: Introduction to Probability and Measure (Corrected Reprint)
34. A. Mukherjee: Topics in Differential Topology
35. K. R. Parthasarathy: Mathematical Foundations of Quantum Mechanics